DATA MINING AND KNOWLEDGE DISCOVERY FOR GEOSCIENTISTS

DATA MINING AND KNOWLEDGE DISCOVERY FOR GEOSCIENTISTS

GUANGREN SHI

Professor of Mathematical Geology,
Research Institute of Petroleum Exploration and Development,
Beijing, China

ELSEVIER

AMSTERDAM • BOSTON • HEIDELBERG • LONDON • NEW YORK • OXFORD
PARIS • SAN DIEGO • SAN FRANCISCO • SINGAPORE • SYDNEY • TOKYO

Elsevier
525 B Street, Suite 1900, San Diego, CA 92101-4495, USA
225 Wyman Street, Waltham, MA 02451, USA

Library of Congress Cataloging-in-Publication Data
Shi, Guangren.
 Data mining and knowledge discovery for geoscientists/Guangren Shi, professor of mathematical geology, Research Institute of Petroleum Exploration and Development, Beijing, China. — First edition.
  pages cm
 Includes bibliographical references.
 ISBN 978-0-12-410437-2 (hardback)
1. Geology—Data processing. 2. Data mining. I. Title.
 QE48.8.S54 2014
 006.3'12—dc23
                                        2013032294

British Library Cataloguing in Publication Data
A catalogue record for this book is available from the British Library

For information on all Elsevier publications visit our web site at store.elsevier.com

ISBN: 978-0-12-410437-2

Contents

Preface

This book is an aggregation of principles, methods, codes, and applications for the data mining and knowledge discovery in geosciences based on the author's studies over the past 17 years.

In the past 20 years, the field of data mining has seen an enormous success in terms of both wide-ranging applications and scientific methodologies. *Data mining* is the computerized process of extracting previously unknown and important actionable information and knowledge from large databases. Such knowledge can then be used to make crucial decisions by incorporating individuals' intuition and experience so as to objectively generate for decision makers informed options that might otherwise go undiscovered. So, data mining is also called *knowledge discovery in database*, and it has been widely applied in many fields of economics, science, and technology. However, data mining applications to geosciences are still at an initial stage, partly due to the multidisciplinary nature and complexity of geosciences and partly due to the fact that many new methods in data mining require time and well-tested case studies in geosciences.

Facing the challenges of large amounts of geosciences databases, geoscientists can use database management systems to conduct conventional applications (such as queries, searches, and simple statistical analysis), but they cannot obtain the available knowledge inherent in data by such methods, leading to a paradoxical scenario of "rich data but poor knowledge." The true solution is to apply data mining techniques in geosciences databases and modify such techniques to suit practical applications in geosciences. This book, *Data Mining and Knowledge Discovery for Geoscientists*, is a timely attempt to summarize the latest developments in data mining for geosciences.

This book introduces some successful applications of data mining in geosciences in recent years for knowledge discovery in geosciences. It systematically introduces to geoscientists the widely used algorithms and discusses their basic principles, conditions of applications, and diversity of case studies as well as describing what algorithm may be suitable for a specific application.

This book focuses on eight categories of algorithm: (1) probability and statistics, (2) artificial neural networks, (3) support vector machines, (4) decision trees, (5) Bayesian classification, (6) cluster analysis, (7) Kriging method, and (8) other soft computing algorithms, including fuzzy mathematics, gray systems, fractal geometry, and linear programming.

This consists of 22 algorithms: probability density function, Monte Carlo method, least-squares method constructing linear function, least-squares constructing exponent function, least-squares constructing polynomial, multiple regression analysis, back-propagation neural network, the classification of support vector machine, the regression of support vector machine, ID3 decision trees, C4.5 decision trees, naïve Bayesian, Bayesian discrimination, Bayesian successive discrimination, Q-mode cluster analysis, R-mode cluster analysis, Kriging, fuzzy integrated decision, gray prediction, gray integrated decision, fractal geometry, and linear programming. For each algorithm, its applying ranges and conditions, basic principles, calculating method, calculation flowchart, and one or more detailed case studies are discussed. The book contains 41 case studies, 38 of which are in the area of geosciences. In each case study, for classification and regression algorithms, the solution accuracy comparison and algorithm selection have been made. Finally, a practical system of data mining and knowledge discovery for geosciences is presented. Moreover, this book also provides some exercises in each chapter; answers to all exercises are provided in a special appendix. Therefore, this book is dedicated to two kinds of people: (1) researchers and programmers in data mining, scientists and engineers in geosciences, and university students and lecturers in geosciences; and (2) scientists and engineers in computer science and information technology and university students and lecturers in information-related subjects such as database management.

Introduction

In the early 21st century, *data mining* (DM) was predicted to be "one of the most revolutionary developments of the next decade" and was chosen as one of 10 emerging technologies that will change the world (Hand et al., 2001; Larose, 2005; Larose, 2006). In fact, in the past 20 years, the field of DM has seen enormous success, both in terms of broad-ranging application achievements and in terms of scientific progress and understanding. DM is the computerized process of extracting previously unknown and important actionable information and knowledge from a database (DB). This knowledge can then be used to make crucial decisions by leveraging the individual's intuition and experience to objectively generate opportunities that might otherwise go undiscovered. So, DM is also called *knowledge discovery in database* (KDD). It has been widely used in some fields of business and sciences (Hand et al, 2001; Tan et al., 2005; Witten and Frank, 2005; Han and Kamber, 2006; Soman et al., 2006), but the DM application to geosciences is still in its initial stage (Wong, 2003; Zangl and Hannerer, 2003; Aminzadeh, 2005; Mohaghegh, 2005; Shi, 2011). This is because geosciences are different from the other fields, with miscellaneous data types, huge quantities, different measuring precision, and lots of uncertainties as to data mining results.

With the establishment of numbers of DB for geosciences, including data banks, data warehouses, and so on, the question of how to search for new important information and knowledge from large amounts of data is becoming an urgent task after the data bank is constructed. Facing such large amounts of geoscientific data, people can use the DB management system to conduct conventional applications (such as query, search, and simple statistical analysis) but cannot obtain the available knowledge inhered in data, falling into a puzzle of "rich data but poor knowledge." The only solution is to develop DM techniques in geoscientific databases.

We need to stress here that attributes and variables mentioned in this book are the same terminology; the term *attribute* refers to data related to datalogy, whereas *variable* refers to data related to mathematics. These two terms are called *parameters* when they are related to applications, so these three terms are absolutely the same. There are two types for these three terminologies; one is the *continuous* or *real* type, referring to lots of unequal real numbers occurring in the sample value, and the other is the *discrete* or *integer* type, referring to the fact that sample values are integer numbers such as 1, 2, 3, and so on. *Continuous* and *discrete* are the words of datalogy, such as *continuous attribute, discrete attribute, continuous variable, discrete variable*; whereas *real type* and *integer type* are terms related to software, such as *real attribute, integer attribute, real variable,* and *integer variable*.

1.1. INTRODUCTION TO DATA MINING

1.1.1. Motivity of Data Mining

Just as its meaning implies, data mining involves digging out the useful information from a large amount of data. With the wider application of computers, large amounts of data have piled up each year. It is possible to mine "gold" from these large amounts of data by applying DM techniques.

We are living in an era in which telecommunications, computers, and network technology are changing human beings and society. However, large amounts of information introduce large numbers of problems while bringing convenience to people. For example, it is hard to digest the excessive amounts of information, to identify true and false information, to ensure information safety, and to deal with inconsistent forms of information.

On the other hand, with the rapid development of DB techniques and the wide application of DB management systems, the amounts of data that people accumulate are growing more and more. A great deal of important information is hidden behind the increased data. It is our hope to analyze this information at a higher level so as to make better use of these data.

The current DB systems can efficiently realize the function of data records, queries, and statistics, but they cannot discover the relationship and rules that exist in the data and cannot predict the future development tendency based on the available data.

The phenomenon of rich data but poor knowledge results from the lack of effective means to mine the hidden knowledge in the data. Facing this challenge, DM techniques have been introduced and appear to be vital. The prediction of DM is the next hotpoint technique following network techniques.

Mass data storage, powerful multiple-processor computers, and DM algorithms are the basis of support for DM techniques. DB is advancing at an unprecedented speed, and data warehouse concepts are being widely applied to various industries.

Moreover, DM algorithms have become a technique that is mature, stable, and prone to be understood and run. Table 1.1 (Tan et al., 2009) outlines four stages of data evolution. Each stage is based on the previous one. Nowadays, the fourth stage is data mining, a revolutionary technique. DM can conduct queries based on past data, but it can also find the potential relationships *among* the past data, thus accelerating the information to transfer.

1.1.2. Objectives and Scope of Data Mining

DM is the process of extracting the unknown but potentially useful information and knowledge that is hidden inside mass, noisy, fuzzy, and random practical applied data.

Here the information mined should be characterized by unknown, effective, and practical data; the knowledge discovered is relative, with a given premise and constraint condition as well as a given field, and it should be prone to be understood by users.

In essence, the differentiation between DM and traditional data analysis, such as queries, report forms, and online application analysis, lies in the fact that DM mines information and discovers knowledge at the premises, without specific assumptions. It is obvious that instead of simple data query, DM is a cross-discipline that helps mine knowledge from data so as to provide the support for decision making. Consequently, researchers from different fields,

TABLE 1.1 Four Stages of Data Evolution

Evolution Stage	Time Period	Techniques Supported	Product Supplier	Product Characteristics
Data acquisition	1960s	Computer, tape, etc.	IBM, CDC	Static data provided
Data access	1980s	Relational database, Structured Query Language (SQL)	Oracle, Sybase, Informix, IBM, Microsoft	Dynamic data provided in records
Data warehouse	1990s	OLAP, multidimensional databases	Pilot, Comshare, Arbor, Cognos, Microstrategy	Traceable dynamic data provided in every hierarchy
Data mining	Now	Advanced algorithms, multiprocessor systems, algorithms of mass data	Pilot, Lockheed, IBM, SGI, etc.	Prediction information provided

Tan et al., 2009.

especially specialists and technical personnel of DB technology, artificial intelligent technology, mathematical statistics, visualized technology, and parallel calculation, come together to devote themselves to this emerging field of DM, forming a new technology hotpoint.

The following are the five classes of knowledge that DM discovers: generalization, association, classification and clustering, prediction, and deviation.

1.1.2.1. Generalization

Generalization refers to the recapitulative and descriptive knowledge to class characteristics. The token, universal, higher level conceptual and macroscopic knowledge is discovered according to the microscopic properties of data. This knowledge reflects common properties of same kind of object, and also is the generalizing, refinery and abstract to the data. There are many methods to find and realize generalization, such as data cube, attribute-oriented induction, and so on.

1.1.2.2. Association

Association is the knowledge that reflects the dependence and association of one subject to the others. If association exists between two or more attributes, the value of one attribute can be predicted based on the values of other attributes. The discovery of the *association rule* can be run in two steps. For the first step, identifying by iteration all frequent item sets, it is required that the support rate of frequent item sets is not lower than the lowest value given by the users. For the second step, set up the rule that the reliability is not lower than the lowest value given by the users from the frequent item sets.

1.1.2.3. Classification and Clustering

Classification and clustering are the characteristic knowledge that reflects the common properties of the same kind of products and the characteristics of differences between different things. The function of classification and clustering is to conduct classification and extract the rules in DB. There are many algorithms to do this, including multiple regression analysis, artificial neural networks, supporting vector machines, decision trees, Bayesian classification, and clustering analysis for clustering algorithms.

1.1.2.4. Prediction

Prediction is the knowledge to predict future data from historical and current data according to time-series data. Prediction also can be believed to be association knowledge by regarding time as the key attribute. The methods are classical statistics, artificial neural networks, gray prediction, machine learning, and so on.

1.1.2.5. Deviation

Deviation is the description of differences and extremely special cases, revealing abnormal phenomena.

1.1.3. Classification of Data Mining Systems

DM is cross-discipline, subjected to multiple disciplines (Figure 1.1), including DB systems, statistics, machine learning, visualization, and information science. Depending on the mined data type or the given DM application, it is possible for a DM system to integrate the techniques of space data analysis, information recall, model recognition, image analysis, signal processing, computer graphics, Web techniques, economics, business, biological information science, psychology, and so on.

Since DM is based on multiple disciplines, large numbers of various kinds of DM systems would be expected to be generated in DM research. In this way, it is necessary to have a clear classification for the DM system, which can help users distinguish DM systems from one another and define the DM system that it most suitable to a given task. In terms of different criteria, DM systems can be classified by four means: according to the mined DB type, according to the mined knowledge type, according to the available techniques type, and according to the application.

1.1.3.1. To Classify According to the Mined DB Type

The DB system itself can be classified according to different criteria, such as data model, data type, or the application covered. Each classification needs its own DM technique. Thus DM systems can be classified accordingly.

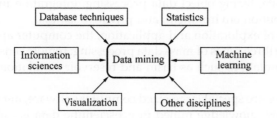

FIGURE 1.1   Data mining is subject to multiple disciplines.

For instance, if the DM system is classified according to data model, there can be a relative, business-like object-related or data warehouse mining system, whereas if it is classified according to the specific type of data to be processed, there can be space, time-series, text, flowing data, multimedia DM system, or Web mining systems.

1.1.3.2. To Classify According to the Mined Knowledge Type

It is possible to classify DM systems according to the mined knowledge type, that is, DM function, such as characterization, differentiation, correlation, association, classification, clustering, prediction, outlier analysis, and evolution analysis. Usually a comprehensive DM system can provide multiple and integrated functions.

In addition, DM systems can conduct differentiation based on the granularity or abstract layer of the mined knowledge, including generalization (high abstract layer), initial layer knowledge (initial data layer), or multiple layer knowledge (in view of number of abstract layers). An advanced DM system should support multiple abstract layers of KDD. DM systems can be classified for the regular to mine data (popular mode) and the singularity to mine data (for example, abnormity or outlier). Generally, the regularity to mine data, such as concept description, correlation, association, classification, clustering, and prediction, will exclude outliers as noise. These methods are useful to detect outliers.

1.1.3.3. To Classify According to the Available Techniques Type

DM systems can be classified according to available DM techniques. These techniques can be described by interaction with users (automatic systems, interactive examination systems, and inquiry drive systems) or the available data analysis method (facing DM techniques, machine learning, statistics, visualization, mode recognition, neural networks). Usually the complicated DM system uses more techniques and integrated techniques.

1.1.3.4. To Classify According to the Application

DM systems can also be classified according to application, such as finance, telecommunications, DNA, stock market, or email applied to some DM system. It is necessary to use special methods for different applications. Therefore, an all-powerful DM system is probably not possible for mining data in a given field.

1.1.4. Major Issues in Data Mining for Geosciences

DM usually faces the following ten problems: super DB and high-dimension data, data loss, variational data and knowledge, understandability for mode, data in nonstandard format, multimedia data, facing object data processing, integration in other systems, KDD in network and distribution environment, and private business.

In the past 50 years of exploration and application, the computer application in geosciences can be divided into three types of mass data processing: remote sense, seismic survey, and logging: numerical simulation, such as basin and reservoir simulation; and some computer science techniques.

The former two types are special large-sized computer software, not in the area of DM. For the third type, any new knowledge mined in geoscientific data is called data mining. The scope of geosciences contains two parts: the Earth itself (underground) and sky (air). All case studies in this book only cover underground.

An old saying goes that it is easy to go up to the sky, whereas it is difficult to go underground, which refers to the difficulty for people or measuring devices in terms of their reach. This pushes geoscientists to use DM techniques as much as possible with the condition that there are less direct methods, such as drilling, and large numbers of indirect methods, such as equipment on surface or in the air, to bring DM techniques into full play in discovery and development of underground resources using the limited and treasure measured data, to extrapolate the whole underground object from point to line, from line to plane, and from plane to volume.

Since the underground resources are distributed based on heterogeneity, randomness, and complexity, some of the DM techniques that people used succeeded, and some failed. The reason for success is that the technique is restricted to a definite application area; the reason for failure is the limitation of an objective condition. Therefore, comparing with DM in the other fields, DM in geosciences is facing the following five special problems, besides the previous ten problems: (1) an encrypted local area network should be used due to expensive data; (2) data cleaning should follow specific physical and chemical rules, since data come from field measuring and laboratory experimentation; (3) the space DM technique and visualization technique should be used from point to line, from line to plane, and from plane to volume, because the study object is in the underground; (4) the corresponding nonlinear DM algorithm should be used, since most correlations between the underground factors are nonlinear, with different strengths; and (5) the knowledge obtained from data mining is of probable risk due to lots of uncertainties underground.

1.2. DATA SYSTEMS USABLE BY DATA MINING

In general, the data systems usable by DM include databases, data warehousing, and data banks as well as file systems or any data set with another organization.

1.2.1. Databases

A DB is a "warehouse" in which data are built up, stored, and managed according to data structure. Databases were first generated 50 years ago. With the development of information technology and markets, particularly after the 1990s, data management came to not only store and manage data but to manage various kinds of data that users need. DB covers lots of types, from the simplest tables that store various kinds of data to the large-scaled DB systems that store mass data and are widely applied in various fields (McDonald et al., 2004; Connoly and Begg, 2005; Kyte, 2005; Silberschatz et al., 2006; Schwartz et al., 2008; Welling and Thomson, 2008).

1.2.1.1. Database types

There are four types of database:

1. *Management database*. For example, a human resources department in a company usually stores the employee number, name, age, sex, place of birth, salary, and résumé of staff members in tables, which can be regarded collectively as a DB. In financial management,

warehouse management, and production management, it is also necessary to set up automatic management for financial, warehouse, and production applications that use computers.

2. *Digital database.* For example, a number that is set up and stored in second-class storage (tape, disk) according to a certain data model is a data set. It is developed from a file management system. This kind of data set is characterized by the fact that multiple applications are served for a specific organization in an optimized way, whereas the data structure is independent of the application program that the data structure uses.

3. *Archive database.* For example, the information in a library and information archive room, whether stored in the form of print, a computer storage unit, or other forms, can be stored by digitization for users to access through a network using computers.

4. *Input/output database.* For large software, large amount of input data and output results are often stored in DB for users to use before and after the software runs, to improve calculation efficiency.

1.2.1.2. Data Properties

Two properties of data:

1. *Integrity.* The data in a DB are set up in view of the whole picture and are organized, described, and stored according to a given data model. The structure is based on natural links between data, thus providing all essential paths for access, and the data are not used for one given application but for the whole.

2. *Share.* The data in DB are set up for multiple users to share the information, getting rid of the limitation and restrict of specific programs. Different users can use the data in a DB as they want and they can access the same data at the same time. Data share meets user requirements for information content and also meets user requirements for communication.

1.2.1.3. Development Phases

DB has experienced four stages of development: manual management, file system, DB system, and advanced DB.

MANUAL MANAGEMENT

Before the mid-1950s, computer software and hardware were not so perfect. Storage equipment was only tape, card, and paper tape for hardware, and there was no operating system for software. The computers at that time were used for scientific calculation. Because no software system was available for data management, programmers not only had to stipulate the logical structure for data in the program, they also had to design the physical structure, including storage structure, access method, and input/output method. Users had to reprogram when the physical data structure or storage equipment changed. Since the organization of data was only aimed at applications, data could not be shared among different programs, which caused large amounts of repeated data between different applications. Consequently, it was difficult to maintain data consistency between the applied programs.

FILE SYSTEM

From the mid-1950s to the mid-1960s, the occurrence of mass storage equipment for computers, such as disks, promoted the development of software techniques, and the development of the operating system marked a new stage for data management. During the file system period, data were stored in the form of files and managed by the operating system. Operating systems provided a friendly interface via which users could access and use files. The disconnection of logical structure from physical structure and the separation of program data made data independent of programs. The program and data could be respectively stored in peripheral devices, and every application program could share a set of data, realizing the data share in the form of files.

However, since data organization was still geared to the needs of programs, large amounts of data redundancy still existed. Furthermore, the logical structure of data could not be modified and expanded conveniently, and every minor change to the logical structure of data had an impact on the application program. Because files were independent of each other, they could not reflect the relationships between things in the real world; the operating system was not responsible for the maintenance of information contact between files. If there appeared some contact between files, the application program would handle it.

DATABASE SYSTEM

Since the 1960s, with wide application of computers in data management, there has been a higher demand for the technique of data management. It is required to serve enterprises, set up data around data, reduce data redundancy, and provide higher data share ability. Meanwhile, it requires program and data to be independent so that when the logical structure of data changes, it will not cover the physical structure of data and will not impact the application program, which will decrease the cost of research and maintenance of the application program. DB techniques were developed based on such an application need.

DB techniques are characterized by (a) working for enterprises by a comprehensive DB that is shared for each application; (b) a given data model adopted to describe data properties and relationships between data; (c) small data redundancy, which is easy to modify and expand so as to keep the data consistent; (d) relatively independent of programs and data; (e) excellent interface for users with which users can develop and use DB easily; and (f) the uniform management and control of data provide safety, integrality, and concentrated control of data.

DB is of landmark significance in the field of information since the file system developed into the DB system. During the period of the file system, the key issue in information processing was system function design. Therefore, program design dominated. However, in the form of DB, data start to be in a central position; data structure design has become the first issue in forming systems, whereas application programs should be designed based on the established data structure.

ADVANCED DATABASE SYSTEM

With the continuous expansion of information management content, many of data models have appeared, such as the hierarchical model, the netlike model, the relationship model, the object model, and the semistructure model, and there has been lots of new technology of data flow, Web data management, DM, and so on.

1.2.1.4. Commonly Used Databases

Technology is getting more mature and perfect thanks to decades of development and practical application. The typical products in this area are DB2, Oracle, Informix, Sybase, SQL Server, PostgreSQL, MySQL, Access, and FoxPro.

1.2.2. Data Warehousing

Facing a large amount of data of various kinds, some with even larger granularity, we find it is not easy to deal with these data using only a database. To solve these problems and increase efficiency for the system backstage, it is necessary to introduce data warehousing (Adelman and Moss, 2000; Mallach, 2000; Kimball and Ross, 2002; Inmon, 2005). Building a data warehouse is a process of construction, maintenance, and inquiry. Here *construction* means to construct a logical model and a physical DB of the data warehouse; *maintenance* means to extract, transfer, and load data—that is, to extract the data from online transaction processing (OLTP) to data warehouse. *Inquiry* involves collecting information from the data warehouse. To make the concept simple, a data warehouse is a structure to obtain information. Finally, we need to distinguish between information and data. *Data* are only simple descriptions; the dispersed data have no significant information. *Information* comes from data; information involves beginning and continuing to raise questions and to take measures to solve them.

1.2.2.1. Data Storage

The following are four essential issues for data storage in data warehousing:

1. *Data storage mode.* There are two data storage modes for the data in a data warehouse: data stored in the relational DB and data stored in multiple-dimension mode, that is, multiple-dimension data sets.
2. *Data to be stored.* Different levels of data exist in a data warehouse. Usually data are divided into four levels: early detail data, current detail data, mild integration, and high integration. The existence of different levels is generally known as *granularity.* The smaller the granularity, the higher the detail, the lower the integration, and the more types of answer to the inquiry; contrarily, the larger the granularity, the lower the detail, the higher the integration, and the fewer the types of answer to the inquiry. The level is divided by granularity. Other metadata in the data warehouse are data about the data. The data dictionary or system category in the traditional DB is metadata. Two types of metadata appear in the data warehouse, one of which is the metadata established in an environment from operation transferred to warehouse. It contains various kinds of attributes of data source and all attributes in the process of transfer. The other type of metadata is used to establish mapping with multidimensional models and frontside tools.
3. *Granularity and separation.* Granularity is used to weigh the integration of the data in a data warehouse. *Separation* involves distributing the data to their own physical units so as to be handled independently and to increase the efficiency of data processing. The data units after data separation are distributed continuously. The criterion for data separation depends. It is possible to perform separation on dates, area or business, or multiple-criteria combinations.

4. *Organization mode of data superaddition.* Here we address a simple case, a combined file. For example, data storage units are day, week, season, and year; store everyday data in a day record, store seven days' data in a weekly record, store every other season's data in the season record, and so on. With this method, the earlier the data are stored, the higher the integration, that is, the larger the granularity.

1.2.2.2. Construction Step

The framework of a data warehouse consists of data source, data source transformation, and data source loading, forming a new DB and online analytical processing (OLAP). OLAP is the main application of a data warehouse system; it supports complex analysis operation but particularly supports decisions and provides intuitionistic and pellucid inquiry results. The aforementioned OLTP is the main application by the traditional relationship DB, processing basic and daily affairs. The data warehouse runs in three stages: project plan, design, and implementation, as well as maintenance and adjustment for data warehousing.

So far as the framework and running of a data warehouse go, a data warehouse will be constructed in the following five steps:

1. *Specific target and plan as a whole.* In light of an enterprise's development target, to stipulate an information framework plan with a strategic foresight, keep the development target consistent with the data needed.
2. *Consolidated plan and implement in steps.* It is a giant project with hefty investment to construct and maintain an enterprise's data warehouse. So it is necessary to construct a whole information framework and work out a plan to be implemented in steps. Emphasis should be laid on a data center that serves an important case need or data transfer mechanism.
3. *To create a technical environment and set up supporting platform.* to the goal is to create a technical environment and select software and hardware resources to realize a data warehouse, including development platform, a database management system (DBMS), network communication, development tools, terminal accessing tools, setup of a service target of availability, loading, maintenance and inquiry, and the like.
4. *To establish models and select the suitable tools.* It is possible to obtain complete and clear description information by constructing data models, which can provide uniform criteria for multiple application data. We can find lots of tools used in the construction of data warehouses, such as tools for model construction, data cleaning, data extract, data warehouse management, OLAP, DM, and so on.
5. *To reinforce management and maintenance.* It is not allowed to neglect the safety of the data warehouse. We must reinforce the management of the data warehouse operation to make backups for the relative data in the data warehouse so as to improve security and usability.

1.2.3. Data Banks

The term *data bank* refers to a data center that is constructed on a storage network distributed at high speed. A data bank cooperates with a large amount of various types of storage equipment in the network by application software, forming safe data storage and an accessing system that is applicable to data storage, backup, and place on file (Dwivedi, 2001a,

2001b). The superiority of a data bank is its capability to transfer the storage product into storage service.

Data banks are characterized by the powerful ability of data storage and backup, more security, high disaster tolerance, and reduction of operating costs for enterprises.

1.3. COMMONLY USED REGRESSION AND CLASSIFICATION ALGORITHMS

This book introduces in detail three regression and three classification algorithms and their application to geosciences. Chapter 2 introduces the multiple regression analysis (MRA), Chapter 3 introduces the error back-propagation neural network (BPNN), Chapter 4 introduces the support vector machine (SVM, which is divided into C-SVM classification algorithm and R-SVM regression algorithm), Chapter 5 introduces decision trees (DTR), and Chapter 6 introduces the Bayesian classification (BAC). These six algorithms use the same known parameters and share the same predicted unknown. The only difference between them is the method and calculation results. It is well known that MRA, BPNN, and R-SVM are regression algorithms, whereas C-SVM, DTR and BAC are classification algorithms; moreover, only MRA is a linear algorithm, whereas the other five are nonlinear algorithms.

Assume that there are n learning samples, each associated with $m + 1$ numbers (x_1, x_2, ..., x_m, y_i^*) and a set of observed values (x_{1i}, x_{2i}, ..., x_{mi}, y_i^*), with $i = 1, 2, ..., n$ for these numbers. In principle, $n > m$, but in actual practice $n >> m$. The n samples associated with $m + 1$ numbers are defined as n vectors:

$$x_i = \left(x_{i1}, x_{i2}, ..., x_{im}, y_i^*\right) \quad (i = 1, 2, ..., n) \tag{1.1}$$

where n is the number of learning samples; m is the number of independent variables in samples; x_i is the ith learning sample vector; x_{ij} is the value of the jth independent variable in the ith learning sample, $j = 1, 2, ..., m$; and y_i^* is the value of the ith learning sample, the observed value.

Equation (1.1) is the expression of learning samples.

Let x_0 be the general form of a vector of (x_{i1}, x_{i2}, ..., x_{im}). The principles of MRA, BPNN, DTR, and BAC are the same, i.e., try to construct an expression, $y = y(x_0)$, such that Equation (1.2) is minimized. Certainly the four different algorithms use different approaches and result in differing accuracy of calculation results.

$$\sum_{i=1}^{n}\left[y(x_{0i}) - y_i^*\right]^2 \tag{1.2}$$

where $y(x_{0i})$ is the calculation result of the dependent variable in the ith learning sample, and the other symbols have been defined in Equation (1.1).

However, the principles of C-SVM and R-SVM algorithms are to try to construct an expression, $y = y(x_0)$, to maximize the margin based on support vector points so as to obtain the optimal separating line.

This $y = y(x_0)$ is called the *fitting formula* obtained in the learning process. The fitting formulas of different algorithms are different. In this book, y is defined as a single variable.

The flowchart is as follows: The first step is the *learning process*, using n learning samples to obtain a fitting formula (Figure 1.2); the second step is the *learning validation*, substituting n learning samples into the fitting formula to get prediction values $(y_1, y_2, ..., y_n)$, respectively, so as to verify the fitness of an algorithm; and the third step is the *prediction process*, substituting k prediction samples expressed with Equation (1.3) into the fitting formula to get prediction values $(y_{n+1}, y_{n+2}, ..., y_{n+k})$, respectively (Figure 1.3).

$$x_i = (x_{i1}, x_{i2}, ..., x_{im}) \quad (i = n+1, n+2..., n+k) \tag{1.3}$$

where k is the number of prediction samples, x_i is the ith prediction sample vector, and the other symbols have been defined in Equation (1.1).

Equation (1.3) is the expression of prediction samples.

1.3.1. Linear and Nonlinear Algorithms

In the aforementioned six algorithms, only MRA is a linear algorithm, whereas the other five are nonlinear algorithms. This is due to the fact that MRA constructs a linear function, whereas the other five construct nonlinear functions, respectively. Because of the complexities of geosciences rules, the correlations between different classes of geoscientific data are nonlinear in most cases. In general, therefore, it is better to use C-SVM and R-SVM when the nonlinearity is very strong, and otherwise use BPNN, DTR, and BAC. As for MRA, it can serve as an auxiliary tool, e.g., a pioneering dimension-reduction tool, cooperating

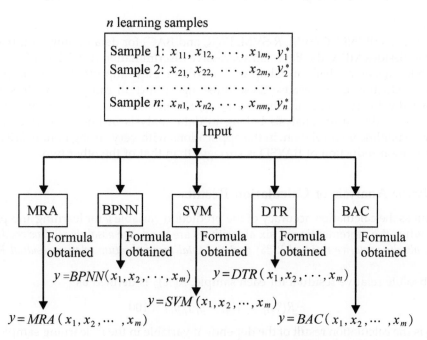

FIGURE 1.2   Sketch map of the learning process for MRA, BPNN, SVM, DTR, and BAC.

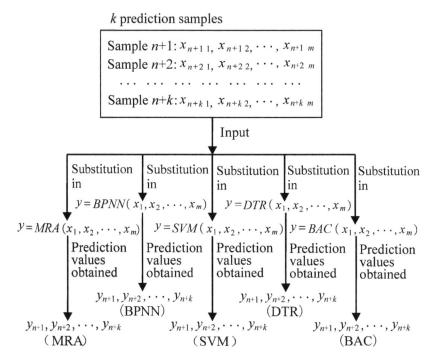

FIGURE 1.3   Sketch map of the prediction process for MRA, BPNN, SVM, DTR, and BAC.

with major tools (BPNN, C-SVM, R-SVM, DTR, and BAC) for data mining, as introduced in Section 4.4. Besides MRA, the Bayesian successive discrimination (BAYSD) in Chapter 6 and the R-mode cluster analysis in Chapter 7 also can play important roles as pioneering dimension-reduction tools because these three algorithms all can give the dependence of the predicted value (y) on parameters ($x_1, x_2, \ldots, x_m$), in decreasing order. However, because MRA and R-mode cluster analysis belong to data analysis in linear correlation whereas BAYSD is in nonlinear correlation, in the applications with very strong nonlinearity the ability of dimension-reduction of BAYSD is stronger than that of the other two.

1.3.2. Error Analysis of Calculation Results

To express the calculation accuracy of the prediction variable y for learning and prediction samples when the aforementioned six algorithms are used, the *absolute relative residual R(%)*, the *mean absolute relative residual $\overline{R}$(%)* and the *total mean absolute relative residual $\overline{R}^*$(%)* are adopted.

The absolute relative residual for each sample, $R(\%)_i$, is defined as

$$R(\%)_i = \left| (y_i - y_i^*) / y_i^* \right| \times 100 \tag{1.4}$$

where y_i is the calculation result of the dependent variable in the ith learning sample, and the other symbols have been defined in Equations (1.1) and (1.3).

It is noted that zero must not be taken as a value of y_i^* to avoid floating-point overflow. Therefore, for a regression algorithm, delete the sample if it is $y_i^* = 0$; and for a classification algorithm, a positive integer is taken as values of y_i^*.

The mean absolute relative residual for all learning samples or prediction samples, $\overline{R}(\%)$, is defined as

$$\overline{R}(\%) = \sum_{i=1}^{N_s} R(\%)_i / N_s \tag{1.5}$$

where $N_s = n$ for learning samples, whereas $N_s = k$ for prediction samples, and the other symbols have been defined in Equations (1.1) and (1.3).

For learning samples, $R(\%)$ and $\overline{R}(\%)$ are called the *fitting residual* to express the fitness of the learning process, and here $\overline{R}(\%)$ is designated as $\overline{R}_1(\%)$. For prediction samples, $R(\%)$ and $\overline{R}(\%)$ are called the *prediction residual* to express the accuracy of the prediction process, and here $\overline{R}(\%)$ is designated as $\overline{R}_2(\%)$.

The total mean absolute relative residual for all samples, $\overline{R}^*(\%)$, is defined as

$$\overline{R}^*(\%) = \left[\overline{R}_1(\%) + \overline{R}_2(\%)\right]/2 \tag{1.6}$$

when there are no prediction samples, $\overline{R}^*(\%) = \overline{R}_1(\%)$.

1.3.3. Differences between Regression and Classification Algorithms

The essential difference between classification and regression algorithms is represented on the data type of calculation results y. For regression algorithms, y is a real-type value and generally differs from y^* given in the corresponding learning sample; whereas for classification algorithms, y is an integer-type value and must be one of y^* defined in the learning samples. In datalogy, the real-type value is called a *continuous attribute*, whereas the integer-type value is called a *discrete attribute*.

To have the comparability of these six algorithms apply to classification examples in Chapters 4, 5, and 6, the results y of these three regression algorithms are converted from real numbers to integer numbers using the round rule. Certainly it is possible that some y after the conversion are not equal to any y^* in all learning samples.

1.3.4. Nonlinearity of a Studied Problem

Since MRA is a linear algorithm, its $\overline{R}^*(\%)$ for a studied problem expresses the nonlinearity of $y = y(x)$ to be solved, i.e., the nonlinearity of the studied problem. This nonlinearity can be divided into five classes: very weak, weak, moderate, strong, and very strong (Table 1.2).

1.3.5. Solution Accuracy of Studied Problem

Whether linear algorithm (MRA) or nonlinear algorithms (BPNN, C-SVM, R-SVM, DTR, and BAC), their $\overline{R}^*(\%)$ of a studied problem expresses the accuracy of $y = y(x)$ obtained by each algorithm, i.e., solution accuracy of the studied problem solved by each algorithm. This solution accuracy can be divided into five classes: very high, high, moderate, low, and very low (Table 1.2).

TABLE 1.2   Two Uses of the Total Mean Absolute Relative Residual $\overline{R}^*(\%)$

Range of $\overline{R}^*(\%)$	Nonlinearity of Studied Problems Based on $\overline{R}^*(\%)$ of MRA	Solution Accuracy of MRA, BPNN, C-SVM, R-SVM, DTR, and BAC Based on Their $\overline{R}^*(\%)$
$\overline{R}^*(\%) \leq 2$	Very weak	Very high
$2 < \overline{R}^*(\%) \leq 5$	Weak	High
$5 < \overline{R}^*(\%) \leq 10$	Moderate	Moderate
$10 < \overline{R}^*(\%) \leq 25$	Strong	Low
$\overline{R}^*(\%) > 25$	Very strong	Very low

1.4. DATA MINING SYSTEM

A data mining system (DMS) is a kind of information system that effectively integrates one or a number of DM algorithms on a software platform and combines corresponding data sources to complete specific mining applications or common mining tasks to extract the modes, rules, and knowledge that are helpful to users (Hand et al, 2001; Tan et al., 2005; Witten and Frank, 2005; Han and Kamber, 2006; Soman et al., 2006). Research and development has been made to a specific DMS for a specific application area.

A complete DM technique should have the following six functions (Larose, 2005, 2006): (1) ability to deal with insufficiency and difference of each data structure; (2) ability to figure out the degree of confidence for the specific technique used; (3) does not force promotion of a certain data structure; (4) determines the subsystem of large data sets under the unbiased condition; (5) brings out effective ways to realize risk evaluation law; and (6) the method generated can be combined with other methods and data.

Three steps for DM are illustrated in Figure 1.4 and described as follows:

1. *Data preprocessing*, such as data selection, cleaning, handling missed ones, misclassification recognition, outlier recognition, transformation, normalization, and so on.
2. *Knowledge discovery*, first making sure which factors are enough to support DM by using dimension-reduction tools, then selecting an appropriate DM algorithm base on the characteristics of learning samples so as to obtain new knowledge.
3. *Knowledge application*, to deduce the target value for each prediction sample according to new knowledge.

1.4.1. System Functions

DM will make the proactive and knowledge-based decision by predicting trends and behavior. The object of DM is to discover hidden significant knowledge from a DB. It has the following five functions:

1. *Automatic prediction of trends and behavior*. DM can automatically look for predictive information in a DB. Now it is possible to rapidly get a solution directly from the data

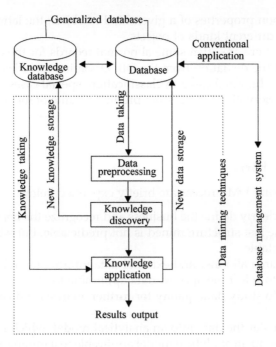

FIGURE 1.4 Flowchart of a data mining system.

instead of doing a large amount of manual analysis, as in the past. A typical example is market prediction. DM uses the data about promotion to look for the user who will get maximum return from his investment. The other prediction includes bankruptcy prediction.

2. *Association analysis.* Data association is a kind of important knowledge to be discovered in DB. If there is a rule of value derivation between two or more variables, this is association. Association can be divided into simple association, time-series association, and consequence association. The purpose of association analysis is to find out the association net hidden in DB. Sometimes the data association function in DB is unknown and is uncertain even if it is known. So the rule resulting from association analysis is one of reliability.

3. *Clustering.* The record in DB can be resolved into a series of significant subset, i.e., clustering. Clustering makes people understand the objective reality and is the premise to conceptual description and deviation analysis. Clustering technique includes traditional mode recognition methods and mathematical classification. Conceptual clustering technique not only can consider the distance between the objects in classification of the objects but also requires that the classified should have a meaningful description, thus avoiding one-sidedness of traditional techniques.

4. *Conceptual description.* Conceptual description provides the meaning of a given kind of object and generalizes relative characteristics of this kind of object. Conceptual description is divided into characteristics of description and distinguishment description. The former

describes the common properties of a given kind of object; the latter describes the difference between different kinds of objects.

5. *Deviation detection.* There are often some abnormal records for the data in a DB. It is significant to detect the deviation from the DB. Deviation includes lots of potential knowledge, such as abnormal cases in classification, special cases that do not meet the rule, deviation of observation results from predicted value of the model, and variation of the values with time.

1.4.2. System Flowcharts

The seven steps during DM process are briefly explained below.

1. *Define the object.* To clearly define the business and recognize the object of DM is one of the important steps. The last structure mined is not predictable, but what is going to be researched is predictable.
2. *Data selection.* To search all inner and external data information related to the business object, from which to select the applied data applicable to DM.
3. *Data preprocessing.* To study data quality for further analysis and confirm the type for mining.
4. *Data transfer.* To transfer the data into an analytical model, which is set up for a mining algorithm. To setting up an analytical model applicable to a mining algorithm is the key to the success of DM.
5. *Data mining.* To mine the data obtained from transfer. Except for the selection of an appropriate mining algorithm, the other work will be accomplished automatically.
6. *Result analysis.* To explain and evaluate the results for which the method used will be determined with the operation of DM, usually using a visualization technique.
7. *Knowledge assimilation.* To integrate the knowledge obtained from analysis to the structure of a business information system.

1.4.3. Data Preprocessing

The data in the real world on the whole are incomplete and inconsistent dirt data, which cannot be DM directly or the mined results are not satisfactory. To improve DM quality, the technique of *data preprocessing* is created. A DB tends to be disturbed by noise, data loss, and data inconsistency, because a DB can be as large as several gigabytes or even more, and most come from multiple isomerous data sources. Data quality is involved with DM process and quality; therefore, data preprocessing is an important step of the KDD process. The purpose of data preprocessing is to improve data quality so as to make DM more effective and easier, to increase the quality of mined results, and to reduce the time that DM requires.

The methods of data preprocessing are various, such as data cleaning, data integration, data transformation, and data reduction. These data processing methods will be used before data mining. Maimon and Rokach (2010) outlined a more detailed summary of data preprocessing, and Han and Kamber (2006) specified more detailed methods related to data preprocessing. Refer to these references; this book provides only a conceptual, not detailed, introduction.

1.4.3.1. Data Cleaning

Realistic data are often noisy, imperfect, and inconsistent. The main job for data cleaning is to fill up the missed data value, make noisy data smooth, identify or eliminate abnormal values, and solve inconsistency problems.

Generally, the process for data cleaning is to handle the missed data first, followed by processing noisy data, and to solve the inconsistent data last. Most methods to clean industrial data are roughly the same; few were effective five years ago.

The practical application shows that (a) data cleaning is a new field to be modified; (b) it is impossible to have common software applicable to every walk of life; (c) users should select commercial software that is applicable to the trade they serve and modify the software according to its own characteristics; and (d) data cleaning is very important because the data are complicated and most data are dirt, and the software is more strictly required.

1.4.3.2. Data Integration

Data integration involves making up an integrated data set by combining the correlative data from multiple data sources. The following should be taken into account in data integration: (a) mode integration, i.e., how to realize a match of the things from multiple data sources—in fact, to identify the things; (b) redundancy, another problem that often occurs in data integration; and (c) detection and elimination of data value interference, the reason for which is possibly that the attribute value in different data sources for one thing has an expression difference, a different scale and different code and so on.

1.4.3.3. Data Transformation

Data transformation involves normalizing data. Data transformation will do the following: (a) smooth the data, i.e., help eliminate noise in the data; (b) aggregate the data; (c) generalize the data by using the concept at a higher level to substitute the data object at a lower level; (d) normalize the data to transform the related attribute data at a given proportion to a specific small area so as to eliminate the deviation of DM resulting from different values of numerical attributes; and (e) construct the attribute to make up a new attribute based on the available attribute set.

1.4.3.4. Data Reduction

It usually takes lots of time to analyze the complicated data in the large-scaled DB content, which often makes this analysis unrealistic and unfeasible, particularly if interactive DM is needed. A data reduction technique is used to obtain a simplified data set from the original huge data set and to keep this simplified data set integration as the original data set. In this way it is obviously more efficient to perform DM on the simplified data set, and the mined results are almost the same as the results from the original data set. The main strategies to do data reduction are: (a) data aggregation, such as building up data cube aggregation; this aggregation is mainly used for the construction of data cube data warehouse operation; (b) data reduction, mainly used for the detection and elimination of those unrelated, weak correlated or redundant attributes or dimensions; (c) data compression, to compress data sets using a coding technique (e.g., minimum coding length or wavelet); and (d) numerosity reduction, to use simpler data expression modes, e.g., parameter models or nonparameter models (clustering, sampling, and histograms) to displace the original data.

1.4.4. Summary of Algorithms and Case Studies

In light of the available applications and future applications of DM, this book introduces the following eight types of algorithms:

1. Probability and statistics, including probability density function, Monte Carlo algorithm, least-squares method, and multiple regression analysis, which are introduced in Chapter 2.
2. An artificial neural network, BPNN, including the conventional prediction algorithm and a special prediction algorithm, i.e., prediction for time-series data, which are introduced in Chapter 3.
3. Supporting vector machine, including C-SVM and R-SVM algorithms, and a dimension-reduction procedure using machine learning (MRA serves as a pioneering dimension-reduction tool), which are introduced in Chapter 4.
4. Decision trees, including ID3 algorithms and C4.5 algorithms, which are introduced Chapter 5.
5. Bayesian classification, including naïve Bayesian, Bayesian discrimination, and Bayesian successive discrimination, which are introduced in Chapter 6.
6. Cluster analysis, including Q-mode cluster analysis and R-mode cluster analysis, where the former can serve as a pioneering sample-reduction tool and the latter can serve as a pioneering dimension-reduction tool; introduced in Chapter 7.
7. The Kriging algorithm, introduced in Chapter 8.
8. Other soft computing algorithms for geosciences, including fuzzy mathematics, gray system, fractal geometry, and linear programming, which are introduced in Chapter 9.

The aforementioned 22 DM algorithms are the keystones of this book. For each algorithm, its applying ranges and conditions, basic principles, complete calculation method, calculation flowchart, and one or more detailed case studies are discussed. There are 41 case studies, 38 of which are in the area of geosciences. In each case study for classification and regression algorithms, the solution accuracy comparison and algorithm selection are made. So this book not only supplies practical application in geosciences but also to all-purpose algorithm study and software development.

A practical software system of data mining and knowledge discovery for geosciences is presented at the end. Moreover, exercises are attached to each chapter, and all the answers are given in the back of the book.

EXERCISES

1-1. With over 20 years of development, data mining has applied to some scientific and technological as well as business fields, but its application in geosciences is still in the initial stage. Why?

1-2. The attributes, variables, and parameters mentioned in this book are the same terminology. But which area, such as datalogy, mathematics, or applications, do the three terminologies fall into?

1-3. Data are divided into two types: continuous and discrete in datalogy, whereas real type and integer type in software. The question is, are continuous data equal to real data and discrete data equal to integer data?

1-4. What are the five most popular types of knowledge discovered by data mining?

1-5. What are the five problems that DM is facing compared with the data mining in other fields?

1-6. What data systems are usable by data mining?

1-7. The absolute relative residual $R(\%)$ is defined in Equation (1.4), where the denominator is an observed value y_i^*. How can we avoid $y_i^* = 0$?

1-8. (a) What is the basic difference between a regression algorithm and a classification algorithm? (b) What is the sameness in application? (c) Is it possible to approximately regard a regression algorithm as a classification algorithm? (d) What are the regression algorithms and classification algorithms introduced in this book? (e) Among them, which one is a linear algorithm?

1-9. What are the two uses of the total mean absolute relative residual $\overline{R}^*(\%)$ expressed in Equation (1.6)?

1-10. What is the definition of a data mining system? What are the three major steps of data mining?

References

Adelman, S., Moss, L.T., 2000. Data Warehouse Project Management. Addison-Wesley, New York, NY, USA.

Aminzadeh, F., 2005. Applications of AI and soft computing for challenging problems in the oil industry. J. Petro. Sci. Eng. 47 (1–2), 5–14.

Connoly, T.M., Begg, C.E., 2005. Database Systems: A Practical Approach to Design, Implementation and Management, fourth ed. Addison-Wesley, San Francisco, CA, USA.

Dwivedi, A.M., 2001a. Data Bank on General Science. Anmol Publications, New Delhi, India.

Dwivedi, A.M., 2001b. Data Bank on Geography. Anmol Publications, New Delhi, India.

Han, J.W., Kamber, M., 2006. Data Mining: Concepts and Techniques, second ed. Morgan Kaufmann, San Francisco, CA, USA.

Hand, D., Mannila, H., Smyth, P., 2001. Principles of Data Mining. MIT Press, Cambridge, MA, USA.

Inmon, W.H., 2005. Building the Data Warehouse, fourth ed. Wiley India, New Delhi, India.

Kimball, R., Ross, M., 2002. The Data Warehouse Toolkit: The Complete Guide to Dimensional Modeling, second ed. Wiley, New York, NY, USA.

Kyte, T., 2005. Expert Oracle Database Architecture: 9i and 10g Programming Techniques and Solutions. A press, Berkeley, CA, USA.

Larose, D.T., 2005. Discovering Knowledge in Data. Wiley, New York, NY, USA.

Larose, D.T., 2006. Data Mining Methods and Models. Wiley, New York, NY, USA.

Maimon, O., Rokach, L., 2010. The Data Mining and Knowledge Discovery Handbook, second ed. Springer, New York, NY, USA.

Mallach, E.G., 2000. Decision Support and Data Warehouse Systems. McGraw-Hill, New York, NY, USA.

McDonald, C., Katz, C., Beck, C., Kallman, J.R., Knox, C.D., 2004. Mastering Oracle PL/SQL: Practical Solutions. Apress, Berkeley, CA, USA.

Mohaghegh, S.D., 2005. A new methodology for the identification of best practices in the oil and gas industry, using intelligent systems. J. Petro. Sci. Eng. 49 (3-4), 239–260.

Schwartz, B., Zaitsev, P., Tkachenko, V., Jeremy, D., Zawodny, D., Lentz, A., Balling, D.J., 2008. High Performance MySQL, second ed. O'Reilly Media, Sebastopol, CA, USA.

Shi, G., 2011. Four classifiers used in data mining and knowledge discovery for petroleum exploration and development. Adv. Petro. Expl. Devel. 2 (2), 12–23.

Silberschatz, A., Korth, H.F., Sudarshan, S., 2006. Database System Concepts, fifth ed. McGraw-Hill, New York, NY, USA.

Soman, K.P., Diwakar, S., Ajay, V., 2006. Insight into Data Mining: Theory and Practice. Prentice Hall of India, New Delhi, India.

Tan, J., Zhang, J., Huang, Y., Hu, Z., 2009. Data Mining Techniques. China WaterPower Press, Beijing, China (in Chinese).

Tan, P., Steinbach, M., Kumar, V., 2005. Introduction to Data Mining. Pearson Education, Boston, MA, USA.

Welling, L., Thomson, L., 2008. PHP and MySQl Web Development, fourth ed. Addison-Wesley, San Francisco, CA, USA.

Witten, I.H., Frank, E., 2005. Data Mining: Practical Machine Learning Tools and Techniques, second ed. Morgan Kaufmann, San Francisco, CA, USA.

Wong, P.M., 2003. A novel technique for modeling fracture intensity: A case study from the Pinedale anticline in Wyoming. AAPG Bull. 87 (11), 1717–1727.

Zangl, G., Hannerer, J., 2003. Data Mining: Applications in the Petroleum Industry. Round Oak Publishing, Katy, TX, USA.

This chapter introduces the calculating techniques of probability and statistics as well as their applications in geosciences. For each technique, the applying ranges and conditions, basic principles, calculation methods, and case studies are provided. Each case study contains the relevant calculation flowchart, calculation results, and analyses. Though each case study is small, it reflects the whole process of calculations to benefit readers in understanding and mastering the techniques applied. Though some flowcharts are drawn for specific case studies, they are actually all-purpose only if the specific case study is replaced by readers' own application samples.

Section 2.1 (probability) introduces two simple case studies. The first is about the probability density function at a more explored area in a basin; this case study is used to explain how to adopt the probability density function for prediction of undiscovered resources. Concretely, using the reserves in place of 40 reservoirs in a basin, a reservoir probability density function of the basin is constructed by the data mining tool of probability density function, and the results coincide with practicality. This function is called *mined knowledge*. This knowledge provides the basis for further exploration decisions related to the basin. Therefore, this method can be spread to the exploration decision of other basins.

The second simple case study is about the pore volume of a trap, whereby we explain how to adopt the Monte Carlo method to calculate an unknown number. Concretely, using the data of values of an area, 25 porosity and 40 relief in a trap, a probability distribution function for pore volume of the trap is constructed by the data mining tool of the Monte Carlo method, and the results coincide with practicality. This function is called *mined knowledge*. This knowledge reflects the oil and gas potential of the trap. Therefore, this method can be spread to the oil and gas potential prediction of other traps.

Section 2.2 (statistics) introduces two simple case studies and one other case study. Simple Case Study 1 is the linear function of porosity with respect to acoustictime in a well, whereby we explain how to adopt the least-squares method to construct a linear function of an unknown number with respect to a parameter. Concretely, using the data of 33 porosity values and corresponding 33 acoustictime values in a well, a formula of porosity with respect to acoustictime is constructed by the data mining tool of constructing linear functions by the least-squares method, and the results coincide with practicality. This formula is called *mined knowledge*. Similarly, using the data of 23 porosity values and corresponding 23 burial depth values in an area, a formula of porosity with respect to burial depth is constructed by the data mining tool of constructing exponent functions by the least-squares method, and the results coincide with practicality. This formula is also called *mined knowledge*. Under the same structural and sedimentary conditions, these two types of knowledge can be borrowed by the neighboring wells or areas. Therefore, the two methods can be spread to the porosity expression of other wells or areas.

Simple Case Study 2 is the polynomial of porosity with respect to acoustictime in the well, whereby we can explain how to adopt the least-squares method to construct a polynomial of an unknown number with respect to a parameter. Concretely, using the data of 33 porosity values and corresponding 33 acoustictime values in the well, a formula of porosity with respect to acoustictime is constructed by the data mining tool of constructing polynomials by the least-squares method, and the results coincide with practicality. This formula is called *mined knowledge*. Under the same structural and sedimentary conditions, this knowledge can be borrowed by the neighboring wells. Therefore, this method can be spread to the formula of porosity with respect to acoustictime in other wells. The case study is about the linear function of proved reserve with respect to some parameters of sags in eastern China, whereby we can explain how to adopt multiple regression analysis (MRA) to construct a function of an unknown number with respect to multiple parameters. Concretely, using data of 15 sags in eastern China, i.e., the five parameters (geothermal gradient, organic carbon content, volume ratio of source rock to sedimentary rock, thickness ratio of sandstone to mudstone in the vicinity of hydrocarbon generation sag, and organic matter transformation rate) and a proved reserve concentration of each sag, an explicit linear function is constructed and the order of dependence between the proved reserve concentration and the five parameters is obtained by the data mining tool of MRA, and the results coincide with practicality. This function and the order are called *mined knowledge*. This knowledge can be applied to the other sags in eastern China. Therefore, this method can be spread to the proved reserve prediction of other sags.

2.1. PROBABILITY

2.1.1. Applying Ranges and Conditions

2.1.1.1. Applying Ranges

The following are the two commonly used projects:

1. *Prediction of undiscovered resources by probability density function.* Assuming that there is a resource of various sizes discovered in an area, on which we base construction of the probability density function of the resource. On this function curve, the probability of each interval on the x-coordinate axis (i.e., Probability density function value × Interval length) can be obtained. Subtracting the probability of a discovered resource in the corresponding interval from the probability on the curve, this difference, multiplied by the total number of discoveries, is the number of undiscovered resources.

2. *Calculation of an unknown number by the Monte Carlo method.* Assuming a formula about an unknown number contains several parameters and each parameter has observed values in different numbers, how do we calculate this unknown number? The most simple and coarse way is to calculate the mean of observed values for each parameter at first, to substitute these means into the formula, and then to obtain the calculated mean of this unknown number. However, since the subsurface parameter is randomly distributed, the parameter should be expressed in probability form, distributed randomly so as to

objectively reflect the parameter. The solution is to calculate the probability distribution function for each parameter and then to calculate the probability distribution function for the unknown number by the Monte Carlo method in conjunction with the formula so that the values of the unknown number at various probabilities are obtained. In general, for the values of unknown numbers corresponding to 0.95, 0.50, and 0.05 three probabilities are output, and the value corresponding to 0.50 is the solution.

2.1.1.2. Applying Conditions

For each parameter, the number of observations is large enough, the observed values are as precise as possible, and the distribution of observation locations is as uniform as possible.

2.1.2. Basic Principles

2.1.2.1. Event

Observing geological phenomena, the following three cases are found.

1. An event that will inevitably occur under definite conditions is called a *certain event*. For example, when the source rock maturity is equal to or greater than the oil generation threshold, kerogens will inevitably generate oil.
2. An event that is impossible to occur under definite conditions is called an *impossible event*. For example, when the source rock maturity is less than the oil generation threshold, it is impossible for kerogens to generate oil.
3. An event that might or might not occur under some conditions is called a *random event*. For example, when some source rock maturities are equal to or greater than the oil generation threshold and some source rock maturities are less than the oil generation threshold, kerogens may or may not generate oil if a source rock is randomly spot-checked. Supposing there are several wells in an area and there are several source rock formations in each well. A steady probability of oil generation appears after a large number of spot-checks, i.e., the probability of oil generation in this area (Number of oil-generated source rocks / Number of source rocks) approaches a stable value. That explains the possibility of the oil generation random event, which is an event-own attribute.

Probability theory is a science to quantitatively study the statistical rules of random phenomena.

2.1.2.2. Probability

1. *Single-event probability.* The probability for the occurrence of event $\mathbf{A}$ is defined by

$$P(\mathbf{A}) = \frac{\text{The number of event } \mathbf{A} \text{ occurrencs}}{\text{The toal number of observations to chek if event } \mathbf{A} \text{ occurs}} \tag{2.1}$$

2. *Multi-event probability.* If we let multi-events $\mathbf{A}_1$, $\mathbf{A}_2$, ..., and $\mathbf{A}_n$ be independent of each other, the probability when they simultaneously occur is defined by

$$P(\mathbf{A}_1\mathbf{A}_2\cdots\mathbf{A}_n) = P(\mathbf{A}_1)\cdot P(\mathbf{A}_2)\cdots P(\mathbf{A}_n) = \prod_{i=1}^{n} P(\mathbf{A}_i) \tag{2.2}$$

2.1.2.3. *Density Function and Distribution Function of Probability*

1. Probability density function $p(x)$:

$$\int_{-\infty}^{+\infty} p(x)\mathrm{d}x = 1 \tag{2.3}$$

For example, a uniform distributed probability density function is

$$p(x) = \begin{cases} 0 & \text{when } x < a \\ \frac{1}{b-a} & \text{when } a \le x \le b \\ 0 & \text{when } x > b \end{cases}$$

It is shown in Figure 2.1 that the probability density function expresses the value of each random variable, and the area enclosed by the probability density function and the random variable axis (x-coordinate axis) is 1 which coincides with the definition of Equation (2.3).

2. Probability distribution function $F(x)$

$$F(x) = \int_{-\infty}^{x} p(x)dx \tag{2.4}$$

$$F'(x) = p(x) \tag{2.5}$$

For example, a uniform distributed probability distribution function is

$$F(x) = \begin{cases} 0 & \text{when } x \le a \\ \frac{x-a}{b-a} & \text{when } a \le x \le b \\ 1 & \text{when } x \ge b \end{cases}$$

It is easy to see that the relationship between the uniform distributed probability density function and the probability distribution function meets the conditions of Equations (2.4) and (2.5).

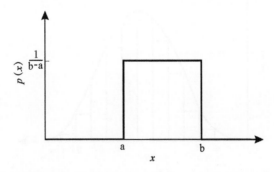

FIGURE 2.1 Uniform distributed probability density function.

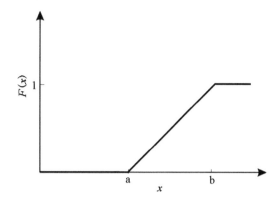

FIGURE 2.2   Uniform distributed probability distribution function.

Figure 2.2 shows that the probability distribution function expresses the cumulative value of the probability along the random variable axis (x-coordinate axis), which is the definition of Equation (2.4).

2.1.3. Prediction of Undiscovered Resources by Probability Density Function

2.1.3.1. Calculation Method

Supposing there are n reservoirs in different sizes at a more explored area. Their reserves are $q_1, q_2, ..., $ and q_n, respectively, based on which we plot a chart of reserve discovery density. On this chart, there are n_i ($i = 1, 2, ..., m$) reservoirs fallen in the intervals (q_i^*, q_{i+1}^*) on the x-coordinate axis, and y-coordinates at each interval are $\overline{p_i} = n_i/[n(q_{i+1}^* - q_i^*)]$, where $q_i^* < q_{i+1}^*$ and $n = \sum_{i=1}^{m} n_i$ (Figure 2.3). In Figure 2.3, there is a jagged discovery density curve, which can be called a *roughcast* of a probability density function defined by Equation (2.3). Based on the shape of this roughcast, an appropriate probability density function, e.g., a logarithmic normal distribution, can be selected.

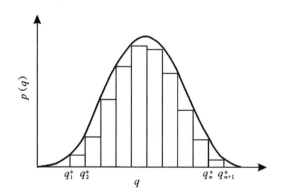

FIGURE 2.3   Discovery density and probability density function of a reservoir at more explored area.

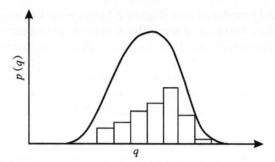

FIGURE 2.4   Discovery density and probability density functions of a reservoir in a less or moderately explored area.

For the probability density function used in reservoir prediction, here a logarithmic normal distribution is introduced as follows:

$$p(q) = \frac{1}{\sigma q \sqrt{2\pi}} \exp\left\{ \frac{-[\ln(q) - \mu]^2}{2\sigma^2} \right\} \tag{2.6}$$

where

$$\mu = \frac{1}{n} \sum_{i=1}^{n} \ln(q_i) \tag{2.7}$$

$$\sigma^2 = \frac{1}{n-1} \sum_{i=1}^{n} |\ln(q_i) - \mu| \tag{2.8}$$

When the probability density function expressed with Equation (2.6) is determined, it is drawn on Figure 2.3.

Some reservoirs have been discovered at a less or moderately explored area. Then how can we predict the size and number of those reservoirs that have not been discovered? Using the preceding logarithmic normal distribution with μ and σ determined by analogy method, the reserve probability density function of the area is drawn on a chart, and the discovery density of discovered reservoirs is also drawn on this chart. Affirmatively, the columns of the discovery density cannot be full of the proportions under the curve of the probability density function. These unfilled proportions are undiscovered reservoirs, and their sizes and the amount of the reservoirs can be determined from the chart (Figure 2.4).

The aforementioned calculation method is described by sample of reservoir but is applicable to any subsurface resources. To predict undiscovered resources by probability density function, the chart of resource discovery density is drawn first, and then the probability density function is constructed by the two following ways, according to the different exploration degrees:

1. For a more explored area (Figure 2.3), since the known data are enough, an appropriate function (e.g., a logarithmic normal distribution) is chosen as the probability density function, and this function is determined by the known data.

2. For a less or moderately explored area (Figure 2.4), since the known data are not enough, a probability density function (e.g., a logarithmic normal distribution) of another area is borrowed by the analogy method. Then a chart of discovery density and probability density function (Figure 2.3 or 2.4) is drawn for resources in this area. Affirmatively, the columns of the discovery density cannot be full of the proportions under the curve of the probability density function. These unfilled proportions are undiscovered reservoirs, and their sizes and the amount of the reservoirs can be determined from the chart (Figure 2.3 or 2.4).

2.1.3.2. Simple Case Study: Discovery Density and Probability Density Functions of a Reservoir at a More Explored Area

Consider a basin in a more explored area where 40 reservoirs are discovered. The reserves in place of these 40 reservoirs are, respectively, estimated to be q_k ($k = 1, 2, ...,$ 40): 1.5, 2.3, 2.7, 3.2, 3.4, 3.6, 3.8, 4.3, 4.4, 4.5, 4.6, 4.7, 4.8, 5.2, 5.3, 5.4, 5.5, 5.6, 5.7, 5.8, 5.9, 6.2, 6.3, 6.4, 6.5, 6.6, 6.7, 6.8, 7.3, 7.4, 7.5, 7.6, 7.7, 7.8, 8.4, 8.5, 8.6, 9.5, 9.6, and 10.5 (10^6t). The values at terminals of each statistical interval are respectively q_i^* ($i = 1, 2, ..., 11$): 1, 2, 3, 4, 5, 6, 7, 8, 9, 10, and 11. What is the reserve probability density function in this basin?

1. Calculation flowchart (Figure 2.5).
2. Calculation results and analyses.

Using the aforementioned calculation method, the discovery density values at each statistical interval are calculated to be 0.025, 0.05, 0.1, 0.15, 0.2, 0.175, 0.15, 0.075, 0.05, and 0.025, respectively. The sum of these 10 numbers is 1. That is a jagged discovery density curve, illustrated in Figure 2.3. Using the logarithmic normal distribution expressed in Equation (2.6), it is calculated by Equations (2.7) and (2.8) that $\mu = 1.713$ and $\sigma = 0.5593$, respectively. Substituting μ and σ values into Equation (2.6), the probability density function values at midpoints of each statistical interval are calculated: 0.03087, 0.1033, 0.1452, 0.1478, 0.1297, 0.1054, 0.08225, 0.06273, 0.04728, and 0.03544. The sum of these 10 numbers is not 1 but 0.88997, which results from ignoring the outside of the total statistical range [1, 11]; otherwise, the sum is 1. However, that does not significantly affect the studied problem. Shaped like Figure 2.3, the reservoir probability density function of this basin is

$$p(q) = \frac{1}{1.4q} \exp\left\{ \frac{-[\ln(q) - 1.713]^2}{0.6256} \right\}$$

The mean-square error between 10 discovery density values and 10 probability density values is 0.04424.

Summarily, using the reserves in place of 40 reservoirs in a basin, a reservoir probability density function of the basin is constructed by the data mining tool of the probability density function, and the results coincide with practicality. This function is called *mined knowledge*. This knowledge provides the basis for the further exploration decision of the basin. Therefore, this method can be spread to the exploration decision of other basins.

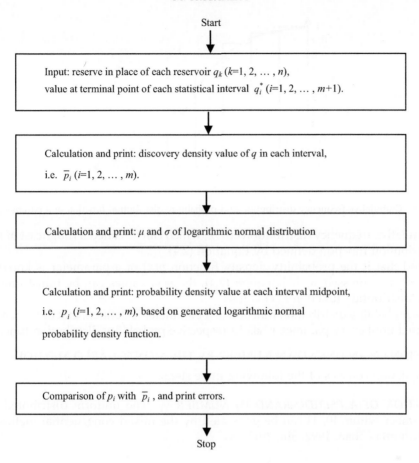

FIGURE 2.5   Calculation flowchart of probability density.

2.1.4. Calculating an Unknown Number by Monte Carlo Method

2.1.4.1. Calculation Method

Assume that a formula about an unknown number contains several parameters, and each parameter has observed values in different numbers. The calculation procedure for the unknown number is successively described here.

CALCULATION OF PROBABILITY DISTRIBUTION FUNCTION FOR EACH PARAMETER

Supposing a parameter has n observed values. Based on these values, we make a chart of cumulative frequency distribution for this parameter. On this chart, there are n_i ($i = 1$, 2, ..., m) observed values fallen in the intervals $[x_i^*, x_{i+1}^*]$ on the x-coordinate axis, and y-coordinates at each interval are n_i/n where $n = \sum_{i=1}^{m} n_i$. This is a chart of noncumulative frequency distribution. If y-coordinates are changed from n_i/n to $\sum_{k=i}^{m} n_k/n$, which is 1 when $i = 1$, the cumulative frequency distribution is obtained (Figure 2.6). In Figure 2.6, there is a

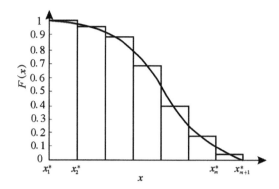

FIGURE 2.6   Cumulative frequency distribution and probability distribution function of a parameter.

jagged cumulative frequency distribution curve, which can be called a roughcast of the probability distribution function defined by Equation (2.4).

It is noted that if the probability density function $p(x)$ of a parameter is known, using $F(x) = \int p(x)dx$, expressed with Equation (2.4), the $p(x)$ values can be transformed to the probability distribution function $F(x)$ values.

Once the probability distribution function of a parameter is defined, that of the next parameter is defined until every parameter has its respective probability distribution function.

CALCULATION OF AN UNKNOWN NUMBER BY THE MONTE CARLO METHOD

This calculation consists of the following three steps.

GENERATION OF A PSEUDO-RANDOM NUMBER γ_k The uniform distributed pseudo-random number within [0, 1] can be generated by the mixed congruential method, and a recursion formula (Zhao, 1992; Shi, 2011) is

$$\left.\begin{array}{l} \gamma_k = X_k/M \\ X_{k+1} = \alpha X_k + \beta \quad (\text{mod } M) \\ (k = 1, 2, \cdots) \end{array}\right\} \tag{2.9}$$

where X_k is the kth pseudo-random number; X_{k+1} is the $(k+1)^{th}$ pseudo-random number; α is a coefficient multiplier that can be taken as $5^5 = 3125$; β is an increment that can be taken as 3, 7, 11, or 17; M is a module that can be taken as $2^{19} = 524288$; mod M is a complementation in term of M, e.g., if $\alpha X_k + \beta = 1572874$, which is divided by M (524288) so that its quotient is 3 and its remainder is 10, thus $X_{k+1} = 10$; and γ_k is the kth pseudo-random number within [0, 1], which is the solution.

X_1 can be initialized as 23, 11, 19, or 37. These four values correspond to the previous four values of β one by one, forming four pairs of values. Any one of the four may be selected in applications. This chapter employs $\beta = 3$ and $X_1 = 23$, i.e., in Equation (2.9) the four numbers are:

$$M = 2^{19} = 524288$$
$$\alpha = 3125$$
$$\beta = 3$$
$$X_1 = 23$$

However, in the case studies of artificial neural networks (ANN) in Chapter 3, it is found that when the number of learning samples is more than 50, each learning sample cannot be randomly chosen at a same iteration for learning, resulting in the failure of the ANN to run. This problem is resulted from the previous four numbers used in the generation of a pseudo-random number. Hence, when the number of learning samples is more than 50, the other four numbers used in the generation of pseudo-random number (Shi, 2005) are:

$$M = 2^{35} = 3.43597 \times 10^{10}$$
$$\alpha = 7$$
$$\beta = 1$$
$$X_1 = 1$$

CALCULATION OF THE VALUE OF PARAMETER x_k CORRESPONDING TO PSEUDO-RANDOM NUMBER γ_k Based on the calculated γ_k of Equation (2.9), the corresponding value of the parameter x_k can be obtained on the chart of probability distribution function for the parameter (Figure 2.6). Since the curve of $F(x)$ is stored in the computer in discrete point form, γ_k falls in two discrete points on the y-coordinate axis, and x_k to be solved also falls in two discrete points on the x-coordinate axis. Therefore, x_k can be calculated by linear interpolation. Once the value of a parameter is calculated, the value of the next parameter is calculated until the value of every parameter has been calculated.

CALCULATION OF AN UNKNOWN NUMBER G_k CORRESPONDING TO PSEUDO-RANDOM NUMBER γ_k Substituting the value of each parameter into the formula about the unknown number, the value of the unknown number corresponding to pseudo-random number γ_k, G_k, is calculated.

When $k = 1$, γ_1 and G_1 are obtained by the aforementioned three-step calculations; when $k = 2$, γ_2 and G_2 are obtained by repeating these three-step calculations. The generated number of pseudo-random numbers γ_k is the number of samplings. Theoretically, the larger the number of samplings , the better the results. Actually, the number of samplings is not necessarily infinite as long as the form of the probability distribution function curve consisting of γ_k and G_k ($k = 1, 2, ...$) reaches stability. In general, the number of samplings can be taken as 500—5,000 when the number of statistical intervals is 100. Denoting γ_k and G_k on the chart, finally the probability distribution function of the unknown number is obtained (Figure 2.7).

In Figure 2.7, $G|_{F=0.95}$ and $G|_{F=0.05}$ are the values of the unknown number at the maximum cumulative probability and minimum cumulative probability, respectively. $[G|_{F=0.95}, G|_{F=0.05}]$ is the possible range of the unknown number. $G|_{F=0.50}$ between $G|_{F=0.95}$ and $G|_{F=0.05}$ is the solution.

2.1.4.2. Simple Case Study: Pore Volume of Trap

In the calculation of trap resources, the pore volume of a trap must be calculated:

$$V_{\varphi} = 10^6 \varphi H S \tag{2.10}$$

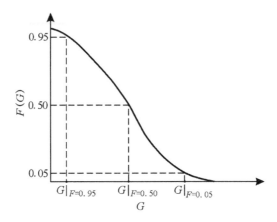

FIGURE 2.7   Probability distribution function of an unknown number.

where V_φ is the pore volume of the trap, m³; φ is the porosity of the trap, a fraction, which will be calculated by the Monte Carlo method; H is the relief of the trap, m, which will be calculated by the Monte Carlo method; and S is the area of the trap, km².

For a trap, its area $S = 1$ km². There are 25 measured porosity φ: 0.21, 0.23, 0.255, 0.26, 0.27, 0.28, 0.29, 0.305, 0.31, 0.315, 0.32, 0.325, 0.33, 0.335, 0.34, 0.345, 0.348, 0.355, 0.36, 0.37, 0.38, 0.39, 0.42, 0.43, and 0.44. And there are 40 measured relief H: 15, 26, 28, 31, 33, 35, 37, 42, 43, 44, 45, 46, 47, 50.5, 51, 52, 53, 54, 55, 56, 57, 58, 59, 62, 63, 64, 65, 66, 67, 68, 73, 74, 75, 76, 77, 84, 85, 86, 95, and 96. What is the pore volume V_φ of the trap?

1. Calculation flowchart (Figure 2.8).
2. Calculation results and analyses.

In calculations, for the known parameters (porosity and relief), the cumulative frequency distribution is not calculated, but the probability distribution function is calculated to calculate the pore volume by the Monte Carlo method. In practical applications, if the probability distribution function for a known parameter cannot be defined, the cumulative frequency distribution can be directly defined through samples for the calculations of the Monte Carlo method.

For porosity, a logarithmic normal distribution expressed in Equation (2.6) is employed. The calculated $\mu = -1.129$ by Equation (2.7), and the calculated $\sigma = 0.3866$ by Equation (2.8). Substituting μ and σ into Equation (2.6) and then using $F(x) = \int p(x)dx$, expressed with Equation (2.4), the value of probability density function is transformed to the value of probability distribution function. Thus, as with the values of probability distribution function, the values of cumulative frequency distribution at midpoints of each statistical interval are calculated: 1, 1, …, 0.04148, and 0.02032, which is a curve as illustrated in Figure 2.6.

For relief, a logarithmic normal distribution expressed with Equation (2.6) is employed. The calculated $\mu = 3.985$ by Equation (2.7), and the calculated $\sigma = 0.5455$ by Equation (2.8). Substituting μ and σ into Equation (2.6) and then using $F(x) = \int p(x)dx$, expressed with Equation (2.4), the value of the probability density function is transformed to the value of the probability distribution function. Thus, as with the values of the probability

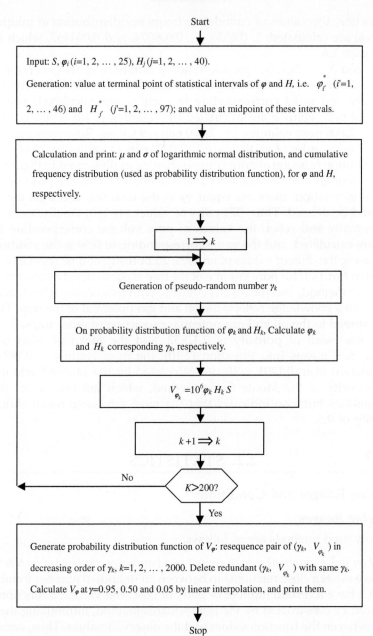

FIGURE 2.8   Calculation flowchart of the Monte Carlo method.

distribution function, the values of cumulative frequency distribution at midpoints of each statistical interval are calculated: 1, 0.8538, ..., 0.009074, and 0.004467, which is a curve as illustrated in Figure 2.6.

Taking $\beta = 3$ and $X_1 = 23$, 2,000 pseudo-random numbers γ_k are generated by Equation (2.9). From the calculated two probability distribution functions for porosity and relief, the values of porosity and relief corresponding to each γ_k are calculated, respectively, and then the values of pore volume are calculated by Equation (2.10). Thus, corresponding to 2,000 γ_k, there are 2,000 pore volumes, i.e., 2,000 pairs of values. Resequencing the 2,000 pairs of values in decreasing order of γ_k, the orderly 2,000 pairs of values are obtained, which is a probability distribution function of pore volume shaping as illustrated in Figure 2.7. It should be indicated that though there are no two equal neighboring γ_k in the procedure of pseudo-random number generation, there are equal γ_k in the total set, and thus these repeatedly equal ones should be deleted. Thus, 374 pairs of values are left. Finally, from the 374 pairs of values for porosity and relief, the values of pore volume corresponding to 0.95, 0.50, and 0.05 of γ_k are calculated, and the value corresponding to 0.50 is the solution.

Summarily, using the data of values of an area, 25 porosity and 40 relief in a trap, a probability distribution function for pore volume of the trap is constructed by the data mining tool of the Monte Carlo method, and the results coincide with practicality. This function is called *mined knowledge*. This knowledge reflects the oil and gas potential of the trap. Therefore, this method can be spread to the oil and gas potential prediction of other traps.

In addition, the mean of porosity is 0.32852, and the mean of relief is 57.3375 m. Substituting the two means into Equation (2.10), the pore volume is 0.188365×10^8 m^3, which is far different from 0.1201×10^8 m^3 calculated by the Monte Carlo method. That shows the superiority of the Monte Carlo method, which not only gives the results at different probabilities but also indicates that the most advisable result should be taken at the probability of 0.5.

2.2. STATISTICS

2.2.1. Applying Ranges and Conditions

2.2.1.1. *Applying Ranges*

The commonly used methods are as follows:

1. *A function of an unknown number with respect to a parameter constructed by least-squares method.* If there exists a close relationship between an unknown number y and a parameter x, and y and x have a certain number of pairs of observed values, a function of y with respect to x can be constructed by the least-squares method, minimizing the deviation square sum between the function values and the observed values. Thus, once a value of x is known, the value of y can be calculated by this function.

2. *A function of an unknown number with respect to multiple parameters constructed by MRA.* As Equation (1.1) described, assume that there are n learning samples, each associated with $m + 1$ numbers $(x_1, x_2, ..., x_m, y_i^*)$ and a set of observed values $(x_{1i}, x_{2i}, ..., x_{mi}, y_i^*)$, with $i = 1$, 2, ..., n for these parameters. In principle, $n > m$, but in actual practice $n \gg m$. As Equation (1.3) described, assume that there are k prediction samples, each associated with

m parameters $(x_1, x_2, ..., x_m)$ and a set of observed values $(x_{1i}, x_{2i}, ..., x_{mi})$, with $i = n + 1$, $n + 2, ..., n + k$ for these parameters.

If there exists a certain relationship between an unknown number y and a set of parameters x_i ($i = 1, 2, ..., m; m \geq 2$), and y and x_i have a certain number of sets of observed values, a function of y with respect to x_i can be constructed by MRA, minimizing the deviation square sum between the function values and the observed values. Thus, once a set of values for x_i is known, the value of y can be calculated by this function, called a *regression equation*. Actually, since the importance of these x_i in a regression equation is different, a practical regression equation is not always expressed with all x_i; usually the unimportant x_i are eliminated, and then only important x_i are left in the regression equation. Thus, once a set of the values for important x_i are known, the value of y can be calculated using this regression equation.

2.2.1.2. *Applying Conditions*

For each parameter, the number of observations is large enough, the observed values are as precise as possible, and the distribution of observation locations is as uniform as possible.

2.2.2. Basic Principles

2.2.2.1. *Least-Squares Method*

Assuming that there are n pairs of observed values (x_i, y_i), where $i = 1, 2, \cdots, n$ for a parameter x and an unknown number y, an expression $y = P(x)$ is constructed, minimizing

$$\sum_{i=1}^{n} [P(x_i) - y_i]^2 \tag{2.11}$$

In general, the expression to be solved is usually the $(m - 1)^{\text{th}}$-order polynomial:

$$P(x) = a_1 + a_2 x + a_3 x^2 + \cdots + a_m x^{m-1} \tag{2.12}$$

where $n > (m - 1)$. In actual practice, $n >> (m - 1)$.

So:

$$y = a_1 + a_2 x + a_3 x^2 + \cdots + a_m x^{m-1} \tag{2.13}$$

2.2.2.2. *Multiple Regression Analysis*

Multiple regression analysis (MRA) has been widely applied in the natural and social sciences since the 1970s (Chatterjee et al., 2000). Successive regression is one of the most popular methods in MRA and is still a very useful tool in many fields, including geosciences (e.g., Shi, 1999, 2009; Lee and Yang, 2002; Shi et al., 2004; Singh et al., 2008; Shi and Yang, 2010; Shi, 2011).

The learning samples and prediction samples for MRA are expressed with Equations (1.1) and (1.3), respectively. The learning process and prediction process of a calculation flowchart are illustrated in Figures 1.2 and 1.3, respectively. In Figure 1.3, if the prediction samples are replaced by the learning samples, Figure 1.3 becomes such a calculation flowchart for learning validation.

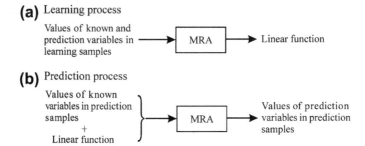

FIGURE 2.9   Learning process (a) and prediction process (b) of MRA.

The expression about y constructed by MRA is a linear combination with respect to m parameters $(x_1, x_2, \ldots, x_m)$, plus a constant term, which is called a *regression equation*:

$$y = b_0 + b_1 x_1 + b_2 x_2 + \cdots + b_m x_m \tag{2.14}$$

where the constants $b_0, b_1, b_2, \ldots,$ and b_m are calculated by the successive regression. Equation (2.14) is called the *fitting formula* and is obtained from the learning process.

The basic idea of MRA is illustrated in Figure 2.9.

In addition, MRA can construct a function of y with respect to m parameters $(x_1, x_2, \ldots, x_m)$. MRA can also indicate the order of dependence between y and $x_1, x_2, \ldots, x_m$, respectively. Based on this order, MRA can serve as a pioneering dimension-reduction tool in data mining, introduced in Chapter 4.

2.2.3. A Function of an Unknown Number with Respect to a Parameter Constructed by the Least-Squares Method

2.2.3.1. Calculation Method

The commonly used approaches are as follows:

1. *Linear fitting.* If $m = 2$ in Equation (2.13), the function to be solved is

$$y = a_1 + a_2 x \tag{2.15}$$

Equation (2.15) is rewritten to a usual expression

$$y = ax + b \tag{2.16}$$

From the principle of the least-squares method,

$$
\left.
\begin{aligned}
a &= \frac{n \sum_{i=1}^{n} x_i y_i - \sum_{i=1}^{n} x_i \sum_{i=1}^{n} y_i}{n \sum_{i=1}^{n} x_i^2 - \left(\sum_{i=1}^{n} x_i \right)^2} \\[2mm]
b &= \frac{\sum_{i=1}^{n} x_i^2 \sum_{i=1}^{n} y_i - \sum_{i=1}^{n} x_i \sum_{i=1}^{n} x_i y_i}{n \sum_{i=1}^{n} x_i^2 - \left(\sum_{i=1}^{n} x_i \right)^2}
\end{aligned}
\right\}
\tag{2.17}
$$

Equation (2.16) is a linear function of y with respect to x.

2. *General polynomial fitting.* The function to be solved is Equation (2.13). From the principle of the least-squares method,

$$a_1 \sum_{k=1}^{n} x_k^j + a_2 \sum_{k=1}^{n} x_k^{j+1} + \cdots + a_m \sum_{k=1}^{n} x_k^{j+m-1} = \sum_{k=1}^{n} y_k x_k^j \qquad (j = 1, 2, \cdots, m) \qquad (2.18)$$

Equation (2.18) is a linear algebraic equation of unknown numbers a_1, a_2, ..., and a_m, and its coefficient matrix is symmetrical and positive-definite. Solutions can be chosen to calculate the m unknown numbers. However, in practical applications this coefficient matrix is sometimes ill conditioned, so a solution of orthogonal polynomial fitting is usually employed.

3. *Orthogonal polynomial fitting.* The function to be solved is still Equation (2.13). The solution employed is not through linear algebraic equations expressed with Equation (2.18) to get a_1, a_2, ..., and a_m, but the solution employed is through constructing m orthogonal polynomials to directly get a_1, a_2, ..., and a_m (Xu, 1997, Shi, 1999).

Let m orthogonal polynomials be

$$\left. \begin{aligned} Q_1(x) &= 1 \\ Q_2(x) &= (x - a_1) \\ Q_{j+1}(x) &= (x - a_j)Q_j(x) - \beta_j Q_{j-1}(x) \qquad (j = 2, 3, \cdots, m-1) \end{aligned} \right\} \qquad (2.19)$$

where

$$a_j = \sum_{i=1}^{n} x_i Q_j^2(x_i)/d_i \qquad (j = 1, 2, \cdots, m-1) \qquad (2.20)$$

$$\beta_j = d_j/d_{j-1} \qquad (j = 2, 3, \cdots, m-1) \qquad (2.21)$$

and

$$d_j = \sum_{i=1}^{n} Q_j^2(x_i) \qquad (j = 1, 2, \cdots, m-1) \qquad (2.22)$$

It is easy to see that $Q_j(x)$ $(j = 1, 2, \cdots, m)$ is the $(j-1)^{\text{th}}$-order polynomial. The orthogonality of each $Q_j(x)$ can be proven by mathematical induction.

Thus, y in Equation (2.13) is written in

$$y = q_1 Q_1(x) + q_2 Q_2(x) + \cdots + q_m Q_m(x) \qquad (2.23)$$

From the principle of the least-squares method,

$$q_j = \frac{\sum_{k=1}^{n} y_k Q_j(x_k)}{\sum_{k=1}^{n} Q_j^2(x_k)} \qquad (j = 1, 2, \cdots, m) \qquad (2.24)$$

Substituting Equations (2.19) and (2.24) into Equation (2.23), Equation (2.23) then becomes Equation (2.13), i.e., how to calculate $a_j (j = 1, 2, \cdots, m)$.

The following series of steps gives the product to calculate $a_j (j = 1, 2, \cdots, m)$:

Step 1. Calculate a_1.

$$b_1 = 1;\ Q_1(x_i) = b_1,\ i = 1, 2, \cdots, n;\ d_1 = n;\ q_1 = \sum_{i=1}^{n} y_i/d_1;\ a_1 = \sum_{i=1}^{n} x_i/d_1;$$

$$\beta_1 = 0(\text{useless}).$$

The result is: $q_1 b_1 \Rightarrow a_1$.

Step 2. Calculate a_2.

$$b_2 = 0(\text{useless, will be assigned to a useful value at step 3});\ t_1 = -a_1;\ t_2 = 1;$$

$$Q_2(x_i) = t_1 + t_2 x_i,\ i = 1, 2, \cdots, n;\ d_2 = \sum_{i=1}^{n} Q_2^2(x_i);\ q_2 = \sum_{i=1}^{n} y_i Q_2(x_i)/d_2;$$

$$a_2 = \sum_{i=1}^{n} x_i Q_2^2(x_i)/d_2;\ \beta_2 = d_2/d_1.$$

The results are: $a_1 + q_2 t_1 \Rightarrow a_1$, and $q_2 t_2 \Rightarrow a_2$.

Step 3. Circularly calculate $a_j (j = 3, 4, \cdots, m)$ through recursion relation:

$$S_1 = -a_{j-1} t_1 - \beta_{j-1} b_1$$
$$S_k = -a_{j-1} t_k + t_{k-1} - \beta_{j-1} b_k \qquad (k = 2, 3, \cdots, j - 2)$$
$$S_{j-1} = -a_{j-1} t_{j-1} + t_{j-2}$$
$$S_j = t_{j-1}$$

$$Q_j(x_i) = \sum_{k=1}^{j} S_k x_i^{k-1} \qquad (i = 1, 2, \cdots, n)$$

$$d_j = \sum_{i=1}^{n} Q_j^2(x_i)$$

$$q_j = \sum_{i=1}^{n} y_i Q_j(x_i)/d_j$$

$$a_j = \sum_{i=1}^{n} x_i Q_j^2(x_i)/d_j$$

$$\beta_j = d_j/d_{j-1}$$

The results are: $q_j S_j \Rightarrow a_j$, $a_k + q_j S_k \Rightarrow a_k$ $(k = 1, 2, \cdots, j - 1)$, and t_k and S_k for the next circle in this step are:

$$t_k \Rightarrow b_k \qquad (k = 1, 2, \cdots, j - 1)$$
$$S_k \Rightarrow t_k \qquad (k = 1, 2, \cdots, j)$$

After $(m-2)$ circles in Step 3, all unknown numbers $a_j (j = 1, 2, \cdots, m)$ in Equation (2.13) are obtained. To avoid the floating-point overflow in practical applications, $x_i - \bar{x}$ substitute for x_i, where $\bar{x} = \sum_{i=1}^{n} x_i/n$. Thus, Equation (2.13) becomes

$$y = a_1 + a_2(x - \bar{x}) + a_3(x - \bar{x})^2 + \cdots + a_m(x - \bar{x})^{m-1} \tag{2.25}$$

Equation (2.25) is a polynomial of y with respect to x. It is noted that $n > (m - 1)$. In practical applications, since the fitting of high-order polynomials may cause numerical instability, $m < 6$ is proposed. Moreover, to ensure the representativeness of fitting, $n >> (m - 1)$.

2.2.3.2. *Simple Case Study 1: Linear Function of Porosity with Respect to Acoustictime; Exponent Function of Porosity with Respect to Burial Depth*

The commonly used function of porosity with respect to burial depth is in two forms, as follows:

1. Linear function of porosity with respect to acoustictime:

$$\phi = a\Delta t + b \tag{2.26}$$

where ϕ is porosity, fraction; Δt is acoustictime, μs/m; and a and b are slope and intercept, respectively.

In a well, there are 33 measured values of Δt that are used as x_i in Equation (2.17): 430, 435, 418, 450, 445, 410, 438, 450, 462, 440, 410, 440, 430, 442, 435, 442, 430, 435, 420, 405, 415, 438, 448, 435, 423, 420, 422, 407, 380, 397, 395, 382, and 180. There are a corresponding 33 measured values of Δt that are used as y_i in Equation (2.17): 0.372, 0.392, 0.375, 0.354, 0.358, 0.338, 0.345, 0.358, 0.367, 0.347, 0.322, 0.369, 0.366, 0.372, 0.368, 0.363, 0.334, 0.355, 0.337, 0.335, 0.326, 0.334, 0.317, 0.341, 0.325, 0.354, 0.349, 0.321, 0.337, 0.330, 0.331, 0.321, and 0.0. Using these data and by Equation (2.17), a and b in Equation (2.26) are obtained.

2. Exponent function of porosity with respect to burial depth (Shi, 2005):

$$\phi = \phi_0 \exp(-cz) \tag{2.27}$$

where ϕ is porosity, fraction; z is burial depth, m; ϕ_0 to be solved is porosity at the Earth's surface, fraction; and c to be solved is the compaction coefficient, 1/m.

Natural logarithm is performed to two handsides of Equation (2.27), resulting in

$$\ln(\phi) = \ln(\phi_0) - cz = -cz + \ln(\phi_0) \tag{2.28}$$

Regarding $\ln(\phi)$ and z in Equation (2.28) as y and x in Equation (2.16), respectively, $-c$ and $\ln(\phi_0)$ in Equation (2.28) are a and b, respectively, i.e.,

$$\begin{cases} \phi_0 = \exp(b) \\ c = -a \end{cases} \tag{2.29}$$

In an area under the condition of normal compaction, there are 23 measured values of z that are used as x_i in Equation (2.17): 0, 50, 100, 200, 300, 400, 500, 600, 700, 800, 1000, 1500, 2000, 2500, 3000, 3500, 4000, 4500, 5000, 5500, 6000, 6500, and 7000. Furthermore, there are a corresponding 23 measured values of ϕ : 0.55, 0.54, 0.53, 0.52, 0.51, 0.50, 0.49, 0.48, 0.47, 0.46, 0.45, 0.44, 0.43, 0.42, 0.41, 0.39, 0.37, 0.35, 0.33, 0.31, 0.28, 0.24, and 0.19. Natural logarithm is performed to these values of ϕ, and the results are used as y_i in Equation (2.17). Using these data and by Equation (2.17), a and b in Equation (2.26) are obtained; substituting a and b into Equation (2.29), ϕ_0 and c to be solved in Equation (2.27) are obtained.

3. Calculation flowchart (Figure 2.10).

4. Calculation results and analyses. The results show that porosity can be expressed with either acoustictime or burial depth, i.e.,

$$\phi = 0.001255\Delta t - 0.1882 \quad \text{(mean square error is 0.0226)}$$

$$\phi = 0.5344 \exp(-0.113 \times 10^{-3}z) \quad \text{(mean square error is 0.07214)}$$

When we attempt to express an unknown number with a parameter, it is required to choose an appropriate function expression such as the preceding linear function or exponent function. Through fitting tests of multiple expressions, the most appropriate function with a minimum mean square error is chosen as a fitting function to be solved. In fitting tests, dealing with nonlinear functions may refer to the preceding exponent function.

Summarily, using the data of 33 porosity values and corresponding 33 acoustictime values in a well, a formula of porosity with respect to acoustictime is constructed by the data mining tool of constructing a linear function by the least-squares method, and the results coincide with practicality. Similarly, using the data of 23 porosity values and corresponding 23 burial depth values in an area, a formula of porosity with respect to burial depth is constructed by the data mining tool of constructing exponent function by the least-squares method, and the results coincide with practicality. Under the same structural and sedimentary conditions, these two kinds of mined knowledge can be borrowed by the neighboring wells or areas. Therefore, the two methods can be spread to the porosity expression of other wells or areas.

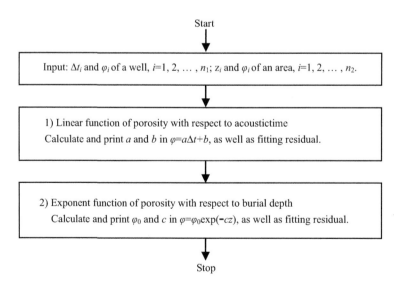

FIGURE 2.10   Calculation flowchart of the least-squares method (linear function and exponent function).

2.2.3.3. Simple Case Study 2: Polynomial of Porosity with Respect to Acoustictime

$$\phi = a_1 + a_2(\Delta t - \overline{\Delta t}) + a_3(\Delta t - \overline{\Delta t})^2 + \cdots + a_m(\Delta t - \overline{\Delta t})^{m-1} \tag{2.30}$$

where ϕ is porosity, fraction; Δt is acoustictime, $\mu s/m$; $\overline{\Delta t}$ is the mean of acoustictime, $\sum_{i=1}^{n} \Delta t_i / n$, $\mu s/m$; and a_j to be solved is a coefficient of polynomial $(j = 1, 2, \cdots, m)$.

The measured values of φ and Δt used are the same as the aforementioned Simple Case Study 1. Using these data and by the above orthogonal polynomial fitting method, a_j $(j = 1, 2, \cdots, m)$ in Equation (2.30) are obtained.

1. Calculation flowchart (Figure 2.11).
2. Calculation results. When m is 2, 3, 4, and 5 in Equation (2.15) or (2.30), respectively, these four calculation results are, respectively:

$$\phi = 0.3368 + 0.1254 \times 10^{-2}(\Delta t - 418.5) \quad \text{(mean square error is 0.02260)};$$

$$\phi = 0.3451 + 0.5200 \times 10^{-3}(\Delta t - 418.5) - 0.3883$$
$$\times 10^{-5}(\Delta t - 418.5)^2 \quad \text{(mean square error is 0.01603)};$$

$$\phi = 0.3444 + 0.5029 \times 10^{-3}(\Delta t - 418.5) - 0.2091 \times 10^{-5}(\Delta t - 418.5)^2 + 0.7782$$
$$\times 10^{-8}(\Delta t - 418.5)^3 \quad \text{(mean square error is 0.01601)};$$

$$\phi = 0.3427 + 0.7349 \times 10^{-3}(\Delta t - 418.5) + 0.1416 \times 10^{-5}(\Delta t - 418.5)^2 - 0.2391$$
$$\times 10^{-6}(\Delta t - 418.5)^3 - 0.1079 \times 10^{-8}(\Delta t - 418.5)^4 \quad \text{(mean square error is 0.01575)}.$$

3. Results analysis. These results show that the larger m is, i.e., the higher the order of polynomial, the smaller the mean square error. But that does not mean that the higher the order of polynomial, the better the results in practical applications. In general, $m = 2$ is taken when the relationship between an unknown number and a parameter is linear; $m > 2$ is taken when the relationship is nonlinear but $m < 6$. In practical applications, when a fitting formula with large m is employed to calculate y from an x, if the value of x is outside the range of n measured values, the calculated y would probably be far from the real value.

If Equation (2.13) is also employed for fitting, comparing the results with that of Equation (2.25) fitting, it is found that (a) when m is the same, a_j $(j = 1, 2, \cdots, m)$ of the two fittings are different, but the mean square error is completely the same; and (b) when $m = 2$, the results of Equation (2.13) fitting are the same as the linear function in the aforementioned Simple Case Study 1, i.e., $\phi = 0.001255\Delta t - 0.1882$, and the mean square error is 0.0226, indicating linear fitting is just a special case of orthogonal polynomial fitting.

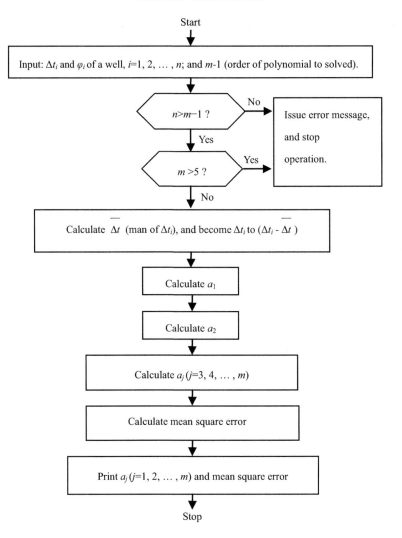

FIGURE 2.11   Calculation flowchart of least-squares method (polynomial).

Summarily, using the data of 33 porosity values and corresponding 33 acoustictime values in a well, a formula of porosity with respect to acoustictime is constructed by the data mining tool of constructing a polynomial by the least-squares method, and the results coincide with practicality. Under the same structural and sedimentary conditions, this mined knowledge can be borrowed by the neighboring wells. Therefore, this method can be spread to the formula of porosity with respect to acoustictime in other wells.

2.2.4. A Function of an Unknown Number with Respect to Multiple Parameters Constructed by MRA

2.2.4.1. Calculation Method

The function to be solved is Equation (2.14). For the sake of easy expression, y is also denoted as x_{m+1}.

1. *Single correlation coefficient matrix.* Letting r_{ij} be the single correlation coefficient between x_i and x_j, the formula is

$$r_{ij} = \sum_{k=1}^{n} \frac{(x_{ik} - \overline{x_i})(x_{jk} - \overline{x_j})}{\sigma_i \sigma_j} \qquad (i, j = 1, 2, \cdots, m, m+1) \tag{2.31}$$

where

$$\overline{x_i} = \frac{1}{n} \sum_{k=1}^{n} x_{ik} \qquad \overline{x_j} = \frac{1}{n} \sum_{k=1}^{n} x_{jk} \tag{2.32}$$

$$\sigma_i = \sqrt{\sum_{k=1}^{n} \left(x_{ik} - \overline{x_i} \right)^2} \qquad \sigma_j = \sqrt{\sum_{k=1}^{n} \left(x_{jk} - \overline{x_j} \right)^2} \tag{2.33}$$

The single correlation coefficient matrix is obtained by calculation of Equation (2.31), and this matrix is also called the 0th-*step regression matrix:*

$$R_{cor}^{(0)} = R_{cor}^{(0)}(r_{ij}) = \begin{bmatrix} r_{11} & r_{12} & \cdots & r_{1m} & r_{1y} \\ r_{21} & r_{22} & \cdots & r_{2m} & r_{2y} \\ \cdots & \cdots & \cdots & \cdots & \cdots \\ r_{m1} & r_{m2} & \cdots & r_{mm} & r_{my} \\ r_{y1} & r_{y2} & \cdots & r_{ym} & r_{yy} \end{bmatrix} \tag{2.34}$$

where $r_{iy} = r_{i,m+1}$, $r_{yi} = r_{m+1,i}$, and $r_{yy} = r_{m+1,m+1}$.

It is easy to see that $|r_{ij}| \leq 1$, especially $r_{ij} = 1$ when $i = j$. The more $|r_{ij}|$ approaches 1, the closer the relationship between x_i and x_j; the more $|r_{ij}|$ approaches 0, the more distant the relationship between x_i and x_j; when $r_{ij} > 0$, x_i increases as x_j increases; and when $r_{ij} < 0$, x_i decreases as x_j increases. Therefore, Equation (2.34) can show the order of dependence between any two numbers among $(m + 1)$ numbers of $(x_1, x_2, \cdots, x_m, y)$.

2. *Successive regression.* Successive regression is performed based on the 0th-step regression matrix (single correlation coefficient matrix).

For the sake of symbol unification, Equation (2.34) is rewritten to

$$R_{cor}^{(0)} = R_{cor}^{(0)}\left(r_{ij}^{(0)}\right) = \begin{bmatrix} r_{11}^{(0)} & r_{12}^{(0)} & \cdots & r_{1m}^{(0)} & r_{1y}^{(0)} \\ r_{21}^{(0)} & r_{22}^{(0)} & \cdots & r_{2m}^{(0)} & r_{2y}^{(0)} \\ \cdots & \cdots & \cdots & \cdots & \cdots \\ r_{m1}^{(0)} & r_{m2}^{(0)} & \cdots & r_{mm}^{(0)} & r_{my}^{(0)} \\ r_{y1}^{(0)} & r_{y2}^{(0)} & \cdots & r_{ym}^{(0)} & r_{yy}^{(0)} \end{bmatrix} \tag{2.35}$$

Successive regression contains the following three contents: regression results of Step 0, regression process of Step 1, and regression process of Step s.

a. Regression results of Step 0:

Introduce or eliminate: None.

Regression equation:

$$y = b_0^{(0)} \tag{2.36}$$

where $b_0^{(0)} = \bar{y} = \frac{1}{n} \sum_{k=1}^{n} x_{m+1,k}$, i.e. $\frac{1}{n} \sum_{k=1}^{n} y_k$.

Residual variance: $Q^{(0)} = r_{yy}^{(0)} = 1$.

Multiple correlation coefficient: $R^{(0)} = \sqrt{1 - \frac{Q^{(0)}}{Q^{(0)}}} = 0$.

The following is Step 1. At first, the filtration criterion of F_1 and F_2 are given, $F_1 \geq F_2$, and they can be assigned to 0, respectively; and the degree of freedom ϕ is given, $\phi = n - 1$.

b. Regression process of Step 1.

This process contains the following five operations.

Operation 1. Calculate:

$$v_i^{(0)} = r_{iy}^{(0)} r_{yi}^{(0)} / r_{ii}^{(0)} \qquad (i = 1, 2, \cdots, m) \tag{2.37}$$

Operation 2. $v_i^{(0)}$ consists of two types: positive and negative. $v_{max}^{(0)}$ in the positive v and its corresponding variable are chosen; and $|v^{(0)}|_{min}$ in the negative v and its corresponding variable are chosen, which does not occur in Step 1 but occurs in every later step.

Operation 3. It is to discriminate if a variable is eliminated, which does not occur in Step 1. Calculate:

$$F_2^* = \frac{|v^{(0)}|_{min} \phi}{Q^{(0)}} \tag{2.38}$$

If $F_2^* < F_2$, the variable x_k corresponding to $|v^{(0)}|_{min}$ is eliminated, $\phi + 1 \Rightarrow \phi$, and turn to Operation 5. If $F_2^* \geq F_2$, turn to Operation 4.

Operation 4. It is to discriminate if a variable is introduced. Calculate:

$$F_1^* = \frac{v_{max}^{(0)}(\phi - 1)}{Q^{(1)}} \tag{2.39}$$

where $Q^{(1)} = Q^{(0)} - v_{max}^{(0)}$.

If $F_1^* > F_1$, the variable x_k corresponding to $v_{max}^{(0)}$ is introduced, $\phi + 1 \Rightarrow \phi$, and turn to Operation 5. If $F_1^* \leq F_1$, the whole regression process ends.

Operation 5. It is the elimination to introduce or eliminate x_k.

The 0th-step regression matrix $R_{\text{cor}}^{(0)} = R_{\text{cor}}^{(0)}(r_{ij}^{(0)})$ becomes the 1st-step regression matrix $R_{\text{cor}}^{(1)} = R_{\text{cor}}^{(1)}(r_{ij}^{(1)})$ using the elimination formula

$$r_{ij}^{(1)} = \begin{cases} r_{ij}^{(0)} - r_{ik}^{(0)} r_{kj}^{(0)} / r_{kk}^{(0)} & (i \neq k, j \neq k) \\ r_{kj}^{(0)} / r_{kk}^{(0)} & (i = k, j \neq k) \\ -r_{ik}^{(0)} / r_{kk}^{(0)} & (i \neq k, j = k) \\ 1 / r_{kk}^{(0)} & (i = k, j = k) \end{cases} \tag{2.40}$$

Thus, the regression results of Step 1 are listed here:
Introduce or eliminate: No elimination occurs in Step 1.
Regression equation:

$$y = b_0^{(1)} + b_k^{(1)} x_k \tag{2.41}$$

where $b_0^{(1)} = \bar{y} - b_k^{(1)} \bar{x}_k$, and $b_k^{(1)} = r_{ky}^{(1)} \frac{\sigma_y}{\sigma_k}$.

Residual variance: $Q^{(1)} = r_{yy}^{(1)}$.

Multiple correlation coefficient: $R^{(1)} = \sqrt{1 - \frac{Q^{(1)}}{Q^{(0)}}}$.

The 2nd, 3rd, ... step regression processes are the same as the 1st-step regression process. To clearly describe, s is denoted as the regression process number, and $s = 1, 2, ...$

  c. Regression process of step s. This process contains the following five operations:
    Operation 1. Calculate:

$$v_i^{(s-1)} = r_{iy}^{(s-1)} r_{yi}^{(s-1)} / r_{ii}^{(s-1)} \qquad (i = 1, 2, \cdots, m) \tag{2.42}$$

Operation 2. $v_i^{(s-1)}$ consists of two types: positive and negative. $v_{\max}^{(s-1)}$ in the positive v and its corresponding variable are chosen; and $|v^{(s-1)}|_{\min}$ in the negative v and its corresponding variable are chosen.
    Operation 3. It is to discriminate if a variable is eliminated. Calculate:

$$F_2^* = \frac{|v^{(s-1)}|_{\min} \phi}{Q^{(s-1)}} \tag{2.43}$$

If $F_2^* < F_2$, the variable x_k corresponding to $|v^{(s-1)}|_{\min}$ is eliminated, $\phi + 1 \Rightarrow \phi$, and turn to Operation 5. If $F_2^* \geq F_2$, turn to Operation 4.
    Operation 4. It is to discriminate if a variable is introduced. Calculate:

$$F_1^* = \frac{v_{\max}^{(s-1)}(\phi - 1)}{Q^{(s)}} \tag{2.44}$$

where $Q^{(s)} = Q^{(s-1)} - v_{\max}^{(s-1)}$.

If $F_1^* > F_1$, the variable x_k corresponding to $v_{max}^{(s-1)}$ is introduced, $\phi + 1 \Rightarrow \phi$, and turn to Operation 5. If $F_1^* \leq F_1$, the whole regression process ends.

Operation 5. This is the elimination step to introduce or eliminate x_k.

The $(s-1)^{th}$-step regression matrix $R_{cor}^{(s-1)} = R_{cor}^{(s-1)}(r_{ij}^{(s-1)})$ becomes the sth-step regression matrix $R_{cor}^{(s)} = R_{cor}^{(s)}(r_{ij}^{(s)})$ using the elimination formula

$$
r_{ij}^{(s)} = \begin{cases}
r_{ij}^{(s-1)} - r_{ik}^{(s-1)} r_{kj}^{(s-1)} / r_{kk}^{(s-1)} & (i \neq k, j \neq k) \\
r_{kj}^{(s-1)} / r_{kk}^{(s-1)} & (i = k, j \neq k) \\
-r_{ik}^{(s-1)} / r_{kk}^{(s-1)} & (i \neq k, j = k) \\
1 / r_{kk}^{(s-1)} & (i = k, j = k)
\end{cases}
\tag{2.45}
$$

Thus, the regression results of step s are listed here:
Introduce or eliminate: x_k.
Regression equation:

$$
y = b_0^{(s)} + b_1^{(s)} x_1 + \cdots + b_m^{(s)} x_m
\tag{2.46}
$$

where $b_0^{(s)} = \bar{y} - \sum_{i=1}^{m} b_i^{(s)} \bar{x}_i$, and $b_i^{(s)} = r_{iy}^{(s)} \frac{\sigma_y}{\sigma_i}$ $(i = 1, 2, \cdots, m)$.
It is noted that the coefficients $b_i^{(s)} = 0$ $(i = 1, 2, \cdots, m)$ for the variables x_i that have not yet introduced or have already been eliminated.

Residual variance: $Q^{(s)} = r_{yy}^{(s)}$.

Multiple correlation coefficient: $R^{(s)} = \sqrt{1 - \frac{Q^{(s)}}{Q^{(0)}}}$.

3. *Some points.* Five points are given here:

Point 1. In each step, an "important" variable x_k is introduced or an "unimportant" variable x_k is eliminated. The elimination refers that x_k had been introduced in the regression equation but now has been eliminated. The "importance" of variable x_k refers to the fact that the correlation coefficient between x_k and y is large, and the correlation coefficients between x_k and those introduced variables are small. The "unimportance" of variable x_k is contrary to the "importance" of variable x_k.

Point 2. Since the parameters are normalized in the regression process, $0 \leq Q \leq 1$ and $0 \leq R \leq 1$ so as to easily analyze the results. In the whole regression process, the residual variance Q is getting smaller, whereas the multiple correlation coefficient R is getting larger.

Point 3. When $F_1 = F_2 = 0$, the successive regression and the classical successive regression are coincident. In the case studies of this book, $F_1 = F_2 = 0$.

Point 4. In real problems, an introduced variable is eliminated in very few cases, and an eliminated variable is introduced in rare cases.

Point 5. To make the residual variance Q as small as possible, the expression of y, Equation (2.14), can be taken as a nonlinear expression, i.e., x_k of the right-hand side can be changed to a function such as logarithm and exponent, and even a combination function of several x_k is taken as a new variable. Thus, the values involved in regression are related to these functions rather than x_k in the functions.

2.2.4.2. Case Study: Linear Function of Proved Reserve with Respect to its Relative Parameters

Using that data of 15 sags in eastern China listed in Table 2.1 and by MRA, a linear function of a proved reserve with respect to some relative parameters is constructed (Tang, Z. et al., 1988; Shi, 1999).

The question is to calculate a regression equation of y with respect to x_k ($k = 1, 2, ..., 5$):

$$y = b_0 + b_1 x_1 + b_2 x_2 + \cdots + b_5 x_5 \tag{2.47}$$

and analyze the order of dependence between y and each x_k.

TABLE 2.1 Proved Reserve Concentration and its Five Relative Parameters of 15 Sags in Eastern China

Sag No.	Relative Parameters for Proved Reserve Concentration Prediction[a]					Proved Reserve Concentration[b]
	x_1 (°C/100m)	x_2 (%)	x_3 (%)	x_4 (%)	x_5 (%)	y^* (10^4t/km^2)
1	3.18	1.15	9.4	17.6	3.00	0.7
2	3.80	0.79	5.1	30.5	3.80	0.7
3	3.60	1.10	9.2	9.1	3.65	1.0
4	2.73	0.73	14.5	12.8	4.68	1.1
5	3.40	1.48	7.6	16.5	4.50	1.5
6	3.29	1.00	10.8	10.1	8.10	2.6
7	2.60	0.61	7.3	16.1	16.16	2.7
8	4.10	2.30	3.7	17.8	6.70	3.1
9	3.72	1.94	9.9	36.1	4.10	6.1
10	4.10	1.66	8.2	29.4	13.00	9.6
11	3.35	1.25	7.8	27.8	10.50	10.9
12	3.31	1.81	10.7	9.3	10.90	11.9
13	3.60	1.40	24.6	12.6	12.76	12.7
14	3.50	1.39	21.3	41.1	10.00	14.7
15	4.75	2.40	26.2	42.5	16.40	21.3

[a]x_1 = geothermal gradient; x_2 = organic carbon content; x_3 = volume ratio of source rock to sedimentary rock; x_4 = thickness ratio of sandstone to mudstone in the vicinity of hydrocarbon generation sag; and x_5 = organic matter transformation rate.
[b]y^* = proved reserve concentration determined by the resources assessment.

1. Calculation flowchart (Figure 2.12).
2. Calculation results and analyses. Since $F_1 = F_2 = 0$, each x_k ($k = 1, 2, ..., 5$) are all introduced, and the obtained regression equation is

$$y = -8.786 - 0.6436x_1 + 4.033x_2 + 0.3845x_3 + 0.1281x_4 + 0.5601x_5 \qquad (2.48)$$

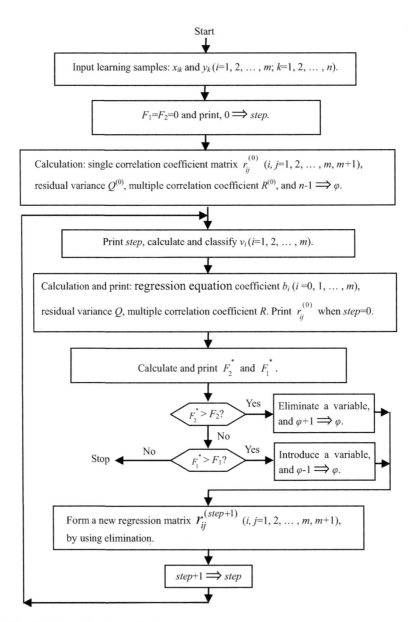

FIGURE 2.12   Calculation flowchart of MRA.

its corresponding residual variance Q is 0.1042 and multiple correlation coefficient R is 0.9465. Since $\overline{R}^*(\%) = 52.42$, calculated from Equation (1.6), the nonlinearity of this case study is very strong, and the solution accuracy of MRA is very low. Therefore, the linear MRA is not applicable to this case study; nonlinear algorithms should be adopted, e.g., the error back-propagation neural network (BPNN) in Chapter 3, the support vector machine (SVM) in Chapter 4, the decision trees (DTR) in Chapter 5, and the Bayesian classification (BAC) in Chapter 6. The main objective of this case study is to help readers understand the whole procedure of MRA.

The order of introducing x_k is x_3, x_5, x_2, x_4, and x_1. In fact, this order can be seen from the correlation coefficient between y and x_k in the single correlation coefficient matrix, i.e., the single correlation coefficient between y and x_3, x_5, x_2, x_4, and x_1 are 0.742, 0.716, 0.579, 0.560, and 0.529, respectively. Certainly, that is not a necessity due to the fact that introducing a variable depends on not only the correlation coefficient between y and this variable but also on the correlation coefficients between those introduced variables and this variable.

In Table 2.2 we see Q and R of introduced x_k. We see that after x_3, x_5, and x_2 are introduced, Q of introduced x_4 or x_1 drops to a very small number, and their R also rise a little bit. So it is enough for the regression equation to be solved to have three variables x_3, x_5, and x_2. This regression equation is the result of the Step 3 regression process:

$$y = -9.499 + 4.461x_2 + 0.4241x_3 + 0.5816x_5 \tag{2.49}$$

Its corresponding $Q = 0.1418$ and $R = 0.9264$. Thus, the 6-D problem (x_1, x_2, x_3, x_4, x_5, y) is reduced to the 4-D problem (x_2, x_3, x_5, y), showing that MRA plays a role in dimension reduction. However, MRA constructs a linear function, whereas BPNN, SVM, DTR, and BAC construct a nonlinear function. Because of the complexities of geoscience rules, the correlations between different classes of geoscientific data are nonlinear in most cases. In general, therefore, it is better to use SVM when the nonlinearity is very strong; otherwise, use BPNN, DTR, and BAC. As for MRA, it can serve as an auxiliary tool, e.g., a pioneering dimension-reduction tool, cooperating with major tools (BPNN, SVM, DTR, and BAC) for data mining.

Proved reserve concentration (y) is shown to depend on the five parameters in decreasing order: volume ratio of source rock to sedimentary rock (x_3), organic matter transformation

TABLE 2.2 Important Results of the Application Example by MRA

Regression Step S	Introduced Variable	Residual Variance $Q^{(s)}$	Multiple Correlation Coefficient $R^{(s)}$
0	N/A	1.0	0
1	x_3	0.4497	0.7418
2	x_5	0.2742	0.8520
3	x_2	0.1418	0.9264
4	x_4	0.1051	0.9460
5	x_1	0.1042	0.9465

rate (x_5), organic carbon content (x_2), thickness ratio of sandstone to mudstone in the vicinity of hydrocarbon generation sag (x_4), and geothermal gradient (x_1). Furthermore, y mainly depends on the former three parameters (x_3, x_5, and x_2), i.e. , the major factor that affects the proved reserve concentration is the hydrocarbon generation. In fact, the major factors that control the proved reserve are the generation, migration, trapping, accumulation, sealing, and preservation of hydrocarbons. If some variables in $x_i(i = 1, 2, \cdots, m)$ can reflect the six factors, it is possible to make a through and multisided study on the factors of proved reserve concentration.

In summary, using data of 15 sags in eastern China, the five parameters (geothermal gradient, organic carbon content, volume ratio of source rock to sedimentary rock, thickness ratio of sandstone to mudstone in the vicinity of hydrocarbon generation sag, organic matter transformation rate) and a proved reserve concentration of each sag, an explicit linear function is constructed and the order of dependence between the proved reserve concentration and the five parameters is obtained by the data mining tool of MRA, and the results coincide with practicality. This function and the order are called *mined knowledge*. This knowledge can be applied to the other sags in eastern China. Therefore, this method can be spread to the proved reserve prediction of other sags.

EXERCISES

2-1. For the various exploration degrees, what are different ways to construct the probability density function?

2-2. In the simple case study for the pore volume of trap calculated by the Monte Carlo method, for parameters φ and H, logarithmic normal distribution is employed as a probability density function $p(x)$ to perform the cumulative frequency distribution as a probability distribution function $F(x)$, i.e., $F(x) = \int p(x)dx$. Can the data of φ and H be directly used to perform the cumulative frequency distribution rather than the logarithmic normal distribution?

2-3. The least-squares method can be adopted to construct a function of an unknown number with respect to a parameter. When the general polynomial fitting is used, does the case that the higher the order of polynomial, the better result in practical applications?

2-4. Multiple regression analysis (MRA) can be adopted to construct a function of an unknown number with respect to multiple parameters. When $F_1 = F_2 = 0$, the successive regression and the classical successive regression are coincident so that two additional benefits are obtained. What are the two benefits?

References

Chatterjee, S., Hadi, A.S., Price, B., 2000. Regression Analysis by Examples, third ed. Wiley, New York, NY, USA.

Lee, J.H., Yang, S.H., 2002. Statistical optimization and assessment of a thermal error model for CNC machine tools. Int. J. Mach. Tool. Manufact. 42 (1), 147–155.

Shi, G., 1999. New Computer Application Technologies in Earth Sciences. Petroleum Industry Press, Beijing, China (in Chinese).

Shi, G., Zhou, X., Zhang, G., Shi, X., Li, H., 2004. The use of artificial neural network analysis and multiple regression for trap quality evaluation: a case study of the Northern Kuqa Depression of Tarim Basin in western China. Mar. Petro. Geol. 21 (3), 411–420.

Shi, G., 2005. Numerical Methods of Petroliferous Basin Modeling, third ed. Petroleum Industry Press, Beijing, China.

Shi, G., 2009. The use of support vector machine for oil and gas identification in low-porosity and low-permeability reservoirs. Int. J. Math. Model. Numer. Optimisa. 1 (1/2), 75–87.

Shi, G., Yang, X., 2010. Optimization and data mining for fracture prediction in geosciences. Procedia Comput. Sci. 1 (1), 1353–1360.

Shi, G., 2011. Four classifiers used in data mining and knowledge discovery for petroleum exploration and development. Adv. Petro. Expl. Devel. 2 (2), 12–23.

Singh, J., Shaik, B., Singh, S., Agrawal, V.K., Khadikar, P.V., Deeb, O., Supuran, C.T., 2008. Comparative QSAR study on para-substituted aromatic sulphonamides as CAII inhibitors: information versus topological (distance-based and connectivity) indices. Chem. Biol. Drug. Design. 71, 244–259.

Tang, Z., et al., 1988. The application of multiple regression to the calculation of oil and gas resources. In: Oil and Gas Resources Assessment. Petroleum Industry Press, Beijing, China (in Chinese).

Xu, S., 1997. Commonly Used Computer Algorithms. Tsinghua University Press, Beijing, China (in Chinese).

Zhao, X., 1992. Conspectus of Mathematical Petroleum Geology. Petroleum Industry Press, Beijing, China (in Chinese).

Artificial neural networks (ANN) constitute a branch of artificial intelligence. This chapter introduces an error back-propagation neural network (BPNN) in ANN as well as its applications in geosciences.

Section 3.1 (methodology) introduces the applying ranges and conditions, basic principles, calculation method, and calculation flowchart of BPNN as well as three simple case studies. These simple case studies give calculation results and analyses. Though each simple case study is small, it reflects the whole process of calculation to benefit readers in understanding and mastering the applied techniques. Let t_{opt} be optimal learning time count (OLTC) at first.

Simple Case Study 1 is an XOR problem, explaining in detail a conventional prediction of BPNN and presenting an OLTC technique. Concretely, using the learning samples of the XOR problem, the structure of BPNN and the final values of W_{ij}, W_{jk}, θ_j, and θ_k at $t_{opt} = 31876$ are obtained by the data mining tool of BPNN, and the results coincide with practicality. This structure and these final values are called *mined knowledge*. In the prediction process, this knowledge can be adopted to perform XOR calculation for any two values consisting of 0 or 1. That proves that the OLTC technique is feasible and available.

Simple Case Study 2 is the prediction of oil production, describing a special time-series prediction of BPNN. Concretely, using the learning samples of the annual oil production for 39 years, from 1952 to 1990, in the Romashkino Oilfield for the prediction problem of oil production, the structure of BPNN and the final values of W_{ij}, W_{jk}, θ_j and θ_k at $t_{opt} = 197636$ are obtained by the data mining tool of BPNN, and the results coincide with practicality. This structure and these final values are called *mined knowledge*. In the prediction process, this knowledge can be adopted to predict the annual oil production for 15 years from 1991 to 2005, and the prediction results basically coincide with practicality. That proves that the special time-series prediction of BPNN is feasible and available. Therefore, this method can be spread to the prediction problem of oil production for other oilfields.

Simple Case Study 3 is the prediction of fracture-acidizing results, explaining a conventional prediction of BPNN. Concretely, using the learning samples of seven wells for the prediction problem of fracture-acidizing results, the structure of BPNN and the final values of W_{ij}, W_{jk}, θ_j and θ_k at $t_{opt} = 55095$ are obtained by the data mining tool of BPNN, and the results coincide with practicality. This structure and these final values are called *mined knowledge*. In the prediction process, this knowledge can be adopted to predict the fracture-acidizing results in the eighth well, and the results are basically correct. Therefore, this method can be spread to the prediction problem of fracture-acidizing results for other structures.

Section 3.2 (integrated evaluation of oil and gas-trap quality) introduces Case Study 1 of BPNN. Using data for the trap evaluation of the Northern Kuqa Depression, i.e., the 14 parameters (unit structure, trap type, petroliferous formation, trap depth, trap relief, trap closed area, formation HC identifier, data reliability, trap coefficient, source rock coefficient, reservoir coefficient, preservation coefficient, configuration coefficient, and resource quantity) and trap quality of 27 traps, the structure of BPNN and the final values of W_{ij}, W_{jk}, θ_j and θ_k at $t_{opt} = 10372$ are obtained by the data mining tool of BPNN, and the results coincide with practicality. This structure and these final values are called *mined knowledge*. In the prediction process, this knowledge can be adopted to predict the trap quality of other two traps, and the results coincide with the real exploration results after this prediction. Therefore, this method can be spread to the trap evaluation of other exploration areas.

Section 3.3 (prediction of fractures using conventional well-logging data) introduces Case Study 2 of BPNN. Using data for fracture prediction of Wells An1 and An2, i.e., the seven well-logging data (acoustictime, compensated neutron density, compensated neutron porosity, microspherically focused resistivity, deep laterolog resistivity, shallow laterolog resistivity, and absolute difference of R_{LLD} and R_{LLS}) and imaging log result of 29 samples, the structure of BPNN and the final values of W_{ij}, W_{jk}, θ_j, and θ_k at $t_{opt} = 6429$ are obtained by the data mining tool of BPNN, and the results coincide with practicality. This structure and these final values are called *mined knowledge*. In the prediction process, this knowledge can be adopted to predict the fractures of other four samples, and the results are consistent with the imaging log results. Therefore, this method can be spread to the fracture prediction of other oilfields.

Moreover, it is found from the aforementioned Case Study 1 and Case Study 2 that (a) since each of the two case studies is a strong nonlinear problem, the major data mining tool can adopt BPNN but not multiple regression analysis (MRA; see Chapter 2); (b) in the two case studies, MRA established the order of dependence between the result and its relative parameter under the condition of linearity; thus MRA can be employed as an auxiliary tool; and (c) certainly, Classification of Support Vector Machine (C-SVM; see Chapter 4) and Regression of Support Vector Machine (R-SVM; see Chapter 4) can be applied to the two case studies and run at a speed as fast as 20 times or more of BPNN; however, it is easy to code the BPNN program, whereas it is very complicated to code the C-SVM or R-SVM program, so BPNN is a good software for this case study when neither the C-SVM nor R-SVM program is available.

3.1. METHODOLOGY

ANN has been widely applied in the natural and social sciences since the 1980s (e.g., Rumelhart et al., 1986; Hecht-Nielsen, 1989). The error back-propagation neural network (BPNN) is one of the most popular neural networks in ANN and is still a very useful tool in some fields, applied widely to geosciences in recent years.

The learning samples and prediction samples for BPNN are expressed with Equations (1.1) and (1.3), respectively. The learning process and prediction process of the calculation flowchart are illustrated in Figures 1.2 and 1.3, respectively. In Figure 1.3, if the prediction samples are replaced by the learning samples, Figure 1.3 becomes such a calculation flowchart for learning validation.

The expression of y created using BPNN is a nonlinear implicit expression with respect to m parameters $(x_1, x_2, ..., x_m)$:

(a) Learning process

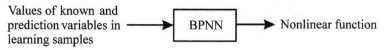

(b) Prediction process

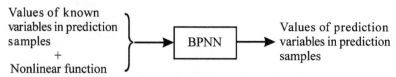

FIGURE 3.1   Learning process (a) and prediction process (b) of BPNN.

$$y = BPNN(x_1, x_2, ..., x_m) \tag{3.1}$$

where *BPNN* is a nonlinear function and calculated out by the BPNN algorithm, which cannot be expressed as a usual mathematical formula and so is an implicit expression. Equation (3.1) is called a *fitting formula* obtained from the learning process.

The basic idea of BPNN is illustrated in Figure 3.1.

3.1.1. Applying Ranges and Conditions

3.1.1.1. *Applying Ranges*

As Equation (1.1) described, assume that there are n learning samples, each associated with $m + 1$ numbers $(x_1, x_2, ..., x_m, y_i^*)$ and a set of observed values $(x_{1i}, x_{2i}, ..., x_{mi}, y_i^*)$, with $i = 1, 2, ..., n$ for these parameters. In principle, $n > m$, but in actual practice $n >> m$. As Equation (1.3) described, assume that there are k prediction samples, each associated with m parameters $(x_1, x_2, ..., x_m)$ and a set of observed values $(x_{1i}, x_{2i}, ..., x_{mi})$, with $i = n + 1, n + 2, ..., n + k$ for these parameters.

3.1.1.2. *Applying Conditions*

The number of learning samples, n, i.e., the number of input layer nodes, must be big enough to ensure the accuracy of prediction by BPNN. The number of known variables, m, must be greater than one; otherwise, $m = 1$. It is a special prediction related to time in general; thus the special processing should be conducted, e.g., Simple Case Study 2 (prediction of oil production), which follows.

3.1.2. Basic Principles

This section introduces BPNN, and the following discussion outlines its basic principles.

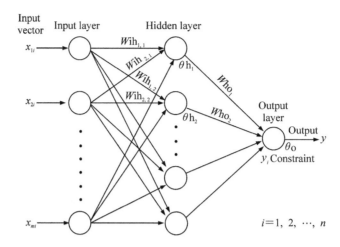

$Wih_{1,1}$, $Wih_{1,2}$, $Wih_{2,1}$, $Wih_{2,2}$, ...   Connection weight value between input layer and hidden layer

Who_1, Who_2, ...   Connection weight value between hidden layer and output layer

θh_1, θh_1, ...   Threshold value at hidden layer node

θo Threshold value at output layer node

FIGURE 3.2   Sketch map of the structure of BPNN.

In general, an error back-propagation neural network (BPNN) consists of one input layer, one or more hidden layers, and one output layer. There is no theory yet to determine how many hidden layers are needed for any given case, but in the case of an output layer with only one node, it is enough to define one hidden layer (Figure 3.2). Moreover, it is also difficult to determine how many nodes a hidden layer should have. For solving a local minima problem, it is suggested to use the large $N_{\text{hidden}} = 2(N_{\text{input}} + N_{\text{output}}) - 1$, where N_{hidden} is the number of hidden nodes, N_{input} is the number of input nodes, and N_{output} is the number of output nodes. The values of the network learning rate for the output layer and the hidden layer are within (0, 1), and in practice they can be the same and taken as 0.6 (Shi, 2009; Shi, 2011). Hence, BPNN, as introduced in this section, is a three-layer neural network (Figure 3.2).

Figure 3.2 shows that BPNN structure has the following characteristics: network nodes (nerve cells) in the same layer are unconnected to each other, and nerve cells between two neighboring layers are connected each other.

Figure 3.2 illustrates the structure frame of BPNN; Table 3.1 lists its structure parameters. By combining Figure 3.2 and Table 3.1, we can lay out the general picture, inherent relationship, and calculation flowchart of BPNN.

The structure parameters listed in Table 3.1 are explained here:

1. *Layers.* BPNN consists of one input layer, one hidden layer, and one output layer.
2. *Node numbering and the number of nodes.* Let i be a node number of input layer, and $i = 1, 2, ..., N_i$, where N_i is the number of known variables. Let j be a node number of hidden layer,

TABLE 3.1   Structure Parameters of BPNN

Layer	Input Layer	Hidden Layer	Output Layer
Node numbering	i ($i = 1, 2, ..., N_i$)	j ($j = 1, 2, ..., N_j$)	k ($k = 1, 2, ..., N_k$)
Number of nodes	N_i (known)	N_j (given by user)	N_k (known)
Learning sample (mainly used in learning process)	Input layer values A_{ui} (known) ($u = 1, 2, ..., N_u$) ($i = 1, 2, ..., N_i$)	Hidden layer values B_{uj} (intermediate results) ($u = 1, 2, ..., N_u$) ($j = 1, 2, ..., N_j$)	Output layer values C^*_{uk} and C_{uk} ($u = 1, 2, ..., N_u$) ($k = 1, 2, ..., N_k$) Note: C^*_{uk} are known, i.e., desired values in learning process, whereas C_{uk} are calculation results.
Connection weight values between nodes of neighboring layers (calculated in learning process and used in prediction process)	Connection weight values between input layer node and hidden layer node W_{ij}[a] ($i = 1, 2, ..., N_i$) ($j = 1, 2, ..., N_j$) Note: Initialized at beginning but later determined by learning process.	Connection weight values between hidden layer node and output layer node W_{jk}[a] ($j = 1, 2, ..., N_j$) ($k = 1, 2, ..., N_k$) Note: Initialized at beginning but later determined by learning process.	N/A
Threshold values for Sigmoid function (calculated in learning process and used in prediction process)	N/A	Threshold values at hidden layer nodes θ_j[a] ($j = 1, 2, ..., N_j$) Note: Initialized at beginning but later determined by learning process.	Threshold values at output layer nodes θ_k[a] ($k = 1, 2, ..., N_k$) Note: Initialized at beginning but later determined by learning process.
Prediction sample (used in prediction process)	Input layer values A_{vi} (known) ($v = 1, 2, ..., N_v$) ($i = 1, 2, ..., N_i$)	Hidden layer values B_{vj} (intermediate results) ($v = 1, 2, ..., N_v$) ($j = 1, 2, ..., N_j$)	Output layer values C_{vk} ($v = 1, 2, ..., N_v$) ($k = 1, 2, ..., N_k$) Note: They are final calculation results, i.e., the solution of BPNN.

[a]W_{ij} and W_{jk} are different weights, even when they have the same subscript. So are θ_j and θ_k.

and $j = 1, 2, ..., N_j$, where N_j is given by users. Let k be a node number of output layer, and $k = 1, 2, ..., N_k$, where N_k is the number of prediction variables.

3. *Learning sample.* The learning sample is used mainly in the learning process.

Let u be a learning sample number and $u = 1, 2, ..., N_u$, where N_u is the number of learning samples. Each learning sample has the values of N_i known variables and N_k prediction variables. The following discusses the values of learning samples on input layer, hidden layer, and output layer, respectively.

a. Values of input layer (A_{ui}). Learning sample number u at node i on the input layer is

$$A_{ui}(u = 1, 2, ..., N_u; \quad i = 1, 2, ..., N_i) \tag{3.2}$$

where A_{ui} is known.

b. Values of hidden layer (B_{uj}). Learning sample number u at node j on the hidden layer is

$$B_{uj}(u = 1, 2, ..., N_u; \quad j = 1, 2, ..., N_j) \tag{3.3}$$

where B_{uj} is intermediate results expressed with Equation (3.15).

c. Values of output layer (C^*_{uk} and C_{uk}). There are two kinds of values of an output layer: the known value and the calculated value. The known value of learning sample number u at node k on the output layer is

$$C^*_{uk}(u = 1, 2, ..., N_u; \quad k = 1, 2, ..., N_k) \tag{3.4}$$

where C^*_{uk} is known and the desired value in learning process.

The calculated value of learning sample u at node k on the output layer is

$$C_{uk}(u = 1, 2, ..., N_u; \quad k = 1, 2, ..., N_k) \tag{3.5}$$

where C_{uk} is a calculation result expressed with Equation (3.17). It will be required to be reverted using Equation (3.25) and thus could become the fitting solution of BPNN.

The calculated C_{uk} approaches the practical value C^*_{uk} as t increases.

The convergence criterion of the learning process is to minimize the difference of C_{uk} and C^*_{uk}, and it should be a global minimum but not a local minimum.

4. *Connection weight values between nodes of neighboring layers.* The connection weight values between node i on the input layer and node j of the hidden layer is

$$W_{ij}(i = 1, 2, ..., N_i; \quad j = 1, 2, ..., N_j) \tag{3.6}$$

The connection weight values between node j on the hidden layer and node k of the output layer is

$$W_{jk}(j = 1, 2, ..., N_j; \quad k = 1, 2, ..., N_k) \tag{3.7}$$

W_{ij} in Equation (3.6) and W_{jk} in Equation (3.7) are initialized when BPNN starts but later are continuously adjusted by using Equations (3.22) and (3.20), respectively.

5. *Threshold value of the Sigmoid function.* The threshold value of hidden node j is

$$\theta_j(j = 1, 2, ..., N_j) \tag{3.8}$$

The threshold value of output node k is

$$\theta_k(k = 1, 2, ..., N_k) \tag{3.9}$$

θ_j in Equation (3.8) and θ_k in Equation (3.9) are initialized when BPNN starts, but later they are continuously adjusted using Equations (3.23) and (3.21), respectively.

6. *Prediction sample.* The prediction sample is used the in prediction process.

Let v be a prediction sample number, and $v = 1, 2, ..., N_v$, where N_v is the number of prediction samples. Each prediction sample has the values of N_i known variables. The following discusses the values of prediction samples on input layer, hidden layer, and output layer, respectively.

a. Values of input layer (A_{vi}). Prediction sample v at node i on the input layer is

$$A_{vi}(v = 1, 2, ..., N_v; \quad i = 1, 2, ..., N_i) \tag{3.10}$$

where A_{vi} is known.

b. Values of hidden layer (B_{vj}). Prediction sample v at node j on the hidden layer is

$$B_{vj}(v = 1, 2, ..., N_v; \quad j = 1, 2, ..., N_j) \tag{3.11}$$

where B_{vj} is intermediate results expressed with Equation (3.15), but in this equation u should be substituted with v.

c. Values of output layer (C_{vk}). The value of prediction sample v at node k on the output layer is

$$C_{vk}(v = 1, 2, ..., N_v; \quad k = 1, 2, ..., N_k) \tag{3.12}$$

where C_{vk} is a calculation result expressed with Equation (3.17), but in this equation u should be substituted with v.

Now the BPNN procedure can be summarized as follows: (a) using the values of known variables and prediction variables in learning samples, a nonlinear function (nonlinear mapping) of the prediction variable with respect to the known variables is constructed; that is the learning process, and (b) using the values of known variables in learning samples and prediction samples and employing this nonlinear function, the values of prediction variables in learning samples and prediction samples are calculated out; that is the prediction process. The obtained values of prediction variables in learning samples and prediction samples are called *fitting solutions* and *prediction solutions* for the BPNN procedure, respectively.

3.1.3. Error Back-Propagation Neural Network (BPNN)

3.1.3.1. *Calculation Method*

BPNN is the most widely applied, most intuitionistic, and most intelligible algorithm in ANN (Wang, 1995; Shi, 1999). The BPNN learning process is illustrated in Figures 3.1a and 3.2, consisting of the following two iterative sweeps:

1. Based on the values of the known variables for learning samples, connection weight values between nodes of neighboring layers (w), and threshold values at hidden layer node and output layer node (θ), and by the prescriptive way to calculate the prediction variable along the direction of input layer to hidden layer and then hidden layer to output layer, the output results (the prediction variable value) are obtained. This is called *sample forward propagation.* Obviously, the error between the calculated and practical values of the prediction variable is big at the beginning.

2. Based on the error and by the prescriptive way to correct w and θ along the direction of the output layer to the hidden layer and then the hidden layer to the input layer, new w and θ are obtained. This is called *error back propagation*. The two sweeps are repeated until the error falls in an allowable range. The major results of the learning process are two sets of w and θ meeting error conditions (Table 3.1), i.e., a created nonlinear relationship from input layer to output layer, which is a basis for the next prediction process. The prediction process is similar to the step a, but a calculation once; prediction samples are involved rather than learning samples. Thus the results are the prediction variable values of prediction samples (Figures 3.1b). Summarily, the term *back propagation* refers to the way (Güler and Übeyli, 2003): The error computed at the output side is propagated backward from the output layer to the hidden layer and finally to the input layer. Each iteration of BPNN constitutes two sweeps: forward to calculate a solution by using a sigmoid activation function, and backward to compute the error and thus to adjust the weights and thresholds for the next iteration. This iteration is performed repeatedly until the solution agrees with the desired value within a required tolerance.

The following discussion introduces the learning and prediction processes of BPNN. The learning process performs in the following 15 steps:

1. Input known parameters and control parameters. The input known parameters are N_u, the number of learning samples; N_v, the number of prediction samples; N_i, the number of nodes on input layer; N_k, the number of nodes on output layer; N_j, the number of nodes on hidden layer; $N_j = 2(N_i + N_k) - 1$, which can be adjusted by users if it is not appropriate; A_{ui} and C_{uk}^*, the known variables and prediction variable of learning samples, respectively; and A_{vi}, the known variables of prediction samples.

   It is noted that the above $N_j = 2(N_i + N_k) - 1$ is appropriate for practical applications in most cases, e.g., all case studies in this book.

   The input control parameters are t_{max}, maximum of t, a number within [10000, 100000], determined by users; α, the network learning rate of output layer, can be taken as 0.6 or so; β, the network learning rate of hidden layer, can be taken as 0.6 or so.

   To have the comparability of BPNN results among various applications in this book, $N_j = 2(N_i + N_k) - 1$, $t_{max} = 100000$, $\alpha = \beta = 0.6$. Through BPNN calculation, the OLTC t_{opt} can be obtained, which is also called the *optimal iteration count*. This OLTC is the learning time count corresponding to the minimum of the root mean square error.

2. Data normalization of samples. At the starting point of BPNN running, samples must be normalized at first. Samples are divided into two types: learning samples and prediction samples. Each learning sample contains known variables and prediction variables shown Equation (1.1); each prediction sample only contains known variables shown in Equation (1.3). Learning samples or prediction samples must be normalized. In learning samples, maximum x_{max} and minimum x_{min} of each variable are calculated, respectively. Letting x be a variable, x can be transformed to the normalized x' by the maximum difference normalization formula (3.13), and $x' \in [0, 1]$.

$$x' = \frac{x - x_{min}}{x_{max} - x_{min}} \tag{3.13}$$

3. Assign a pseudo-random number within $[-0.1, 0.1]$ to W_{ij}, W_{jk}, θ_j and θ_k. Using Equation (2.9) to generate a uniform distributed pseudo-random number γ_k within $[0, 1]$, and then using Equation (3.14) to transform γ_k to a uniform distributed pseudo-random number γ_k' within $[-1, 1]$,

$$\gamma_k' = 2\gamma_k - 1 \quad (k = 1, 2, ...) \tag{3.14}$$

Those γ_k' that are within $[-0.1, 0.1]$ are orderly assigned to W_{ij}, W_{jk}, θ_j, and θ_k which are respectively expressed with Equations (3.6), (3.7), (3.8), and (3.9) as their initial values.

It is noted that only one set of pseudo-random numbers is generated and assigned in an orderly manner to the preceding four parameters so as to ensure that no same values exist in the initial values of these four parameters.

4. t sets to 1:

$$1 \Rightarrow t$$

5. Randomly select one sample from N_u learning samples for learning. Reusing Equation (2.9) to generate a uniform distributed pseudo-random number γ_k within $[0, 1]$, and equally dividing the interval $[0, 1]$ into N_u small intervals, γ_k must fall in a small interval. The number of this small interval is the number of the selected learning sample, designated as u, i.e., the uth learning sample. The implementation is

$$u = \text{INT}[\gamma_k \cdot N_u] + 1$$

where INT is integral function.

6. Calculate the value of the uth learning sample at each node on the hidden layer. The value at each node on the hidden layer equals the output value at each node on the input layer. The value of the uth learning sample at the jth node on hidden layer is

$$B_{uj} = f\left(\sum_{i=1}^{N_i} W_{ij} A_{ui} + \theta_j \right) \quad \left(j = 1, 2, ..., N_j \right) \tag{3.15}$$

where θ_j is the threshold value at the jth node on the hidden layer, its initial value expressed with Equation (3.14) is a pseudo-random number within $[-0.1, 0.1]$ but later calculated by Equation (3.23); $f(x)$ is a Sigmoid function expressed with Equation (3.16), used as an activation function; and the other symbols have been defined in Equations (3.6) and (3.2).

$$f(x) = \frac{1}{1 + \exp(-x)} \tag{3.16}$$

To accelerate the convergence of BPNN and to avoid the false convergence, a limit control should be put on $f(x)$. Concretely, if $f(x) < 0.01$, then $f(x) = 0.01$; if $f(x) > 0.99$, then $f(x) = 0.99$. This limit control is used in the learning process but not in the prediction process.

7. Calculate the value of the uth learning sample at each node on the output layer. The value at each node on the output layer equals the output value at each node on the hidden layer.

The value of the uth learning sample at the kth node on output layer is

$$C_{uk} = f\left(\sum_{j=1}^{N_j} W_{jk}B_{uj} + \theta_k\right) \quad (k = 1, 2, ..., N_k) \tag{3.17}$$

where θ_k is the threshold value at the k^h node on the output layer; its initial value expressed with Equation (3.14) is a pseudo-random number within $[-0.1, 0.1]$ but later calculated by Equation (3.21); and the other symbols have been defined in Equations (3.7) and (3.15).

W_{ij} in Equation (3.15) and W_{jk} in Equation (3.17) are different connection weight values between nodes of neighboring layers, even when their subscripts are the same; and θ_j in Equation (3.15) and θ_k in Equation (3.17) are also different threshold values, even when their subscripts are the same.

8. Calculate the error at each node on output layer. The error of the uth learning sample at the kth node on the output layer is

$$E_{uk} = C_{uk}(1 - C_{uk})(C^*_{uk} - C_{uk}) \quad (k = 1, 2, ..., N_k) \tag{3.18}$$

where all symbols have been defined in Equations (3.17) and (3.4).

9. Calculate the error at each node on the hidden layer. The error of the uth learning sample at the jth node on the hidden layer is

$$E_{uj} = B_{uj}(1 - B_{uj})\sum_{k=1}^{N_k} W_{jk}E_{uk} \tag{3.19}$$

where all symbols have been defined in Equations (3.15), (3.7) and (3.18).

E_{uk} expressed with Equation (3.18) and E_{vj} expressed with Equation (3.19) are different errors, even when their subscripts are the same.

10. Adjust W_{jk} and θ_k. For next learning, W_{jk} is adjusted to

$$W_{jk} + \alpha E_{uk}B_{uj} \Rightarrow W_{jk} \quad (j = 1, 2, ..., N_j; \quad k = 1, 2, ..., N_k) \tag{3.20}$$

where α is the network learning rate of the output layer, $0 < \alpha < 1$, can be taken as 0.6, and the other symbols have been defined in Equations (3.7), (3.18), and (3.15).

For the next learning, θ_k is adjusted to

$$\theta_k + \alpha E_{uk} \Rightarrow \theta_k \quad (k = 1, 2, ..., N_k) \tag{3.21}$$

where all symbols have been defined in Equations (3.9), (3.20), and (3.18).

11. Adjust W_{ij} and θ_j. For the next learning, W_{ij} is adjusted to

$$W_{ij} + \beta E_{uj}A_{ui} \Rightarrow W_{ij} \quad (i = 1, 2, ..., N_i; \quad j = 1, 2, ..., N_j) \tag{3.22}$$

where β is the network learning rate of the hidden layer, $0 < \beta < 1$, can be taken as 0.6, and the other symbols have been defined in Equations (3.6), (3.19), and (3.2).

For the next learning, θ_j is adjusted to

$$\theta_j + \beta E_{uj} \Rightarrow \theta_j \quad (j = 1, 2, ..., N_j) \tag{3.23}$$

where all symbols have been defined in Equations (3.8), (3.22), and (3.19).

12. Check whether every learning sample is randomly selected for learning under the condition of the same t. If yes, go to Step 13; otherwise, return to Step 5.
13. Calculate the root mean square error. The root mean square error is defined as

$$RMSE(\%) = \sqrt{\frac{1}{n} \sum_{i=1}^{n} (y_i - y_i^*)^2} \times 100 \qquad (3.24)$$

where all symbols of the right-hand side have been defined in Equation (1.4).
14. Calculate t_{opt}. If the current $RMSE(\%)$ is less than the last $RMSE(\%)$,

$$t \Rightarrow t_{opt}$$

15. Check the size of t. If $t < t_{max}$, $t + 1 \Rightarrow t$, and return to Step 5. If $t = t_{max}$, print a warning message, and a discrimination is required as follows: if $t_{opt} < t_{max}$, the learning process ends and we turn to the prediction process; if $t_{opt} = t_{max}$, we need to enlarge t_{maxt} by taking a number that is larger than 100000 (e.g., $t_{max} = 200000$ in Simple Case Study 2), and return to Step 5, but this phenomenon occurs in few cases.

In the aforementioned learning process, Steps 6 and 7 constitute sample forward propagation, and Steps 8 to 11 constitute error back propagation. Finally, W_{ij}, W_{jk}, θ_j and θ_k are knowledge obtained by the learning process.

The aforementioned Steps 14 and 15 constitute a technique of optimal OLTC as well as a technique of optimal iteration count.

The prediction process performs based on obtained W_{ij}, W_{jk}, θ_j, and θ_k, in the following five steps:

1. Calculate the values at each node on the hidden layer and the output layer for all learning samples and prediction samples. For learning samples, B_{uj} is calculated with Equation (3.15), C_{uk} is calculated with Equation (3.17), for N_u times in total. For prediction samples, B_{vj} is calculated with Equation (3.15), C_{vk} is calculated with Equation (3.17), but in this case u should be substituted by v in the two formulas, and N_u should be substituted by N_v, so for N_v times in total. It is noted that the limit control should not be put on the calculation of the Sigmoid function $f(x)$.
2. Revert and print the known value (desired value) at each node on the output layer for learning samples. From the normalization formula expressed with Equation (3.13), its corresponding reverting formula is as follows:

$$x = x_{min} + x'(x_{max} - x_{min}) \qquad (3.25)$$

where x_{min} and x_{max} are the minimum and maximum of the known value C_{uk}^* at each node on output layer for learning samples, respectively, referring to Equation (3.4). x' is the C_{uk}^* value after data normalization using Equation (3.13) at the starting point of the learning process, whereas x is C_{uk}^* itself. It is noted that when C_{uk} and C_{vk} are to be reverted, x' are calculated C_{uk} and C_{vk}, whereas x are reverted C_{uk} and C_{vk}, respectively.

Using Equation (3.25) to revert the normalized C_{uk}^*, the calculation is the original value expressed with Equation (3.4). Print it, and print $RMSE(\%)$ expressed with Equation (3.24).

3. Print W_{ij} and W_{jk}. W_{ij}, W_{jk}, θ_j, and θ_k are knowledge obtained by the learning process, which is worth printing out to refer to later.

4. Revert and print the value at each node on the output layer for learning samples. Use Equation (3.25) to revert the normalized calculation result C_{uk} and print it. The reverted C_{uk} are fitting results of the learning process. Comparing C_{uk} with C_{uk}^*, the fitness of the learning process is obtained, and print $RMSE(\%)$, expressed with Equation (3.24).

5. Revert and print the value at each node on the output layer for prediction samples. Use Equation (3.25) to revert the normalized calculation result C_{vk} and print it. The reverted C_{vk} are prediction results of the prediction process. If C_{vk}^* is given, comparing C_{vk} with C_{vk}^*, the accuracy of the prediction process is obtained; print $RMSE(\%)$ expressed with Equation (3.24).

BPNN converges very slowly, i.e., t is up to several 10^3–10^4 and even 10^5 for exceptional cases. How can we enhance the convergence speed? There are plenty of proposed accelerating techniques, e.g., to multiply γ_y in Equation (3.16) by a coefficient that is unequal to one. However, these techniques mostly need some new control parameters, which are determined by trial and error. In practical applications, since seeking the optimal control parameters is very time consuming, it wise to enlarge t so as to use the aforementioned BPNN procedure once.

3.1.3.2. Calculation Flowchart

The calculation flowchart for BPNN appears in Figure 3.3.

In the aforementioned calculation method and calculation flowchart, the symbols C_{uk} and C_{uk}^* under the condition of $k = 1$ are y_i and y_i^* ($i = 1, 2, ..., n, n + 1, n + 2, ..., n + k$), defined in Equation (1.4), respectively; and the symbols N_u and N_v are n and k, defined in Equation (1.4), respectively. Equation (1.4) is commonly used throughout the book.

3.1.4. Simple Case Study 1: XOR Problem

To better understand the aforementioned BPNN, here an XOR problem is introduced.

XOR is a logical operator of exclusive-OR, including two x_1 and x_2 operands. Each operand is assigned to 0 or 1. The XOR result is 0 if the two operands are the same; otherwise, it is 1. This operation can be expressed as

$$\begin{bmatrix} 0 \\ 0 \\ 1 \\ 1 \end{bmatrix} \cdot XOR \cdot \begin{bmatrix} 0 \\ 1 \\ 0 \\ 1 \end{bmatrix} = \begin{bmatrix} 0 \\ 1 \\ 1 \\ 0 \end{bmatrix} \tag{3.26}$$

The problem to be studied is to test whether BPNN can simulate the function of the XOR operator. Though Equation (3.26) is more easily proved by hand computation or programming, it is often adopted as a basic sample for testing BPNN.

3.1.4.1. Input Known Parameters

$N_u = 4$, $N_v = 4$, $N_i = 2$, $N_k = 1$, $N_j = 5$, i.e., four learning samples, four prediction samples, two input layer nodes, one output layer node, and five hidden layer nodes.

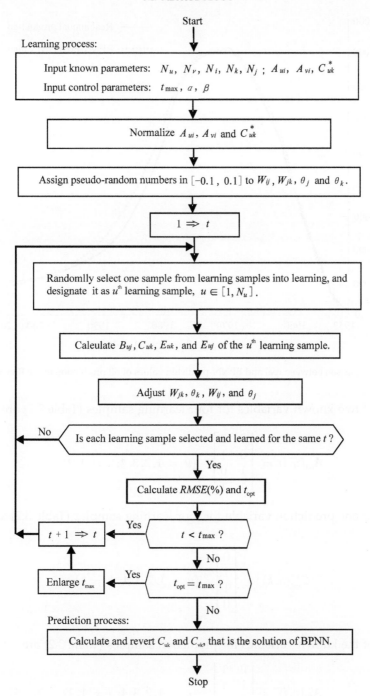

FIGURE 3.3 Calculation flowchart of BPNN.

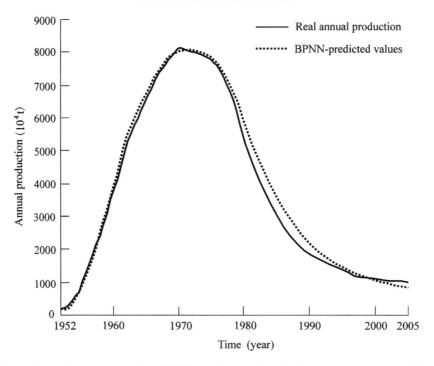

FIGURE 3.4   Comparison between real and BPNN-predicted values of oil production in the Romashkino Oilfield.

The values of two known variables for four learning samples (Table 3.2) are

$$A_{ui}(u, i) = \begin{bmatrix} 0 & 0 \\ 0 & 1 \\ 1 & 0 \\ 1 & 1 \end{bmatrix} \quad (u = 1, 2, 3, 4; \ i = 1, 2) \tag{3.27}$$

The values of one prediction variable for four learning samples (Table 3.2) are

$$C_{uk}^*(u, k) = \begin{bmatrix} 0 \\ 1 \\ 1 \\ 0 \end{bmatrix} \quad (u = 1, 2, 3, 4; \ k = 1) \tag{3.28}$$

The values of the known variables for prediction samples (Table 3.2) are

$$A_{vi}(v, i) = \begin{bmatrix} 0 & 1 \\ 1 & 1 \\ 0 & 1 \\ 1 & 1 \end{bmatrix} \quad (v = 1, 2, 3, 4; \ i = 1, 2) \tag{3.29}$$

TABLE 3.2   Input Data and Prediction Results of the XOR Problem

Sample Type	Sample No.	Two Operands		XOR Results from Definition of Equation (3.26)	BPNN Results	Ra (%)[a]
		x_1	x_2	$C_{uk}^*(u,k)$ ($u = 1, 2, 3, 4; k = 1$)	$C_{uk}(u,k)$ ($u = 1, 2, 3, 4; k = 1$)	
Learning samples	1	0	0	0	0.01	1
	2	0	1	1	0.99	1
	3	1	0	1	0.99	1
	4	1	1	0	0.01	1
Prediction samples	5	0	1	1	0.99	1
	6	1	1	0	0.01	1
	7	0	1	1	0.99	1
	8	1	1	0	0.01	1

[a]R_a(%)—absolute residual between y and y*.

3.1.4.2. Initialization of W_{ij}, W_{jk}, θ_j, and θ_k by Automatic Random Generation

$$W_{ij}(i,j) = \begin{bmatrix} 0.01269531 & 0.07714844 & -0.06347656 & 0.01464844 & 0.05761719 \\ 0.05371094 & -0.04003906 & 0.00097656 & -0.02539063 & -0.08007813 \end{bmatrix}$$

$$(i = 1,2; j = 1,2,3,4,5) \tag{3.30}$$

$$W_{jk}(j,k) = \begin{bmatrix} 0.06347656 \\ 0.03027344 \\ 0.01856995 \\ 0.03100586 \\ 0.09472656 \end{bmatrix} \quad (j = 1,2,3,4,5; \ k = 1) \tag{3.31}$$

$$\theta_j(j) = \begin{bmatrix} 0.02050781 \\ 0.08691406 \\ 0.07519531 \\ 0.01269531 \\ 0.07714844 \end{bmatrix} \quad (j = 1,2,3,4,5) \tag{3.32}$$

$$\theta_k(k) = -0.06347656 \quad (k = 1) \tag{3.33}$$

3.1.4.3. Calculation Results and Analyses

Using four learning samples [Table 3.2 or Equation (3.26)], the calculated $t_{opt} = 31876$. The result

$$y = BPNN(x_1, x_2) \tag{3.34}$$

TABLE 3.3   RMSE(%) and Predicted Results at the Different *LTC* in the Learning Process for the XOR Problem

LTC	1	10	100	1000	10000	31875	31876 (t_{opt})	31877	40000	70000	100000
RMSE(%)	0.5222	0.5420	0.5189	0.1490	0.1067	0.010001	0.010000	0.010001	0.010000	0.010109	0.010146
Predicted results C_{uk} (u = 1, 2, 3, 4; k = 1)	0.4841	0.3107	0.5767	0.1211	0.0100	0.010000	0.010000	0.010000	0.010000	0.010430	0.010572
	0.4899	0.3170	0.5448	0.8875	0.9899	0.990000	0.990000	0.990000	0.990000	0.990000	0.990000
	0.4424	0.2745	0.5397	0.8831	0.9897	0.990000	0.990000	0.990000	0.990000	0.990000	0.990000
	0.5343	0.2927	0.5705	0.2186	0.0122	0.010002	0.010000	0.010005	0.010000	0.010000	0.010000

obtained is an implicit expression corresponding to Equation (3.1), with $RMSE(\%) = 0.010000$, whereas $RMSE(\%) = 0.010146$ at $t_{max} = 100000$.

For learning samples, because $R(\%)$ expressed with Equation (1.4) could not be calculated out due to $y^* = 0$ at Samples 1 and 4 listed in Table 3.2 and Equation (3.26), respectively, it is required to use the absolute residual $R_a(\%)$ rather than $R(\%)$ (Table 3.2), and thus $\overline{R}_a^*(\%) = 1$. Similarly, $\overline{R}_a^*(\%) = 1$ for prediction samples. Table 3.3 shows that the minimum $RMSE(\%)$ and the optimal predicted results are obtained at $t = 31876$ (t_{opt}), indicating that OLTC is feasible and available.

The varying rule of $RMSE(\%)$ is shown in Table 3.3. In detail, (a) $RMSE(\%)$ decreases as LTC increases in general, but $RMSE(\%)$ increases sometimes (e.g., at $t = 10$, 31877, 70000, 100000); and (b) $RMSE(\%)$ and the predicted results at t_{opt} (31876) can be the same as that at $t > t_{opt}$ (e.g., $t = 40000$).

It is shown in Table 3.2 that whether learning process or perdition process, the results of the BPNN approach the real results of XOR, proving that BPNN can simulate the function of the XOR operator.

Summarily, using the learning samples expressed with Equations (3.27) and (3.28) of the XOR problem, the structure of BPNN and the final values of W_{ij}, W_{jk}, θ_j, and θ_k at $t_{opt} = 31876$ are obtained by the data mining tool of BPNN, and the results coincide with practicality. This structure and these final values are called *mined knowledge*. In the prediction process, this knowledge can be adopted to perform XOR calculations for any two values consisting of 0 or 1. That proves that the OLTC technique is feasible and available.

3.1.5. Simple Case Study 2: Oil Production Prediction

Through this simple case study, a special time-series prediction of BPNN is described.

Using the annual oil production y^* at time x (for 39 years, from 1952 to 1990) in the Romashkino Oilfield of Russia (Wen and Lu, 2008), BPNN is adopted to predict y^* at x (for 15 years, from 1991 to 2005).

If y^* and x are directly introduced in BPNN, this is a very simple network structure: Only one node (x) on the input layer, and only one node (y) on the output layer. It is proven from practice that this network structure is not easily converged, mainly due to the fact that there is only one node on the input layer, so there is not enough information for network learning. To solve this problem, a way of changing one-dimensional input space to multidimensional input space was proposed (Wu and Ge, 1994; Shi, 1999). Concretely, for this simple case study, at the first node on the input layer, the known variable of learning samples is A_{ui} $(u, 1)$, with $u = 1, 2, \ldots, 39$; the known variable of prediction samples is A_{vi} $(v, 1)$, with $v = 1, 2, \ldots, 15$; as A_{ui} $(u, 2)$ and A_{vi} $(v, 2)$, A_{ui}^2 $(u, 1)$, and A_{vi}^2 $(v, 1)$ are put at the second node on the input layer, respectively; as A_{ui} $(u, 3)$ and A_{vi} $(v, 3)$, A_{ui}^3 $(u, 1)$, and A_{vi}^3 $(v, 1)$ are put at the third node on the input layer, respectively; $\ldots$; as A_{ui} $(u, 6)$ and A_{vi} $(v, 6)$, A_{ui}^6 $(u, 1)$, and A_{vi}^6 $(v, 1)$ are put at the sixth node on the input layer, respectively. Now, this simple case study employs six input nodes, i.e., the number of nodes on the input layer is increased from 1 to 6, the number of known variables for samples is increased from 1 to 6, and thus one-dimensional input space is changed to multidimensional input space.

Excepting the aforementioned special way of changing one-dimensional input space to multidimensional input space, there is no change in BPNN.

3.1.5.1. Calculation Flowchart

Figure 3.3 shows the calculation flowchart.

Here $i = 1$, A_{ui} and A_{vi} are input data; $i = 2, 3, \ldots, N_i$, A_{ui} and A_{vi} are calculated out by program. That is only difference from the conventional BPNN.

The input known parameters are $N_u = 39$, $N_v = 15$, $N_i = 6$, $N_k = 1$, and $N_j = 2(N_i + N_k) - 1 = 13$; that is, 39 learning samples, 15 prediction samples, six input layer nodes, one output layer node, and 13 hidden layer nodes, respectively.

The values of the first known variable (time with units of year) for learning samples are

$$A_{ui}(u, i) = (1952, 1953, \cdots, 1990)^{\mathrm{T}} \quad (u = 1, 2, \cdots, 39;\ i = 1) \tag{3.35}$$

The values of the prediction variable (annual oil production with units of $10^4 t$) for learning samples are

$$C_{uk}^*(u, k) = (200, 300, \cdots, 1900)^{\mathrm{T}} \quad (u = 1, 2, \cdots, 39;\ k = 1) \tag{3.36}$$

The values of the first known variable (time with units of year) for prediction samples are

$$A_{vi}(v, i) = (1991, 1992, \cdots, 2005)^{\mathrm{T}} \quad (v = 1, 2, \cdots, 15;\ i = 1) \tag{3.37}$$

The elliptical contents in Equations (3.35), (3.36), and (3.37) are shown in Table 3.4. The values of the second to the sixth known variables for learning samples and prediction samples are calculated out by program in the aforementioned way.

3.1.5.2. Calculation Results and Analyses

Using 39 learning samples (Table 3.4), we calculate $t_{\mathrm{opt}} = 197636$. The result

$$y = BPNN\left(x, x^2, x^3, x^4, x^5, x^6\right) \tag{3.38}$$

is an implicit expression corresponding to Equation (3.1), with $RMSE(\%) = 0.005488$, whereas $RMSE(\%) = 0.005489$ at $t_{\mathrm{max}} = 200000$.

Table 3.4 shows that $\overline{R}1(\%) = 7.11$, $\overline{R}2(\%) = 7.54$, and $\overline{R}^*(\%) = 7.32$ (<10), proving that the results predicted by BPNN are basically available.

In summary, using the learning samples of the annual oil production for 39 years, from 1952 to 1990, in the Romashkino Oilfield of Russia (Table 3.4) for the prediction problem of

TABLE 3.4 Predicted Oil Production of the Romashkino Oilfield Using BPNN

Sample Type	Sample No.	x (Year)	Annual Oil Production $(10^4 t)^a$		Sample Type	Sample No.	x (Year)	Annual Oil Production $(10^4 t)^a$	
			y^*	y				y^*	y
Learning samples	1	1952	200	204	Learning samples	28	1979	6090	6490
	2	1953	300	244		29	1980	5450	5970
	3	1954	500	426		30	1981	4900	5441
	4	1955	800	803		31	1982	4350	4927
	5	1956	1400	1295		32	1983	3900	4445
	6	1957	1900	1848		33	1984	3450	4005
	7	1958	2400	2465		34	1985	3100	3609
	8	1959	3050	3154		35	1986	2700	3256
	9	1960	3800	3892		36	1987	2450	2943
	10	1961	4400	4616		37	1988	2200	2667
	11	1962	5000	5264		38	1989	2000	2424
	12	1963	5600	5806		39	1990	1900	2211
	13	1964	6000	6255	Prediction samples	40	1991	(1770)	2023
	14	1965	6500	6654		41	1992	(1680)	1858
	15	1966	6900	7041		42	1993	(1590)	1712
	16	1967	7350	7416		43	1994	(1500)	1584
	17	1968	7600	7729		44	1995	(1410)	1472
	18	1969	7900	7931		45	1996	(1320)	1373
	19	1970	8150	8030		46	1997	(1230)	1285
	20	1971	8100	8068		47	1998	(1180)	1207
	21	1972	8050	8072		48	1999	(1150)	1139
	22	1973	8000	8049		49	2000	(1110)	1078
	23	1974	7900	7990		50	2001	(1090)	1024
	24	1975	7800	7871		51	2002	(1070)	975
	25	1976	7600	7670		52	2003	(1050)	932
	26	1977	7250	7367		53	2004	(1040)	894
	27	1978	6750	6967		54	2005	(1020)	860

^{a}In y^*, the numbers in parenthesis are not input data but are used for calculating $R(\%)$.

oil production, the structure of BPNN and the final values of W_{ij}, W_{jk}, θ_j, and θ_k at $t_{opt} = 197636$ are obtained by the data mining tool of BPNN, and the results coincide with practicality. This structure and these final values are called *mined knowledge*. In the prediction process, this knowledge can be adopted to predict the annual oil production for 15 years from 1991 to 2005, and the prediction results basically coincide with practicality. That proves the special time-series prediction of BPNN is feasible and available. Therefore, this method can be spread to the prediction problem of oil production for other oilfields.

3.1.6. Simple Case Study 3: Prediction of Fracture-Acidizing Results

Through this simple case study, a conventional prediction of BPNN is explained.

The fracture acidizing for increasing oil production was performed in a layer of seven wells in a structure of an oilfield in eastern China, and the results were good from the two-month liquid production after this fracture acidizing. The issue to be studied is whether it is worth performing the fracture acidizing in the layer of the eighth well (Fan et al., 1998). Therefore, we'll try to adopt BPNN to predict the two-month liquid production of the layer of the eighth well if the fracture acidizing is performed on this well (Table 3.5).

This is a standard BPNN problem, without changing one-dimensional space to multidimensional space as we did in Simple Case Study 2.

3.1.6.1. Calculation Flowchart

The calculation flowchart is shown in Figure 3.3.

The known input parameters are the values of the known variables x_i ($i = 1, 2, ..., 8$) for seven learning samples and one prediction sample and the value of the prediction variable y^* for seven learning samples.

TABLE 3.5 Predicted Liquid Production After Fracture Acidizing Using BPNN

Sample Type	Sample No.	Relative Parameters for Fracture-Acidizing Results Prediction[a]								Well Test[b]		
		x_1	x_2	x_3	x_4	x_5	x_6	x_7	x_8	y^* (t)	y (t)	$R(\%)$
Learning samples	1	27.4	45.2	16.9	9.0	11.48	12.1	100	197.14	126	125.6	0.34
	2	34.7	39.0	29.2	11.0	11.57	5.6	70	197.14	196	198.9	1.48
	3	18.3	22.9	45.0	12.1	10.13	11.5	100	197.14	54	54.00	0.00
	4	10.4	23.9	99.0	21.6	10.17	2.5	85	197.14	193	192.8	0.12
	5	32.5	51.6	41.4	11.7	11.65	6.7	100	197.14	278	278.0	0.01
	6	21.2	45.7	44.0	12.3	11.25	1.3	85	197.14	121	123.5	2.03
	7	26.7	46.8	44.7	12.3	11.38	1.8	92.5	197.14	248	250.2	0.89
Prediction samples	8	35.4	36.9	166.8	11.0	10.50	42.4	85	197.14	(272)	278.0	2.21

[a]x_1 = effective thickness of production layer (m); x_2 = oil saturation (%); x_3 = permeability (mD); x_4 = effective porosity (%); x_5 = oil layer pressure (MPa); x_6 = mud content (%)); x_7 = cementing quality (centesimal system); and x_8 = oil viscosity (mPa·s).
[b]y^* = the two-month liquid production after the fracture acidizing is determined by the well test. In y^*, numbers in parentheses are not input data but are used for calculating $R(\%)$.

$N_u = 7$, $N_v = 1$, $N_i = 8$, $N_k = 1$, $N_j = 17$, i.e., seven learning samples, one prediction sample, eight input layer nodes, one output layer node, and 17 hidden layer nodes.

The values of eight known variables for seven learning samples (Table 3.5) are

$$A_{ui}(u,i) = \begin{bmatrix} 27.4 & 45.2 & 16.9 & 9.0 & 11.48 & 12.1 & 100 & 197.14 \\ 34.7 & 39.0 & 29.2 & 11.0 & 11.57 & 5.6 & 70 & 197.14 \\ 18.3 & 22.9 & 45.0 & 12.1 & 10.13 & 11.5 & 100 & 197.14 \\ 10.4 & 23.9 & 99.0 & 21.6 & 10.17 & 2.5 & 85 & 197.14 \\ 32.5 & 51.6 & 41.4 & 11.7 & 11.65 & 6.7 & 100 & 197.14 \\ 21.2 & 45.7 & 44.0 & 12.3 & 11.25 & 1.3 & 85 & 197.14 \\ 26.7 & 46.8 & 44.7 & 12.3 & 11.38 & 1.8 & 92.5 & 197.14 \end{bmatrix}$$
$$(u = 1, 2, \cdots, 7; \; i = 1, 2, \cdots, 8)$$
(3.39)

The values of one prediction variable for seven learning samples (Table 3.5) are

$$C_{uk}^*(u,k) = (126, 196, 54, 193, 278, 121, 248)^\mathsf{T}$$
$$(u = 1, 2, \cdots, 7; \; k = 1)$$
(3.40)

The values of eight known variables for one prediction sample (Table 3.5) are

$$A_{vi}(v,i) = [35.4, 36.9, 166.8, 11.0, 10.50, 42.4, 85, 197.14]$$
$$(v = 1; \; i = 1, 2, \cdots, 8)$$
(3.41)

3.1.6.2. Calculation Results and Analyses

Using seven learning samples (Table 3.5), the calculated $t_{opt} = 55095$. The result

$$y = BPNN(x_1, x_2, x_3, x_4, x_5, x_6, x_7, x_8)$$
(3.42)

is an implicit expression corresponding to Equation (3.1), with $RMSE(\%) = 0.005350$, whereas $RMSE(\%) = 0.005825$ at $t_{max} = 100000$.

Table 3.5 shows that $\overline{R}1(\%) = 0.70$, $\overline{R}2(\%) = 2.21$, and $\overline{R}^*(\%) = 1.46$ (<5), proving the results predicted by BPNN are available.

In summary, using the learning samples of seven wells (Table 3.5) for the prediction problem of fracture-acidizing results, the structure of BPNN and the final values of W_{ij}, W_{jk}, θ_j, and θ_k at $t_{opt} = 55095$ are obtained by the data mining tool of BPNN, and the results coincide with practicality. This structure and these final values are called *mined knowledge*. In the prediction process, this knowledge can be adopted to predicted the fracture-acidizing results in the eighth well, and the results are basically correct. Therefore, this method can be spread to the prediction problem of fracture-acidizing results for other structures.

3.2. CASE STUDY 1: INTEGRATED EVALUATION OF OIL AND GAS-TRAP QUALITY

3.2.1. Studied Problem

The objective of this case study is to conduct an optimal selection of traps using multigeological factors of oil-gas-pool forming, which has practical value in the stage of rolling exploration.

Located north of the Tarim Basin in western China, the Kuqa Depression covers about 40,000 km^2, stretching from the mountainous southern Tianshan fold belt in the north to the Tabei uplift in the south. It is about 400 km long (E-W) and 50–140 km wide (N-S), wide to the west and narrowing to the east. Over 10 oil and gas fields have been discovered in this depression. It is one of the richest areas of natural gas accumulation in China, of which the large gas field Kela2 has become the major gas supplier in the state project "gas in the west delivered to the east." In light of the differences in the structural features and migration mechanisms, the Kuqa Depression can be divided into two separate, north and south petroleum systems. The gas-rich Northern Kuqa Depression comprises about half the whole depression and includes the northern monocline belt, the linear Keyi anticline belt, the Qiulitake anticline belt, the Baicheng sag, and the Yangxia sag. Since it has experienced stronger tectonic movements, the geological conditions are more complicated than in the Southern Kuqa Depression (Shi et al., 2004; Shi et al., 2010). Therefore, it is a challenge to study the traps in the Northern Kuqa Depression under such complicated conditions, and it is a new trial to apply BPNN and MRA to the evaluation of trap quality.

We used data from 29 traps in the Northern Kuqa Depression, of which 27 were taken as learning samples and 2 as prediction samples. Each trap has 14 parameters and the trap quality (Table 3.6). However, the trap quality of each prediction samples was judged to be less reliable. BPNN and MRA were then applied to validate them (Shi et al., 2004; Shi, 2011).

3.2.2. Input Data

The known input data are x_i ($i = 1, 2, \ldots, 14$) of 27 learning samples and three prediction samples and y^* of 27 learning samples. Among them, 21 traps, including numbers 1–18, 25, 26, and 28, are located in the Keyi structural belt; 6 traps, including numbers 19–22, 29, and 30, are located in the Yangxia sag, and 3 traps, including numbers 23, 24, and 27, are located in the Qiulitake structural belt.

It is important to conduct trap evaluation based on a combination of time and space analysis of oil/gas generation, migration, trapping, accumulation, sealing, and preservation. Thus geological factors selected for the evaluation should be able to describe these six processes adequately. In Table 3.7, all relative attributes of each parameter are normalized within [0, 1], their corresponding weights are also normalized within [0, 1], and thus each parameter (coefficient) is within [0, 1]. Taking trap No. 2 as an example, the derivation of trap coefficient $x_9 = 1$ is explained: (a) Trap reliability $= 1$ since this trap is reliable; (b) Trap type $= 1$ since the trap is an anticline; (c) Trap closed area $= 1$ since the area of the trap is larger than 30 km^2; and (d) Trap relief $= 1$ since the relief of the trap is larger than 300 m. Multiplying these four attributes with their corresponding weights (Table 3.7), the cumulating sum is trap coefficient x_9 as follows:

$$\text{Trap coefficient}(x_9) = \text{Trap reliability} \times \text{Its weight} + \text{Trap type} \times \text{Its weight}$$
$$+ \text{Trap closed area} \times \text{Its weight} + \text{Trap relief} \times \text{Its weight}$$
$$= 1 \times 0.35 + 1 \times 0.45 + 1 \times 0.1 + 1 \times 0.1 = 1$$

TABLE 3.6   Input Data for Trap Evaluation of the Northern Kuqa Depression

Sample Type	Trap No.	Relative Parameters for Trap Evaluation[a]														Trap Quality[b]
		x_1	x_2	x_3	x_4	x_5	x_6	x_7	x_8	x_9	x_{10}	x_{11}	x_{12}	x_{13}	x_{14}	y^*
Learning samples	1	1	1	2	2362	300	58	2	0.45	0.753	0.960	0.935	0.808	0.900	6.6	2
	2	1	2	1.5	3150	350	42	2	0.85	1.000	1.000	0.935	0.921	0.900	210.5	1
	3	1	2	2	3650	350	12	2	0.51	0.975	1.000	0.935	0.763	0.900	8.3	1
	4	1	2	2	2630	150	17	2	0.51	0.818	0.898	0.935	0.763	0.900	1.9	2
	5	1	3	2	5950	750	135	2	0.45	0.895	0.940	0.820	0.808	0.900	171.9	1
	6	1	2	2	3970	300	28	2	0.75	0.950	0.868	0.820	0.763	0.900	5.5	2
	7	1	2	2	4680	300	27	2	0.75	0.828	0.868	0.820	0.808	0.900	12.6	2
	8	1	1	2	1450	700	54	1	0.45	0.778	0.898	0.935	0.751	0.900	7.13	2
	9	1	1	3	1450	1000	74	1	0.45	0.778	0.898	0.935	0.808	0.900	9.8	2
	10	1	2	3	1200	750	23	1	0.45	0.888	0.970	0.840	0.681	0.900	1.4	2
	11	1	1	2	1550	1780	34	2	0.45	0.778	0.860	0.820	0.856	0.900	43.2	2
	12	1	1	2	6700	250	11	2	0.45	0.693	0.930	0.935	0.936	0.900	13.6	1
	13	1	1	2	5500	500	16	2	0.45	0.693	1.000	0.935	0.936	0.900	20.3	1
	14	1	1	2	5500	200	11	2	0.45	0.753	1.000	0.935	0.936	0.900	13.6	1
	15	1	2	1	850	550	50	1	0.45	1.000	0.868	0.820	0.681	0.900	1.82	2
	16	1	2	3	1510	750	57	1	0.45	1.000	1.000	0.840	0.794	0.930	3.48	1
	17	1	1	3	3510	1150	161	2	0.75	0.865	1.000	0.900	0.794	0.930	179.9	1
	18	1	1	3	2700	300	56	1	0.45	0.888	0.970	0.840	0.714	0.900	3.41	2
	19	2	3	1	4220	460	66	2	0.51	0.808	0.882	0.840	0.756	0.930	9.8	2
	20	2	2	3	5600	300	27	2	0.51	0.828	0.882	0.840	0.756	0.930	29.2	2
	21	2	1	3	8580	300	17	2	0.51	0.780	0.898	0.840	0.748	0.930	18.5	2
	22	2	3	1	4940	260	46	2	0.51	0.808	0.798	0.840	0.756	0.930	3.9	2
	23	3	1	1	1855	1800	42	2	0.85	0.865	0.798	0.840	0.909	0.900	4	1
	24	3	3	2	4755	700	83	2	0.85	0.808	1.000	0.850	0.920	0.900	169.4	1
	25	1	2	3	1000	400	82	1	0.45	0.913	0.970	0.885	0.673	0.930	2.61	2
	26	1	1	4	3670	300	133	2	0.45	0.753	0.970	0.885	0.673	0.900	37.9	2
	27	3	3	1	2750	1100	118	2	0.85	0.955	0.898	0.780	0.763	0.900	10.2	2
Prediction samples	28	2	3	3	5660	340	56	2	0.51	0.808	0.860	0.850	0.748	0.930	71.8	(2)
	29	2	1	3	5850	180	17	2	0.51	0.753	0.758	0.840	0.728	0.930	19.1	(2)

[a] x_1 = unit structure (1—linear anticline belt, 2—Yangxia sag, 3—Qiulitake anticline belt); x_2 = trap type (1—faulted nose, 2—anticline, 3—faulted anticline); x_3 = petroliferous formation (1—E, 1.5—E+K, 2—K, 3—J, 4—T); x_4 = trap depth (m); x_5 = trap relief (m); x_6 = trap closed area (km^2); x_7 = formation HC identifier (1—oil, 2—gas); x_8 = data reliability (0—1); x_9 = trap coefficient (0—1); x_{10} = source rock coefficient (0—1); x_{11} = reservoir coefficient (0—1); x_{12} = preservation coefficient (0—1); x_{13} = configuration coefficient (0—1); and x_{14} = resource quantity (Mt, million ton oil-equivalent).

[b] y^* = trap quality (1—high,2—low) assigned by geologists. In y^*, numbers in parentheses are not input data but are used for calculating R(%).

TABLE 3.7 Dependent Attribute Weight for Five Parameters with 0−1 Values for Trap Reservoir Forming of the Northern Kuqa Depression

Parameter	Dependent Attribute	Weight
x_9 Trap coefficient	Trap reliability	0.35
	Trap type	0.45
	Trap closed area	0.1
	Trap relief	0.1
x_{10} Source rock coefficient	Distance from trap to source	0.2
	Migration direction and path	0.1
	Migration pathway	0.3
	Hydrocarbon expulsion concentration[a]	0.2
	Hydrocarbon migration-accumulation concentration[b]	0.2
x_{11} Reservoir coefficient	Lithology	0.3
	Sedimentary facies	0.3
	Reservoir thickness	0.05
	Reservoir type	0.1
	Porosity	0.1
	Permeability	0.15
x_{12} Preservation coefficient	Cap rock lithology	0.15
	Cap rock area	0.2
	Cap rock thickness	0.2
	Hydrodynamics	0.125
	Water type	0.05
	Mineralization	0.025
	Cap rock sedimentary facies	0.04
	Igneous rock damage	0.02
	Fault damage	0.15
	Fault type	0.04
x_{13} Configuration coefficient	Time configuration	0.7
	Space configuration	0.3

[a]Calculated using the hydrocarbon expulsion model.
[b]Calculated using the hydrocarbon migration-accumulation model (Shi, 2005; Shi et al., 2010).

3.2.3. Application Comparisons between BPNN and MRA

3.2.3.1. Learning Process

Using the 27 learning samples (Table 3.6) and by BPNN and MRA, the two functions of trap quality (y) with respect to 14 parameters (x_1, x_2, ..., x_{14}) were constructed, i.e., Equation (3.43) corresponding to BPNN formula (3.1) and Equation (3.44) corresponding to MRA formula (2.14), respectively.

Using BPNN, there are 14 input layer nodes, one output layer node, and 29 hidden layer nodes. Setting $t_{max} = 40000$, the calculated $t_{opt} = 10372$. The result (Shi et al., 2004; Shi, 2011)

$$y = BPNN(x_1, x_2, ..., x_{14}) \tag{3.43}$$

is an implicit expression corresponding to Equation (3.1), with $RMSE(\%) = 0.009999$, whereas $RMSE(\%) = 0.009999$ at $t_{max} = 40000$.

Using MRA, the result is an explicit linear function (Shi et al., 2004; Shi, 2011):

$$\begin{aligned} y = {}& 13.766 - 0.026405x_1 - 0.038781x_2 - 0.0016605x_3 - 0.00012343x_4 \\ & -0.00038344x_5 - 0.002442x_6 + 0.045162x_7 + 0.58229x_8 - 3.3236x_9 \\ & -2.1313x_{10} - 3.3651x_{11} - 4.1977x_{12} - 0.67296x_{13} + 0.0010516x_{14} \end{aligned} \tag{3.44}$$

Equation (3.44) yields a residual variance of 0.14157 and a multiple-correlation coefficient of 0.92652. From the regression process, trap quality (y) is shown to depend on the 14 parameters in decreasing order: preservation coefficient (x_{12}), source rock coefficient (x_{10}), trap coefficient (x_9), trap depth (x_4), trap relief (x_5), reservoir coefficient (x_{11}), data reliability (x_8), trap closed area (x_6), resource quantity (x_{14}), unit structure (x_1), trap type (x_2), formation HC identifier (x_7), configuration coefficient (x_{13}), and petroliferous formation (x_3).

Substituting 14 parameters of 27 learning samples (Table 3.6) in Equations (3.43) and (3.44) respectively, the trap quality (y) of each learning sample is obtained. Table 3.8 shows the results of learning process by BPNN and MRA.

3.2.3.2. Prediction Process

Substituting 14 parameters of two prediction samples (Table 3.6) in Equations (3.43) and (3.44), respectively, the trap quality (y) of each prediction sample is obtained. Table 3.8 shows the results of the prediction process by BPNN and MRA. For the predicted quality values of these two traps (Nos. 28 and 29), the results calculated by BPNN are 2 and 2, respectively, whereas the results calculated by MRA are 1.9444 and 2.6174, respectively. From the real exploration discovery after this prediction, no commercial gas flow has yet been found at traps No. 28 and 29.

Table 3.9 shows that the results calculated by BPNN are quite accurate, i.e., not only the fitting residual $\overline{R}_1(\%) = 0.06$ but also the prediction residual $\overline{R}_2(\%) = 0$, and thus the total mean absolute relative residual $\overline{R}^*(\%) = 0.03$, which is almost completely consistent with the real exploration discovery; whereas for the results calculated by MRA, not only $\overline{R}_1(\%) = 9.93$ but also $\overline{R}_2(\%) = 16.83$, and thus $\overline{R}^*(\%) = 13.38$.

3.2.3.3. Application Comparisons between BPNN and MRA

As shown in $\overline{R}^*(\%) = 13.38$ of MRA (Table 3.9), the nonlinearity of the relationship between the predicted value y and its relative parameters (x_1, x_2, ..., x_{14}) is strong from Table 1.2. Since BPNN can describe complex nonlinear relationships, its advantage is very high

TABLE 3.8   Prediction Results from Trap Evaluation of the Northern Kuqa Depression

Sample Type	Trap No.	y^*	BPNN		MRA	
			y	$R(\%)$	y	$R(\%)$
Learning samples	1	2	2	0	1.8160	9.20
	2	1	1	0	0.7675	23.25
	3	1	1.007	0.6977	1.1139	11.39
	4	2	2	0	2.0368	1.84
	5	1	1	0	1.0665	6.65
	6	2	2	0	1.9427	2.86
	7	2	2	0	2.0816	4.08
	8	2	2	0	2.0287	1.43
	9	2	2	0	1.6267	18.67
	10	2	2	0	2.1640	8.20
	11	2	2	0	1.7614	11.93
	12	1	1	0	1.1479	14.79
	13	1	1	0	1.0458	4.58
	14	1	1	0	0.9666	3.34
	15	2	2	0	2.1342	6.71
	16	1	1.009	0.9326	1.1142	11.42
	17	1	1	0	1.1510	15.10
	18	2	2	0	1.9732	1.34
	19	2	2	0	1.9430	2.85
	20	2	2	0	1.9186	4.07
	21	2	2	0	1.7618	11.91
	22	2	2	0	2.1525	7.62
	23	1	1	0	1.3903	39.03
	24	1	1	0	1.1278	12.78
	25	2	2	0	1.9590	2.05
	26	2	2	0	2.2146	10.73
	27	2	2	0	1.5941	20.30
Prediction samples	29	2	2	0	1.9444	2.78
	30	2	2	0	2.6174	30.87

[a]y^* = trap quality (1—high, 2—low) assigned by geologists.

TABLE 3.9   Comparison Between the Applications of BPNN and MRA to trap Evaluation of the Northern Kuqa Depression

Algorithm	Fitting Formula	Mean Absolute Relative Residual			Dependence of the Predicted Value (y) on Parameters ($x_1, x_2, ..., x_{14}$), in Decreasing Order	Time Consumed on PC (Intel Core 2)	Solution Accuracy
		$\overline{R}1(\%)$	$\overline{R}2(\%)$	$\overline{R}^*(\%)$			
BPNN	Nonlinear, implicit	0.06	0	0.03	N/A	1 min 20 s	Very high
MRA	Linear, explicit	9.93	16.83	13.38	$x_{12}, x_{10}, x_9, x_4, x_5, x_{11}, x_8,$ $x_6, x_{14}, x_1, x_2, x_7, x_{13}, x_3$	<1 s	Low

solution accuracy; since MRA only can describe simple linear relationships, its solution accuracy is low, but its advantage is fast and can establish the order of dependence between the predicted value y and its relative parameters ($x_1, x_2, ..., x_{14}$), e.g., MRA can indicate that the trap quality depends on the preservation coefficient first, the source rock coefficient next, the trap coefficient, ..., and the petroliferous formation last.

3.2.4. Summary and Conclusions

In summary, using data for the trap evaluation of the Northern Kuqa Depression, i.e., the 14 parameters (unit structure, trap type, petroliferous formation, trap depth, trap relief, trap closed area, formation HC identifier, data reliability, trap coefficient, source rock coefficient, reservoir coefficient, preservation coefficient, configuration coefficient, and resource quantity) and trap quality of 27 traps, the structure of BPNN and the final values of W_{ij}, W_{jk}, θ_j, and θ_k at $t_{opt} = 10372$ are obtained by the data mining tool of BPNN, and the results coincide with practicality. This structure and these final values are called *mined knowledge*. In the prediction process, this knowledge can be adopted to predict the trap quality of the other two traps, and the results coincide with the real exploration discovery after this prediction. Therefore, this method can be spread to the trap evaluation of other exploration areas.

Moreover, it is found that (a) since this case study is a strong nonlinear problem, the major data mining tool can adopt BPNN but not MRA; (b) in this case study, MRA established the order of dependence between the trap quality and its relative 14 parameters under the condition of linearity; thus MRA can be employed as an auxiliary tool; and (c) certainly, C-SVM and R-SVM can be applied to this case study and run at a speed as fast as 20 times or more of BPNN; however, it is easy to code the BPNN program, whereas it is very complicated to code the C-SVM or R-SVM program, so BPNN is a good software for this case study when neither the C-SVM nor R-SVM program is available.

3.3. CASE STUDY 2: FRACTURES PREDICTION USING CONVENTIONAL WELL-LOGGING DATA

3.3.1. Studied Problem

The objective of this case study is to predict fractures using conventional well-logging data, which has practical value when the data of imaging log and core samples are limited.

Located southeast of the Biyang Sag in Nanxiang Basin in central China, the Anpeng Oilfield covers an area of about 17.5 km², close to Tanghe-zaoyuan in the northwest-west, striking a large boundary fault in the south, and close to a deep sag in the east. As an inherited nose structure plunging from northwest to southeast, this oilfield is a simple structure without faults, where commercial oil and gas flows have been discovered (Ming et al., 2005; Wang et al., 2006). One of its favorable pool-forming conditions is that the fractures are found to be well developed at formations as deep as 2800 m or more. These fractures provide favorable oil-gas migration pathways and enlarged the accumulation space.

We used data from 33 samples in Wells An1 and An2, of which 29 were taken as learning samples and four prediction samples. Each sample contains seven well-logging data and one imaging log result (Table 3.10). However, the imaging log results of the four prediction

TABLE 3.10   Input Data for Fracture Prediction of Wells An1 and An2

Sample Type	Sample No.	Well No.	Depth (m)	$x_1 \Delta t$ (0~1)	$x_2 \rho$ (0~1)	$x_3 \varphi_N$ (0~1)	$x_4 R_{xo}$ (0~1)	x_5 R_{LLD} (0~1)	x_6 R_{LLS} (0~1)	x_7 R_{DS} (0~1)	$y^* IL$
										Fracture[b]	
Learning samples	1	An1	3065.13	0.5557	0.2516	0.8795	0.3548	0.6857	0.6688	0.0169	1
	2		3089.68	0.9908	0.0110	0.8999	0.6792	0.5421	0.4071	0.1350	1
	3		3098.21	0.4444	0.1961	0.5211	0.7160	0.7304	0.6879	0.0425	1
	4		3102.33	0.4028	0.3506	0.5875	0.6218	0.6127	0.5840	0.0287	1
	5		3173.25	0.3995	0.3853	0.0845	0.5074	0.8920	0.8410	0.0510	1
	6		3180.37	0.6117	0.6420	0.0993	0.6478	0.9029	0.8511	0.0518	1
	7		3202.00	0.6463	0.5205	0.5351	0.7744	0.2919	0.3870	0.0951	2
	8		3265.37	0.4154	0.9545	0.4397	0.6763	0.2906	0.5173	0.2267	2
	9		3269.87	0.7901	0.6601	0.1487	0.8994	0.9257	0.9325	0.0068	1
	10		3307.87	0.7162	0.1475	0.4481	0.9164	0.7827	0.7992	0.0165	1
	11		3357.37	0.5546	0.4778	0.0741	0.7725	0.9756	0.9237	0.0519	1
	12		3377.03	0.4909	0.3654	0.1816	0.7625	0.8520	0.8237	0.0283	1
	13		3416.48	0.2567	0.5843	0.2043	0.3412	0.7369	0.7454	0.0085	2
	14		3445.37	0.0944	0.9818	0.5124	0.7614	0.5943	0.6321	0.0378	2
	15		3446.12	0.5215	0.8091	0.7594	0.6924	0.7186	0.7572	0.0386	2
	16		3485.25	0.9443	0.2647	0.9904	0.4794	0.4189	0.4776	0.0587	2
	17		3575.00	0.2078	0.0000	0.0358	0.8246	0.9872	0.9800	0.0072	1
	18		3645.00	0.1193	0.6953	0.8879	0.7839	0.8323	0.8409	0.0086	1
	19		3789.37	0.0579	0.6889	0.9418	0.7261	0.8902	0.8947	0.0045	1

(Continued)

TABLE 3.10   Input Data for Fracture Prediction of Wells An1 and An2—cont'd

Sample Type	Sample No.	Well No.	Depth (m)	x_1 Δt (0~1)	x_2 ρ (0~1)	x_3 φ_N (0~1)	x_4 R_{xo} (0~1)	x_5 R_{LLD} (0~1)	x_6 R_{LLS} (0~1)	x_7 R_{DS} (0~1)	y^* IL
						Relative Parameters for Fracture Prediction[a]						**Fracture**[b]
	20	An2	992.795	0.3471	0.9624	0.3848	0.6115	0.8245	0.8388	0.0143	2
	21		1525.37	0.5256	0.3256	0.0821	0.7450	0.9888	0.9234	0.0654	1
	22		1527.25	0.0753	0.5441	0.1345	0.6750	0.8468	0.9255	0.0787	1
	23		1867.12	0.3145	0.1325	0.0368	0.5744	0.9425	0.8547	0.0878	1
	24		1880.00	0.7755	0.8347	0.5546	0.4578	0.1894	0.4265	0.2371	2
	25		2045.87	0.4928	0.2110	0.5977	0.6892	0.7411	0.6071	0.1340	1
	26		2085.25	0.8678	0.0833	0.9997	0.4085	0.1973	0.4117	0.2144	2
	27		2112.13	0.5467	0.2961	0.8235	0.7250	0.6328	0.6825	0.0497	1
	28		2355.37	0.4524	0.3426	0.6005	0.7658	0.8992	0.8346	0.0646	1
	29		2358.00	0.6463	0.5205	0.5351	0.7744	0.2919	0.3870	0.0951	2
Prediction samples	30	An1	3164.00	0.5300	0.3333	0.0758	0.8939	0.9918	0.9863	0.0055	(1)
	31		3166.50	0.5282	0.4589	0.0459	0.7140	1.0000	1.0000	0.0000	(1)
	32	An2	980.485	0.2024	0.4288	0.2149	0.5581	0.8489	0.8504	0.0015	(1)
	33		987.018	0.0631	0.5278	0.3450	0.7403	0.7368	0.7295	0.0073	(1)

[a]x_1 = acoustictime (Δt), x_2 = compensated neutron density (ρ), x_3 = compensated neutron porosity (φ_N), x_4 = microspherically focused resistivity (R_{xo}), x_5 = deep laterolog resistivity (R_{LLD}), x_6 = shallow laterolog resistivity (R_{LLS}), x_7 = absolute difference of R_{LLD} and R_{LLS} (R_{DS}); and the data of these seven well-logging are normalized over the interval [0, 1].
[b]y^* = fracture identification (1—fracture, 2—nonfracture) determined by the imaging log (IL). In y^*, numbers in parentheses are not input data but are used for calculating R(%).

samples need to be further validated. BPNN and MRA were then applied to confirm the fractures (Shen and Gao, 2007).

3.3.2. Input Data

The known input parameters are the values of the known variables x_i ($i = 1, 2, ..., 7$) for 29 learning samples and four prediction samples, and the value of the prediction variable y^* for 29 learning samples.

3.3.3. Application Comparisons between BPNN and MRA

3.3.3.1. Learning Process

Using the 29 learning samples (Table 3.10) and by BPNN and MRA, the two functions of fracture identification (y) with respect to seven well-logging data ($x_1, x_2, ..., x_7$) have been constructed, i.e., Equation (3.45) corresponding to BPNN formula (3.1) and Equation (3.46) corresponding to MRA formula (2.14), respectively.

Using BPNN, there are seven input layer nodes, one output layer node, and 15 hidden layer nodes. Setting $t_{max} = 20000$, the calculated $t_{opt} = 6429$. The result

$$y = BPNN(x_1, x_2, ..., x_7) \tag{3.45}$$

is an implicit expression corresponding to Equation (3.1), with $RMSE(\%) = 0.009999$, whereas $RMSE(\%) = 0.009999$ at $t_{max} = 20000$.

Using MRA, the result is an explicit linear function:

$$\begin{aligned} y = {}& 2.759 + 0.099491x_1 + 0.64368x_2 - 0.2826x_3 - 0.59405x_4 - 2.005644x_5 \\ & + 0.43176x_6 - 1.9752x_7 \end{aligned} \tag{3.46}$$

Equation (3.46) yields a residual variance of 0.25054 and a multiple correlation coefficient of 0.86571. From the regression process, fracture identification (y) is shown to depend on the seven well-logging data in decreasing order: deep laterolog resistivity (x_5), compensated neutron density (x_2), microspherically focused resistivity (x_4), absolute difference of R_{LLD} and R_{LLS} (x_7), compensated neutron porosity x_3), shallow laterolog resistivity (x_6), and acoustictime (x_1).

Substituting seven well-logging data of 29 learning samples (Table 3.10) in Equations (3.45) and (3.46), respectively, the fracture identification (y) of each learning sample is obtained. Table 3.11 shows the results of the learning process by BPNN and MRA.

TABLE 3.11   Predicted Fracture Results of Wells An1 and An2

Sample Type	Sample No.	Well No.	Depth (m)	y^* IL	Fracture Identification[a]			
					BPNN		MRA	
					y	$R(\%)$	y	$R(\%)$
Learning samples	1	An1	3065.13	1	1	0	1.40	39.71
	2		3089.68	1	1	0	1.03	2.88
	3		3098.21	1	1	0	1.11	10.50
	4		3102.33	1	1	0	1.46	45.60
	5		3173.25	1	1	0	1.19	19.49
	6		3180.37	1	1	0	1.27	27.45
	7		3202.00	2	2	0	1.94	2.96
	8		3265.37	2	2	0	2.08	4.07
	9		3269.87	1	1	0	1.22	21.88
	10		3307.87	1	1	0	1.00	0.31

(Continued)

TABLE 3.11   Predicted Fracture Results of Wells An1 and An2—cont'd

Sample Type	Sample No.	Well No.	Depth (m)	y^* IL	BPNN		MRA	
					y	$R(\%)$	y	$R(\%)$
	11		3357.37	1	1	0	0.98	1.85
	12		3377.03	1	1	0	1.13	12.97
	13		3416.48	2	1.990	0.5132	1.73	13.63
	14		3445.37	2	2	0	1.81	9.52
	15		3446.12	2	1.993	0.3613	1.52	24.24
	16		3485.25	2	2	0	1.71	14.56
	17		3575.00	1	1	0	0.71	29.13
	18		3645.00	1	1	0	1.18	17.87
	19		3789.37	1	1	0	1.10	10.27
	20	An2	992.795	2	2	0	1.62	18.93
	21		1525.37	1	1	0	0.84	15.85
	22		1527.25	1	1	0	1.22	22.35
	23		1867.12	1	1	0	0.83	17.07
	24		1880.00	2	2	0	2.28	14.04
	25		2045.87	1	1	0	0.88	12.34
	26		2085.25	2	2	0	1.73	13.38
	27		2112.13	1	1	0	1.27	26.80
	28		2355.37	1	1	0	0.83	17.08
	29		2358.00	2	2	0	1.94	2.96
Prediction samples	30	An1	3164.00	1	1	0	0.90	10.03
	31		3166.50	1	1	0	1.10	9.60
	32	An2	980.485	1	1	0	1.32	32.45
	33		987.018	1	1	0	1.39	39.06

[a]y^* = fracture identification (1—fracture, 2—nonfracture) determined by the imaging log (IL).

3.3.3.2. Prediction Process

Substituting seven well-logging data of four prediction samples (Table 3.10) in Equations (3.45) and (3.46), respectively, the fracture identification (y) of each prediction sample is obtained. Table 3.11 shows the results of the prediction process by BPNN and MRA.

Table 3.12 shows that the results calculated by BPNN are quite accurate, i.e., not only $\overline{R}_1(\%) = 0$, but also $\overline{R}_2(\%) = 0.03$, and thus $\overline{R}^*(\%) = 0.015$, which is almost completely

TABLE 3.12   Comparison between the Applications of BPNN and MRA to Fracture Prediction of Wells An1 and An2

Algorithm	Fitting Formula	Mean Absolute Relative Residual			Dependence of the Predicted Value (y) on Parameters ($x_1, x_2, ..., x_7$), in Decreasing Order	Time Consumed on PC (Intel Core 2)	Solution Accuracy
		$\overline{R}1(\%)$	$\overline{R}2(\%)$	$\overline{R}^*(\%)$			
BPNN	Nonlinear, implicit	0	0.03	0.015	N/A	50 s	Very high
MRA	Linear, explicit	16.20	22.79	19.50	$x_5, x_2, x_4, x_7, x_3, x_6, x_1$	<1 s	Low

consistent with the real exploration results; whereas for the results calculated by MRA, not only $\overline{R}_1(\%) = 16.20$, but also $\overline{R}_2(\%) = 22.79$, and thus $\overline{R}^*(\%) = 19.50$.

3.3.3.3. Application Comparisons between BPNN and MRA

From $\overline{R}^*(\%) = 19.50$ of MRA (Table 3.12) we can see that the nonlinearity of the relationship between the predicted value y and its relative parameters ($x_1, x_2, ..., x_7$) is strong from Table 1.2. Since BPNN can describe complex nonlinear relationships, its advantage is very high solution accuracy; since MRA only can describe simple linear relationships, its solution accuracy is low, but its advantage is speed and it can establish the order of dependence between the predicted value y and its relative parameters ($x_1, x_2, ..., x_7$), e.g., MRA can indicate that the fracture identification depends on the deep laterolog resistivity first, the compensated neutron density next, the microspherically focused resistivity, ..., and the acoustictime last.

3.3.4. Summary and Conclusions

In summary, using data for fracture prediction of Wells An1 and An2, i.e., the seven well-logging data (acoustictime, compensated neutron density, compensated neutron porosity, microspherically focused resistivity, deep laterolog resistivity, shallow laterolog resistivity, and absolute difference of R_{LLD} and R_{LLS}) and imaging log result of 29 samples, the structure of BPNN and the final values of W_{ij}, W_{jk}, θ_j, and θ_k at $t_{opt} = 6429$ are obtained by the data mining tool of BPNN, and the results coincide with practicality. This structure and these final values are called *mined knowledge*. In the prediction process, this knowledge can be adopted to predict the fractures of another four samples and the results are consistent with the imaging log results. Therefore, this method can be spread to the fracture prediction of other oilfields.

Moreover, it is found that (a) since this case study is a strong nonlinear problem, the major data mining tool can adopt BPNN but not MRA; (b) in this case study, MRA established the order of dependence between the fracture identification and its relative seven well-logging data under the condition of linearity, and thus MRA can be employed as an auxiliary tool; and (c) certainly, C-SVM and R-SVM can be applied to this case study and run at a speed as fast as 20 times or more than BPNN; however, it is easy to code the BPNN program, whereas it is very complicated to code the C-SVM or R-SVM program, so BPNN is a good software for this case study when neither the C-SVM nor R-SVM program is available.

EXERCISES

3-1. In applying the conditions of BPNN, the number of known variables for samples must be greater than one, i.e., the number of input layer nodes must be greater than one. Why?

3-2. Why does BPNN as introduced in this book use only one hidden layer?

3-3. Why is the formula employed for the number of hidden nodes $N_{hidden} = 2(N_{input} + N_{output}) - 1$?

3-4. Since each of Case Studies 1 and 2 is a strong nonlinear problem, the data mining tool can adopt BPNN but not MRA. How can we determine whether these two case studies are strong nonlinear problems?

References

Fan, X., Wu, H., Zhang, H., 1998. Nerve network expert system and its application to a strategic decision for fracturing and acidizing. Petro. Expl. Devel. 25 (1), 73−75 (in Chinese with English abstract).

Güler, I., Übeyli, E.D., 2003. Detection of ophthalmic artery stenosis by least-mean-squares back-propagation neural network. Comput. Biol. Med. 33 (4), 333−343.

Hecht-Nielsen, R., 1989. Theory of the backpropagation neural network. In: Proceedings of the Int. Joint Conf. on Neural Networks. Washington, DC, USA, pp. 593−605.

Ming, H., Jin, Z., Li, Q., Qu, Y., Chen, X., 2005. Sedimentary facies of deep sequences of Anpeng Oilfield in Biyang Depression and its control over oil-gas distribution. J. Earth Sci. Env. 27 (2), 48−51 (in Chinese with English abstract).

Rumelhart, D.E., Hinton, G.E., Williams, R.J., 1986. Learning internal representations by error propagation. In: Rumelhart, D.E., McClelland, J.L. (Eds.), Parallel Distributed Processing, vol. 1. MIT Press, Cambridge, MA, USA, pp. 318−362.

Shen, H., Gao, S., 2007. Research on fracture identification based on BP neural network. Fault-Block Oil Gas Field 14 (2), 60−62 (in Chinese with English abstract).

Shi, G., 1999. New Computer Application Technologies in Earth Sciences. Petroleum Industry Press, Beijing, China (in Chinese).

Shi, G., Zhou, X., Zhang, G., Shi, X., Li, H., 2004. The use of artificial neural network analysis and multiple regression for trap quality evaluation: a case study of the Northern Kuqa Depression of Tarim Basin in western China. Mar. Petro. Geol. 21 (3), 411−420.

Shi, G., 2005. Numerical Methods of Petroliferous Basin Modeling, third ed. Petroleum Industry Press, Beijing, China.

Shi, G., 2009. The use of support vector machine for oil and gas identification in low-porosity and low-permeability reservoirs. Int. J. Math. Model. Numer. Optimisa. 1 (1/2), 75−87.

Shi, G., Zhang, Q., Yang, X., Mi, S., 2010. Oil and gas assessment of the Kuqa Depression of Tarim Basin in western China by simple fluid flow models of primary and secondary migrations of hydrocarbons. J. Petro. Sci. Eng. 75 (1−2), 77−90.

Shi, G., 2011. Four classifiers used in data mining and knowledge discovery for petroleum exploration and development. Adv. Petro. Expl. Devel. 2 (2), 12−23.

Wang, G., Guo, Q., Zhao, Q., Gong, Y., Zhao, L., 2006. Effective fracturing cut-off values of Extremely-low permeability reservoir—take the Anpeng Oilfield of Biyang Depression as an example. J. Southwest Petro. Univ. 28 (4), 40−43 (in Chinese with English abstract).

Wang, W., 1995. Principles of Artificial Neural Networks. Beijing University of Aeronautics and Astronautics Press, Beijing, China (in Chinese).

Wen, X., Lu, C., 2008. Characteristics of geological modeling of the Romashkino Oilfield in the later stage of production. Foreign Oilfield Eng. 24 (4), 1−2 (in Chinese).

Wu, X., Ge, J., 1994. The application of artificial neural network in predicting output of oil fields. Petro. Expl. Devel. 21 (3), 75−78 (in Chinese with English abstract).

4

Support Vector Machines

A *support vector machine* (SVM) is a new machine-learning approach based on statistical learning theory. This chapter introduces the *classification* of SVM (*C*-SVM) and *regression* of SVM (*R*-SVM) as well as their applications in geosciences.

Section 4.1 (methodology) introduces the applying ranges and conditions, basic principles, calculation methods, and calculation flowcharts of *C*-SVM and *R*-SVM.

Section 4.2 (gas layer classification based on porosity, permeability, and gas saturation) introduces Case Study 1 of *C*-SVM. Using data for the gas layer classification based on three parameters of the Tabamiao area, that is, the three parameters (porosity, permeability, gas saturation) and gas test result of 38 oil layers, an explicit nonlinear function is constructed by the data mining tool of *C*-SVM, and the results are consistent with the gas test. This function is called *mined knowledge*. In the prediction process, applying this knowledge, that is, substituting the values of porosity, permeability, and gas saturation in this nonlinear function, respectively, the results are also consistent with the gas test. Therefore, this method can be spread to the gas layer classification based on porosity, permeability, and gas saturation of other areas. Moreover, it is found that (a) since this case study is a very strong nonlinear problem, the major data mining tool can adopt *C*-SVM but not multiple regression analysis (MRA; see Chapter 2), *R*-SVM, or error back-propagation neural network (BPNN; see Chapter 3); and (b) in this case study, MRA established the order of dependence between the gas layer classification and its relative porosity, permeability, and gas saturation under the condition of linearity, thus MRA can be employed as an auxiliary tool.

Section 4.3 (oil layer classification based on well-logging interpretation) introduces Case Study 2 of *C*-SVM. Using data for the oil layer classification based on well-logging interpretation of the Xiefengqiao Anticline, i.e., the five parameters (true resistivity, acoustictime, porosity, oil saturation, permeability) and an oil test result of 24 oil layers, an explicit nonlinear function is constructed by the data mining tool of *C*-SVM, and the results are consistent with the oil test. Again, this function is called *mined knowledge*. In the prediction process, applying this knowledge, that is, substituting the values of true resistivity, acoustictime, porosity, oil saturation, and permeability in this nonlinear function, respectively, the results are also consistent with the oil test. Therefore, this method can be spread to the oil layer classification based on well-logging interpretation of other areas. Moreover, it is found that (a) since this case study is a strong nonlinear problem, the major data mining tool can adopt *C*-SVM but not MRA, *R*-SVM, or BPNN; and (b) in this case study, MRA established the order of dependence between the oil layer classification and its relative true resistivity, acoustictime, porosity, oil saturation, and permeability under the condition of linearity, thus MRA can be employed as an auxiliary tool.

Section 4.4 (dimension-reduction procedure using machine learning) presents a dimension-reduction procedure using MRA as a pioneering dimension-reduction tool and

C-SVM or BPNN as a validating dimension-reduction tool. Taking Case Study 2 (fractures prediction using conventional well-logging data) of Section 3.3 as a case study, this dimension-reduction procedure can reduce the original 8-D problem to a 4-D problem. Through this case study, the following three conclusions are reached:

1. MRA is a good pioneering dimension-reduction tool. It can indicate the order of dimension reduction and the possible final number of dimensions. However, since MRA is a linear data mining tool, the results of dimension reduction are obtained under linear condition, and thus a nonlinear data mining tool (C-SVM or BPNN) is required to validate that the results are feasible.
2. The action of dimension reduction is to reduce the amount of data so as to speed up the data mining process and especially to extend up the applying ranges. For instance, the preceding case study can be reduced to the 4-D problem so that the fracture identification (y) only depends on three well-logging data of deep laterolog resistivity (x_5), compensated neutron density (x_2), and microspherically focused resistivity (x_4), i.e., C-SVM or BPNN can be adopted to the fracture identification without four well-logging data of absolute difference of R_{LLD} and R_{LLS} (x_7), compensated neutron porosity (x_3), shallow laterolog resistivity (x_6), and acoustictime (x_1).
3. Because of the complexities of geosciences rules, the correlations among various classes of geosciences data are nonlinear in most cases. In general, therefore, it is better to use C-SVM when the nonlinearity is very strong; otherwise, use BPNN. For the sake of unification and safety, it is suggested to just adopt C-SVM as a validating dimension-reduction tool, since C-SVM is not only applicable to either strongly or weakly nonlinear problems but also runs at a speed as fast as 20 or more times that of BPNN. Certainly, it is easy to code the BPNN program, whereas it is very complicated to code the C-SVM program, so BPNN is a good validating dimension-reduction tool in case no C-SVM program exists.

4.1. METHODOLOGY

Since the 1990s, SVM has been gradually applied in natural and social sciences, especially widely in the 21st century. SVM includes two principal algorithms: (1) C-SVM, such as the binary classification (e.g., Boser et al., 1992; Cortes and Vapnik, 1995; Cristianini and Shawe-Taylor, 2000; Chang and Lin, 2011) and the ν-binary classification (Crisp and Burges, 2000; Schölkopf et al., 2000; Chang and Lin, 2001); and (2) R-SVM, such as the ε-regression (e.g., Vapnik, 1998; Chang and Lin, 2011) and the ν-regression (e.g., Schölkopf et al., 2000; Chang and Lin, 2002). In this book, the binary classification for C-SVM and the ε-regression for R-SVM are employed.

The learning samples and prediction samples for C-SVM and R-SVM are expressed in Equations (1.1) and (1.3), respectively. The learning process and prediction process of calculation flowchart are illustrated by Figures 1.2 and 1.3, respectively. If the prediction samples are replaced by the learning samples, Figure 1.3 becomes such a calculation flowchart for learning validation.

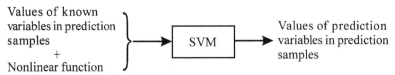

FIGURE 4.1   Learning process (a) and prediction process (b) of SVM (C-SVM, R-SVM).

The expression of y created using C-SVM and R-SVM is a nonlinear explicit expression with respect to m parameters $(x_1, x_2, ..., x_m)$:

$$y = SVM(x_1, x_2, ..., x_m) \tag{4.1}$$

where SVM (C-SVM or R-SVM) is a nonlinear function and calculated out by C-SVM algorithm or R-SVM algorithm, which can be expressed as a usual mathematical formula and so is an explicit expression. Equation (4.1) is called the *fitting formula* and is obtained from the learning process.

The basic idea of SVM is illustrated by Figure 4.1.

4.1.1. Applying Ranges and Conditions

4.1.1.1. Applying Ranges

As Equation (1.1) described, assume that there are n learning samples, each associated with $m + 1$ numbers $(x_1, x_2, ..., x_m, y_i^*)$ and a set of observed values $(x_{1i}, x_{2i}, ..., x_{mi}, y_i^*)$, with $i = 1, 2, ..., n$ for these parameters. In principle, $n > m$, but in actual practice $n >> m$. As Equation (1.3) described, assume that there are k prediction samples, each associated with m parameters $(x_1, x_2, ..., x_m)$ and a set of observed values $(x_{1i}, x_{2i}, ..., x_{mi})$, with $i = n + 1, n + 2, ..., n + k$ for these parameters.

4.1.1.2. Applying Conditions

The number of learning samples, n, must be big enough to ensure the accuracy of prediction by C-SVM or R-SVM. The number of known variables, m, must be greater than one.

4.1.2. Basic Principles

SVM is essentially performed by converting a real-world problem (the original space) into a new higher-dimensional feature space using the kernel function and then constructing a linear discriminate function in the new space to replace the nonlinear discriminate function.

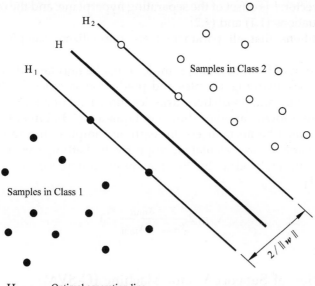

FIGURE 4.2 Optimal separating line under the linear separable condition.

This nonlinear transformation is implemented by defining a proper integral operator kernel function; then the optimal separating hyperplane is obtained in the new feature space.

SVM is developed through the optimal separating hyperplane under the linear separable condition, and its basic idea is illustrated by Figure 4.2 in a two-dimensional plane. In Figure 4.2, the solid dot denotes Sample Class 1, whereas the hollow dot denotes Sample Class 2; H is the optimal separating line; H_1 is the line linking two samples in Class 1 and is the nearest and parallel to H; H_2 is the line linking two samples in Class 2 and is the nearest and parallel to H; the distance between H_1 and H_2 is called the *margin* and is equal to $2/\|w\|$, where w will be defined in Equation (4.2). The optimal separating line H is required to separate the two sample classes correctly, so that the misclassification ratio is zero, and to maximize the margin. Spreading the two-dimensional plane to the higher-dimensional space, this optimal separating line becomes the optimal separating hyperplane.

To separate the two sample classes by the optimal separating hyperplane, a separating hyperplane

$$wx + b = 0 \tag{4.2}$$

must exist between the two sample classes, subject to

$$y_i(wx_i + b) \geq 1 \quad (i = 1, 2, \ldots, n) \tag{4.3}$$

where w is weight vector; b is offset of the separating hyperplane; and the other symbols have been defined in Equations (1.1) and (1.2).

Under the conditions that all parameters are normalized, samples are called *linear separable*.

At the starting point of SVM running, samples must be normalized at first. Samples are divided into two types: learning samples and prediction samples. Each learning sample contains known variables and prediction variables shown in Equation (1.1); each prediction sample only contains known variables shown in Equation (1.3). Either learning samples or prediction samples must be normalized. In learning samples, maximum x_{max} and minimum x_{min} of each variable are calculated, respectively. Letting x be a variable, x can be transformed to the normalized x' by the maximum difference normalization formula (4.4), and $x' \in [-1, 1]$.

$$x = 2\frac{x - x_{min}}{x_{max} - x_{min}} - 1 \tag{4.4}$$

4.1.3. Classification of Support Vector Machine (C-SVM)

Using the binary classifier and taking the radial basis function (RBF) as a kernel function, solving the optimal separating hyperplane is finally transformed to the following optimization.

Object function:

$$\max_{\alpha}\left\{\sum_{i=1}^{n}\alpha_i - \frac{1}{2}\sum_{i,j=1}^{n}\left[\alpha_i\alpha_j y_i y_j \exp\left(-\gamma\|x_i - x_j\|^2\right)\right]\right\} \tag{4.5}$$

Constraint conditions:

$$0 \leq \alpha_i \leq C, \sum_{i=1}^{n} y_i\alpha_i = 0 \quad (i = 1, 2, ..., n) \tag{4.6}$$

where α is the vector of Lagrange multipliers, $\alpha = (\alpha_1, \alpha_2, ..., \alpha_n)$; γ is the regularization parameter, $\gamma > 0$; $\exp(-\gamma\|x_i - x_j\|^2)$ is a RBF kernel function; C is the penalty factor; and the other symbols have been defined in Equations (1.1) and (1.2).

This is a *dual quadratic optimization* for solving α_i, C, and γ.

After accomplishing the optimization, the y expression corresponding to C-SVM formula (4.1) is a nonlinear expression with respect to vector x:

$$y = \sum_{i=1}^{n}\left[y_i\alpha_i \exp\left(-\gamma\|x - x_i\|^2\right)\right] + b \tag{4.7}$$

where b has been defined in Equations (4.2) and (4.3), which can be calculated using the free vectors x_i. These free x_i are those vectors corresponding to $\alpha_i > 0$ on which the final C-SVM model depends, and the other symbols have been defined in Equations (1.1), (1.2), (4.5), and (4.6).

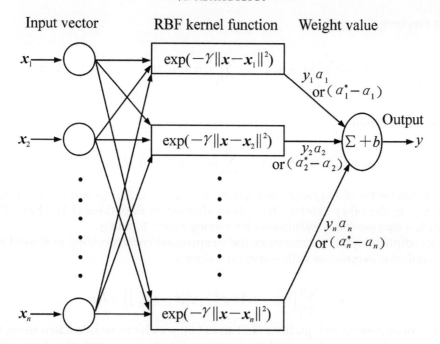

Input vector   **RBF kernel function**   **Weight value**

FIGURE 4.3   Sketch map of the learning process for C-SVM or R-SVM.

Moreover, w in Equation (4.2) is expressed as

$$w = \sum_{i=1}^{n} \alpha_i y_i x_i \tag{4.8}$$

where all symbols have been defined in Equations (1.1), (1.2), (4.2), (4.3), (4.5), and (4.6).

Moreover, it is better to take RBF as a kernel function than to take the linear, polynomial, and sigmoid kernel functions under strong nonlinear conditions (Chang and Lin, 2011). Because the fact that the correlations between different subsurface factors are strongly or weakly nonlinear in most cases, it is suggested to simply employ RBF for the sake of unification and safety.

To have the comparability of SVM results among different applications in this book, RBF is taken as a kernel function, the termination of calculation accuracy (TCA) is fixed to 0.001, and the insensitive function ε in R-SVM is fixed to 0.1.

Figure 4.3 illustrates the calculation flowchart of the learning process for C-SVM or R-SVM.

4.1.4. Regression of Support Vector Machine (R-SVM)

Using the ε-regression and taking the radial basis function (RBF) as a kernel function, solving the optimal separating hyperplane is finally transformed to the following optimization.

Object function:

$$\max_{\alpha,\alpha^*}\left\{\sum_{i=1}^{n}(\alpha_i^* - \alpha_i)y_i - \varepsilon\sum_{i=1}^{n}(\alpha_i + \alpha_i^*) - \frac{1}{2}\sum_{i,j=1}^{n}\left[(\alpha_i - \alpha_i^*)(\alpha_j - \alpha_j^*)\exp\left(-\gamma||x_i - x_j||^2\right)\right]\right\}$$

(4.9)

Constraint conditions:

$$0 \le \alpha_i \le C, \quad 0 \le \alpha_i^* \le C, \quad \sum_{i=1}^{n}(\alpha_i^* - \alpha_i) = 0 \ (i = 1, 2, ..., n)$$

(4.10)

where α^* is the vector of Lagrange multipliers, $\alpha^* = (\alpha_1^*, \alpha_2^*, ..., \alpha_n^*)$ and ε is the insensitive function, $\varepsilon > 0$; the other symbols have been defined in Equations (1.1), (1.2), (4.5), and (4.6). This is a *dual quadratic optimization* for solving α_i, α_i^*, C, and γ.

After accomplishing the optimization, the y expression corresponding to R-SVM formula (4.1) is a nonlinear expression with respect to vector x:

$$y = \sum_{i=1}^{n}\left[(\alpha_i^* - \alpha_i)\exp\left(-\gamma||x - x_i||^2\right)\right] + b$$

(4.11)

where b has been defined in Equations (4.5) and (4.6), which can be calculated using the free vectors x_i. These free x_i are those vectors corresponding to $\alpha_i > 0$ and $\alpha_i^* > 0$, on which the final R-SVM model depends; the other symbols have been defined in Equations (1.1), (1.2), (4.9), and (4.10).

Moreover, w in Equation (4.8) is expressed as

$$w = \sum_{i=1}^{n}(\alpha_i^* - \alpha_i)x_i$$

(4.12)

where all symbols have been defined in Equations (1.1), (1.2), (4.2), (4.3), (4.9), and (4.10).

Note that in case studies the formulas corresponding to Equations (4.7) and (4.11) are not concretely written out due to their large size.

Now the SVM procedure can be summarized as follows: (a) using the values of known variables and a prediction variable in learning samples, a nonlinear function (nonlinear mapping) of the prediction variable with respect to the known variables is constructed, that is, a learning process; and (b) using the values of known variables in learning samples and prediction samples, and employing this nonlinear function, the values of a prediction variable in learning samples and prediction samples are calculated out, that is, a prediction process. The obtained values of prediction variable in learning samples and prediction samples are called a *fitting solution* and a *prediction solution* for the SVM procedure, respectively.

The XOR problem and simple Case Study 2 (prediction of fracture-acidizing results) were successfully solved by BPNN, but SVM cannot, which results from too few learning samples and then the failure of a learning process. Simple Case Study 1 (prediction of oil production) was also successfully solved by BPNN, but SVM cannot, which results from only one known variable. Certainly, to solve this special problem, BPNN uses a technique: 1-D input space is changed to 6-D input space so as to increase the number of variables from 1 to 6. Similarly, SVM can also use this technique, but additional coding is needed.

SVM procedure basically does not refer to the definition of probability measure and the law of great numbers, so SVM defers from existing methods of statistics. The final decision-making function of SVM [e.g., Equation (4.7) for *C*-SVM, Equation (4.11) for *R*-SVM] is defined by only a few support vectors (free vectors), so the complexity of calculation depends on the number of support vectors but not on the number of dimensions of sample space, which can avoid the dimension disaster to some extent and has better robustness. That is a distinguishing advantage of SVM.

Though both SVM and BPNN are based on learning mechanisms, SVM employs mathematical methods and optimization techniques. Theoretically, SVM can obtain the global optimal solution and avoid converging to a local optimal solution, as can possibly occur in BPNN, though this problem in BPNN is rare if BPNN is properly coded (Shi, 2009; Shi, 2011). Comparing with BPNN, the major advantages of SVM are higher accuracy and faster calculation speed, which are validated by the following case studies.

4.2. CASE STUDY 1: GAS LAYER CLASSIFICATION BASED ON POROSITY, PERMEABILITY, AND GAS SATURATION

4.2.1. Studied Problem

The objective of this case study is to conduct a gas layer classification in tight sandstones using porosity, permeability, and gas saturation, which has practical value when gas test data are limited.

Located on the northeast side of the Yishan Slope in the Ordos Basin in western China, the Tabamiao area covers about 2005 km^2. As a large and gentle west-dipping monocline, this slope is structurally simple, with extremely undeveloped faults, and there exist multiple northeast−southwest gently plunging nose structures. Petrophysical statistics indicate a mean porosity of 5−10 (%), with mean permeability of 0.3−0.8 ($\times 10^{-3}$ μm^2) for sandstones in the upper Paleozoic layers, indicating that the gas-bearing layers are tight sandstones with medium-low porosity and low permeability (Chen et al., 2006)

Researchers used data from 40 gas layers in tight sandstones in the Tabamiao area, of which 38 were taken as learning samples and 2 as prediction samples. Each gas layer has data of porosity, permeability, gas saturation, and gas test result (Table 4.1). However, the gas test result of each prediction sample needs to be further validated. BPNN was then applied to confirm the gas layer classification (Chen et al., 2006).

4.2.2. Input Data

The known input parameters are the values of the known variables x_i ($i = 1, 2, 3$) for 38 learning samples and 2 prediction samples, and the value of the prediction variable y^* for 38 learning samples (Table 4.1).

TABLE 4.1   Input Data for Gas Layer Classification of the Tabamiao Area

Sample Type	Sample No.	Well No.	Depth Interval (m)	x_1 Porosity (%)	x_2 Permeability (mD)	x_3 Gas Saturation (%)	y^* Gas Test[a]
Learning samples	1	D3	2701.0–2708.4	10.10	0.8652	74.10	1
	2		2708.4–2725.4	6.17	0.4782	50.30	3
	3		2819.5–2832.0	7.03	0.3526	66.00	1
	4	D4	2856.0–2872.0	5.57	0.3312	51.40	1
	5	D10	2509.0–2514.4	11.06	2.0749	74.90	1
	6		2522.6–2525.0	11.65	3.9939	59.30	1
	7		2600.5–2603.8	4.43	0.1740	45.00	2
	8		2603.8–2606.3	7.05	0.4284	60.00	1
	9		2672.7–2676.2	7.68	1.6651	50.00	1
	10		2676.2–2685.2	7.68	1.5102	60.00	1
	11		2727.3–2730.0	11.17	1.0088	49.00	2
	12		2730.0–2747.0	11.08	2.2951	73.70	1
	13	D14	2683.4–2687.4	5.91	0.3582	60.20	1
	14	D15	2840.6–2849.6	9.60	0.9093	54.20	1
	15		2849.6–2857.0	2.73	0.1429	0.00	4
	16	D16	2647.2–2653.8	7.98	0.4096	70.00	1
	17		2696.8–2703.8	6.48	0.3184	77.00	1
	18		2704.5–2717.7	3.44	0.1184	58.50	4
	19		2852.9–2853.5	6.40	0.3315	47.10	2
	20		2853.5–2858.0	10.46	1.1226	64.90	1
	21		2861.8–2868.5	4.42	0.1976	39.10	3
	22		2868.5–2871.0	7.17	0.4033	41.80	2
	23	D18	2771.0–2777.8	8.94	1.6147	47.80	2
	24		2778.4–2788.1	8.65	1.5373	65.40	1
	25	D22	2763.6–2766.8	6.89	0.5337	39.20	3
	26		2766.8–2768.3	9.11	1.4718	45.70	2
	27		2768.3–2773.0	7.71	0.8055	65.10	1
	28		2773.0–2774.5	8.78	2.7089	47.60	2
	29	DK2	2656.8–2660.0	7.22	0.5379	88.20	1

(Continued)

TABLE 4.1   Input Data for Gas Layer Classification of the Tabamiao Area—cont'd

Sample Type	Sample No.	Well No.	Depth Interval (m)	x_1 Porosity (%)	x_2 Permeability (mD)	x_3 Gas Saturation (%)	y^* Gas Test[a]
	30		2666.5–2669.4	7.86	1.1193	86.20	1
	31		2839.1–2844.0	4.74	0.1501	42.00	2
	32		2867.6–2872.4	5.06	0.1769	63.00	1
	33	DK4	2666.1–2675.1	11.52	4.6680	87.90	1
	34		2676.1–2680.4	10.57	3.7996	69.30	1
	35	DT1	2737.3–2750.0	9.79	0.8721	58.36	1
	36		2829.1–2838.4	7.49	0.4017	59.88	1
	37		2838.4–2842.0	3.42	0.1703	14.90	3
	38		2842.0–2846.1	8.31	0.3762	37.00	2
Prediction samples	39	D10	2607.0–2614.5	8.30	3.0923	65.00	(1)
	40	DK2	2660.0–2666.5	7.60	0.6991	87.30	(1)

[a] y^* = gas layer classification (1—gas layer, 2—poor gas layer, 3—gas-bearing layer, 4—dry layer) determined by the well test; in y^*, numbers in parenthesis are not input data but are used for calculating R(%).

4.2.3. Application Comparisons Among C-SVM, R-SVM, BPNN, and MRA

4.2.3.1. Learning Process

Using the 38 learning samples (Table 4.1) and via C-SVM, R-SVM, BPNN, and MRA, the four functions of gas layer classification (y) with respect to porosity (x_1), permeability (x_2), and gas saturation (x_3) have been constructed, i.e., Equations (4.13) and (4.14) corresponding to SVM formula (4.1), Equation (4.15) corresponding to BPNN formula (3.1), and Equation (4.16) corresponding to MRA formula (2.14), respectively.

Using C-SVM, the result is an explicit nonlinear function (Shi, 2008):

$$y = C\text{-}SVM(x_1, x_2, x_3) \tag{4.13}$$

with $C = 512$, $\gamma = 0.5$, 18 free vectors x_i, and the cross-validation accuracy CVA = 81.6%.

Using R-SVM, the result is an explicit nonlinear function:

$$y = R\text{-}SVM(x_1, x_2, x_3) \tag{4.14}$$

with $C = 1$, $\gamma = 0.333333$, and 29 free vectors x_i.

The BPNN used consists of three input layer nodes, one output layer node, and seven hidden layer nodes. Setting the control parameter of t maximum $t_{max} = 100000$, the calculated optimal learning time count $t_{opt} = 49642$. The result is an implicit nonlinear function (Shi, 2008):

$$y = BPNN(x_1, x_2, x_3) \tag{4.15}$$

with the root mean-square error $RMSE(\%) = 0.1231 \times 10^{-1}$.

Using MRA, the result is an explicit linear function (Shi, 2008):

$$y = 4.2195 - 0.1515x_1 + 0.1363x_2 - 0.0283x_3 \tag{4.16}$$

Equation (4.16) yields a residual variance of 0.3939 and a multiple correlation coefficient of 0.7785. From the regression process, gas layer classification (y) is shown to depend on the three parameters in decreasing order: gas saturation (x_3), porosity (x_1), and permeability (x_2).

Substituting the values of porosity, permeability, and gas saturation given by the former 38 gas layers (learning samples) in the aforementioned 40 gas layers (Table 4.1) in Equations (4.13), (4.14), (4.15), and (4.16), respectively, the gas layer classification (y) of each learning sample is obtained. Table 4.2 shows the results of the learning process by C-SVM, R-SVM, BPNN, and MRA.

TABLE 4.2   Prediction Results from Gas Layer Classification of the Tabamiao Area

					Gas Layer Classification							
					C-SVM		R-SVM		BPNN		MRA	
Sample Type	Sample No.	Well No.	Depth Interval (m)	y^* Gas Test[a]	y	$R(\%)$	y	$R(\%)$	y	$R(\%)$	y	$R(\%)$
Learning samples	1	D3	2701.0−2708.4	1	1	0	1	0	1	0	1	0
	2		2708.4−2725.4	3	3	0	2	33.33	4	33.33	2	33.33
	3		2819.5−2832.0	1	1	0	1	0	1	0	1	0
	4	D4	2856.0−2872.0	1	1	0	2	100	1	0	2	100
	5	D10	2509.0−2514.4	1	1	0	1	0	1	0	1	0
	6		2522.6−2525.0	1	1	0	1	0	1	0	1	0
	7		2600.5−2603.8	2	2	0	2	0	3	50	2	0
	8		2603.8−2606.3	1	1	0	1	0	1	0	2	100
	9		2672.7−2676.2	1	1	0	2	100	1	0	2	100
	10		2676.2−2685.2	1	1	0	1	0	1	0	2	100
	11		2727.3−2730.0	2	2	0	1	50	2	0	1	50
	12		2730.0−2747.0	1	1	0	1	0	1	0	1	0
	13	D14	2683.4−2687.4	1	1	0	1	0	1	0	2	100
	14	D15	2840.6−2849.6	1	1	0	1	0	1	0	1	0
	15		2849.6−2857.0	4	4	0	3	25	4	0	4	0
	16	D16	2647.2−2653.8	1	1	0	1	0	1	0	1	0
	17		2696.8−2703.8	1	1	0	1	0	1	0	1	0
	18		2704.5−2717.7	4	4	0	2	50	4	0	2	50

(Continued)

TABLE 4.2   Prediction Results from Gas Layer Classification of the Tabamiao Area—cont'd

Sample Type	Sample No.	Well No.	Depth Interval (m)	y^{*} Gas Test[a]	C-SVM		R-SVM		BPNN		MRA	
					y	$R(\%)$	y	$R(\%)$	y	$R(\%)$	y	$R(\%)$
	19		2852.9–2853.5	2	2	0	2	0	3	50	2	0
	20		2853.5–2858.0	1	1	0	1	0	1	0	1	0
	21		2861.8–2868.5	3	3	0	2	33.33	4	33.33	2	33.33
	22		2868.5–2871.0	2	2	0	2	0	3	50	2	0
	23	D18	2771.0–2777.8	2	2	0	2	0	2	0	2	0
	24		2778.4–2788.1	1	1	0	1	0	1	0	1	0
	25	D22	2763.6–2766.8	3	3	0	2	33.33	4	33.33	2	33.33
	26		2766.8–2768.3	2	2	0	2	0	3	50	2	0
	27		2768.3–2773.0	1	1	0	1	0	1	0	1	0
	28		2773.0–2774.5	2	2	0	2	0	2	0	2	0
	29	DK2	2656.8–2660.0	1	1	0	1	0	1	0	1	0
	30		2666.5–2669.4	1	1	0	1	0	1	0	1	0
	31		2839.1–2844.0	2	2	0	2	0	3	50	2	0
	32		2867.6–2872.4	1	1	0	1	0	1	0	2	100
	33	DK4	2666.1–2675.1	1	1	0	1	0	1	0	1	0
	34		2676.1–2680.4	1	1	0	1	0	1	0	1	0
	35	DT1	2737.3–2750.0	1	1	0	1	0	1	0	1	0
	36		2829.1–2838.4	1	1	0	1	0	1	0	1	0
	37		2838.4–2842.0	3	3	0	3	0	4	33.33	3	0
	38		2842.0–2846.1	2	2	0	2	0	3	50	2	0
Prediction samples	39	D10	2607.0–2614.5	1	1	0	2	100	1	0	2	100
	40	DK2	2660.0–2666.5	1	1	0	2	100	1	0	1	0

[a]y^{*} = gas layer classification (1—gas layer, 2—poor gas layer, 3—gas-bearing layer, 4—dry layer) determined by the well test.

4.2.3.2. Prediction Process

Substituting the values of porosity, permeability, and gas saturation given by two prediction samples (Table 4.1) in Equations (4.13), (4.14), (4.15), and (4.16), respectively, the gas layer classification (y) of each prediction sample is obtained. Table 4.2 shows the results of prediction process by C-SVM, R-SVM, BPNN, and MRA.

As shown in Table 4.3, the results calculated by C-SVM are completely accurate, i.e., not only the fitting residual $\overline{R}_1(\%) = 0$ but also the prediction residual $\overline{R}_2(\%) = 0$, and thus the

TABLE 4.3   Comparison Among the Applications of C-SVM, R-SVM, BPNN and MRA to Gas Layer Classification of the Tabamiao Area

Algorithm	Fitting Formula	Mean Absolute Relative Residual			Dependence of the Predicted Value (y) on Parameters (x_1, x_2, x_3), in Decreasing Order	Time Consumed on PC (Intel Core 2)	Solution Accuracy
		$\overline{R}1(\%)$	$\overline{R}2(\%)$	$\overline{R}^*(\%)$			
C-SVM	Nonlinear, explicit	0	0	0	N/A	2 s	Very high
R-SVM	Nonlinear, explicit	11.18	100	55.59	N/A	3 s	Very low
BPNN	Nonlinear, implicit	11.40	0	5.70	N/A	55 s	Moderate
MRA	Linear, explicit	21.05	50	35.53	x_3, x_1, x_2	<1 s	Very low

total mean absolute relative residual $\overline{R}^*(\%) = 0$, which is consistent with the real exploration results. For the results calculated by BPNN, though, $\overline{R}_2(\%) = 0$, $\overline{R}_1(\%) = 11.40$, which could not ensure correct prediction of other new prediction samples. As for R-SVM and MRA, the solution accuracy is very low, and $\overline{R}_2(\%) > \overline{R}_1(\%)$.

4.2.3.3. Application Comparisons Among C-SVM, R-SVM, BPNN, and MRA

We can see from $\overline{R}^*(\%) = 35.53$ of MRA (Table 4.3) that the nonlinearity of the relationship between the predicted value y and its relative parameters (x_1, x_2, x_3) is strong from Table 1.2. Since C-SVM can describe very strong nonlinear relationships, its advantage is very high solution accuracy. Since R-SVM can only describe not very strong nonlinear relationships, its solution accuracy is very low. Since BPNN can only describe not very strong nonlinear relationships, its solution accuracy is moderate. Since MRA can only describe simple linear relationships, its solution accuracy is very low, but its advantage is fast and can establish the order of dependence between the predicted value y and its relative parameters (x_1, x_2, x_3), e.g., MRA can indicate that the gas layer classification in tight sandstones depends on gas saturation first, the porosity next, and permeability last.

4.2.4. Summary and Conclusions

Summarily, using data for the gas layer classification based on three parameters of the Tabamiao area, i.e., the three parameters (porosity, permeability, gas saturation) and the gas test result of 38 oil layers, an explicit nonlinear function (4.13) is constructed by the data mining tool of C-SVM, and the results are consistent with the gas test. This function is called mined knowledge. In the prediction process, applying this knowledge, i.e., substituting the values of porosity, permeability, and gas saturation in Equation (4.13), respectively, the results are also consistent with the gas test. Therefore, this method can be spread to the gas layer classification based on porosity, permeability, and gas saturation of other areas.

Moreover, it is found that (a) since this case study is a very strong nonlinear problem, the major data mining tool can adopt C-SVM but not MRA, R-SVM, or BPNN; and (b) in this case study, MRA established the order of dependence between the gas layer classification and its relative porosity, permeability, and gas saturation under the condition of linearity; thus MRA can be employed as an auxiliary tool.

4.3. CASE STUDY 2: OIL LAYER CLASSIFICATION BASED ON WELL-LOGGING INTERPRETATION

4.3.1. Studied Problem

The objective of this case study is to conduct an oil layer classification in tight sandstones using true resistivity, acoustictime, porosity, oil saturation, and permeability, which has practical value when oil test data are limited.

Located on the southwestern margin of the Jianghan Basin in central China, the Xiefengqiao Anticline is a litho-structure complex oil layer with low porosity and low permeability lying at depths of 3100—3600 m.

Researchers used data from 27 oil layers in tight sandstones in the Xiefengqiao Anticline, of which 24 were taken as learning samples and 3 as prediction samples. Each oil layer has data of true resistivity, acoustictime, porosity, oil saturation, and permeability and an oil test result (Table 4.4). However, the oil test result of each prediction sample needs to be further validated. BPNN was then applied to confirm the oil layer classification (Yang et al., 2002).

4.3.2. Input Data

The known input parameters are the values of the known variables x_i ($i = 1, 2, 3, 4, 5$) for 24 learning samples and 3 prediction sample as well as the value of the prediction variable y^* for 24 learning samples (Table 4.4).

4.3.3. Application Comparisons Among C-SVM, R-SVM, BPNN, and MRA

4.3.3.1. Learning Process

Using the 24 learning samples (Table 4.4) and by C-SVM, R-SVM, BPNN, and MRA, the four functions of oil layer classification (y) with respect to true resistivity (x_1), acoustictime (x_2), porosity (x_3), oil saturation (x_4), and permeability (x_5) have been constructed, i.e., Equations (4.17) and (4.18) corresponding to SVM formula (4.1), Equation (4.19) corresponding to BPNN formula (3.1), and Equation (4.20) corresponding to MRA formula (2.14), respectively.

Using C-SVM, the result is an explicit nonlinear function (Shi, 2009; Shi, 2011):

$$y = C\text{-}SVM(x_1, x_2, x_3, x_4, x_5) \tag{4.17}$$

with $C = 8192$, $\gamma = 0.007813$, nine free vectors x_i, and CVA = 91.7%.

Using R-SVM, the result is an explicit nonlinear function:

$$y = R\text{-}SVM(x_1, x_2, x_3, x_4, x_5) \tag{4.18}$$

TABLE 4.4   Input Data for oil Layer Classification of the Xiefengqiao Anticline

Sample Type	Sample No.	Well No.	Layer No.	x_1 True Resistivity ($\Omega \cdot$m)	x_2 Acoustictime (μs/m)	x_3 Porosity (%)	x_4 Oil Saturation (%)	x_5 Permeability (mD)	y^* Oil Test[a]
Learning samples	1	ES4	5	64	206	6.0	48.6	1.1	2
	2		6	140	208	6.5	41.5	1.3	3
	3		7	63	206	6.0	36.4	1.1	2
	4		8	116	196	3.8	0.7	0.5	3
	5		9	17	267	19.6	44.0	32.4	1
	6		10	49	226	10.6	57.2	5.2	2
	7		11	44	208	6.6	36.1	1.4	2
	8		12	90	208	6.5	29.7	1.3	3
	9		13	69	260	18.1	81.7	16.8	1
	10	ES6	4	49	207	6.2	67.5	0.9	2
	11		5	80	207	6.3	50.9	1.0	3
	12		6	95	218	8.7	77.5	2.2	3
	13		8	164	212	7.5	67.5	1.5	3
	14	ES8	5_1	21	202	5.1	22.2	1.0	2
	15		5_2	56	192	2.9	24.2	0.6	3
	16		5_3	36	198	4.1	28.8	0.8	2
	17		6	128	196	3.4	19.2	0.6	3
	18		11	34	197	3.9	28.4	0.7	2
	19		12_1	10	208	6.4	42.4	1.7	1
	20		12_2	6	226	10.4	45.6	5.8	1
	21		12_3	6	225	10.3	50.4	6.1	1
	22		12_4	10	206	6.0	44.0	1.7	1
	23		13_1	7	224	9.9	44.4	5.2	1
	24		13_2	15	197	3.8	34.2	0.6	1
Prediction samples	25	ES8	13_4	11	201	4.8	39.3	0.8	(1)
	26		13_5	25	197	3.8	16.9	0.6	(2)
	27		7_2	109	199	4.4	17.8	0.8	(3)

[a]y^* = oil layer classification (1—oil layer, 2—poor oil layer, 3—dry layer) determined by the oil test; in y^*, numbers in parenthesis are not input data but are used for calculating R(%).

with $C = 1$, $\gamma = 0.2$, and 19 free vectors x_i.

The BPNN used consists of 5 input layer nodes, 1 output layer node and 11 hidden layer nodes. Setting $t_{max} = 100000$, the calculated $t_{opt} = 41467$. The result is an implicit nonlinear function (Shi, 2008):

$$y = BPNN(x_1, x_2, x_3, x_4, x_5) \tag{4.19}$$

with $RMSE(\%) = 0.8714 \times 10^{-2}$.

Using MRA, the result is an explicit linear function (Shi, 2009; Shi, 2011):

$$y = 62.641 + 0.0134x_1 - 0.3378x_2 + 1.3781x_3 - 0.0007x_4 + 0.0386x_5 \tag{4.20}$$

Equation (4.20) yields a residual variance of 0.16649 and a multiple-correlation coefficient of 0.91297. From the regression process, oil layer classification (y) is shown to depend on the five parameters in decreasing order: true resistivity (x_1), acoustictime (x_2), porosity (x_3), permeability (x_5), and oil saturation (x_4).

Substituting the values of true resistivity, acoustictime, porosity, oil saturation, and permeability given by the former 24 oil layers (learning samples) in the aforementioned 27 oil layers (Table 4.4) in Equations (4.17), (4.18), (4.19), and (4.20), respectively, the oil layer classification (y) of each learning sample is obtained. Table 4.5 shows the results of learning process by C-SVM, R-SVM, BPNN, and MRA.

4.3.3.2. Prediction Process

Substituting the values of true resistivity, acoustictime, porosity, oil saturation, and permeability given by three prediction samples (Table 4.4) in Equations (4.17), (4.18), (4.19), and (4.20), respectively, the oil layer classification (y) of each prediction sample is obtained. Table 4.5 shows the results of prediction process by C-SVM, R-SVM, BPNN, and MRA.

Table 4.3 shows that the results calculated by C-SVM are completely accurate, i.e., not only $\overline{R}_1(\%) = 0$ but also $\overline{R}_2(\%) = 0$, and thus $\overline{R}^*(\%) = 0$, which is consistent with the real exploration results. For the results calculated by R-SVM, though, $\overline{R}_2(\%) = 0$, $\overline{R}_1(\%) = 8.33$, which could not ensure correct prediction of other new prediction samples. As for BPNN and MRA, the solution accuracy is low and very low, respectively, and $\overline{R}_2(\%) > \overline{R}_1(\%)$.

4.3.3.3. Application Comparisons Among C-SVM, R-SVM, BPNN, and MRA

We can see from $\overline{R}^*(\%) = 20.83$ of MRA (Table 4.6) that the nonlinearity of the relationship between the predicted value y and its relative parameters (x_1, x_2, x_3, x_4, x_5) is strong from Table 1.2. Since C-SVM can describe very strong nonlinear relationships, its advantage is very high solution accuracy. Since R-SVM can describe strong nonlinear relationships, its solution accuracy is high. Since BPNN can only describe not very strong nonlinear relationships, its solution accuracy is low. Since MRA can only describe simple linear relationships, its solution accuracy is very low, but its advantage is fast and can establish the order of dependence between the predicted value y and its relative parameters (x_1, x_2, x_3, x_4, x_5), e.g., MRA can indicate that the oil layer classification in tight sandstones depends on true resistivity first, acoustictime next, then porosity and permeability, and oil saturation last.

TABLE 4.5　Prediction Results from Oil Layer Classification of the Xiefengqiao Anticline

Sample Type	Sample No.	Well No.	Layer No.	y^* Oil Test[a]	C-SVM		R-SVM		BPNN		MRA	
					y	$R(\%)$	y	$R(\%)$	y	$R(\%)$	y	$R(\%)$
Learning samples	1	ES4	5	2	2	0	2	0	3	50	2	0
	2		6	3	3	0	3	0	3	0	3	0
	3		7	2	2	0	2	0	3	50	2	0
	4		8	3	3	0	3	0	3	0	3	0
	5		9	1	1	0	1	0	1	0	1	0
	6		10	2	2	0	2	0	2	0	2	0
	7		11	2	2	0	2	0	2	0	2	0
	8		12	3	3	0	3	0	3	0	3	0
	9		13	1	1	0	1	0	1	0	1	0
	10	ES6	4	2	2	0	2	0	2	0	2	0
	11		5	3	3	0	2	33.33	3	0	2	33.33
	12		6	3	3	0	2	33.33	3	0	2	33.33
	13		8	3	3	0	3	0	3	0	4	33.33
	14	ES8	5_1	2	2	0	2	0	2	0	2	0
	15		5_2	3	3	0	2	33.33	3	0	3	0
	16		5_3	2	2	0	2	0	2	0	2	0
	17		6	3	3	0	3	0	3	0	3	0
	18		11	2	2	0	2	0	2	0	2	0
	19		12_1	1	1	0	1	0	1	0	1	0
	20		12_2	1	1	0	1	0	1	0	1	0
	21		12_3	1	1	0	1	0	1	0	1	0
	22		12_4	1	1	0	1	0	1	0	1	0
	23		13_1	1	1	0	1	0	1	0	1	0
	24		13_2	1	1	0	2	100	1	0	2	100
Prediction samples	25	ES8	13_4	1	1	0	1	0	1	0	2	100
	26		13_5	2	2	0	2	0	3	50	2	0
	27		7_2	3	3	0	3	0	3	0	3	0

[a]y^* = oil layer classification (1—oil layer, 2—poor oil layer, 3—dry layer) determined by the oil test.

TABLE 4.6 Comparison Among the Applications of C-SVM, R-SVM, BPNN, and MRA to Oil Layer Classification of the Xiefengqiao Anticline

Algorithm	Fitting Formula	Mean Absolute Relative Residual			Dependence of the Predicted Value (y) on Parameters (x_1, x_2, x_3, x_4, x_5), in Decreasing Order	Time Consumed on PC (Intel Core 2)	Solution Accuracy
		$\overline{R}1(\%)$	$\overline{R}2(\%)$	$\overline{R}^*(\%)$			
C-SVM	Nonlinear, explicit	0	0	0	N/A	3 s	Very high
R-SVM	Nonlinear, explicit	8.33	0	4.17	N/A	3 s	High
BPNN	Nonlinear, implicit	4.17	16.67	10.42	N/A	30 s	Low
MRA	Linear, explicit	8.33	33.33	20.83	x_1, x_2, x_3, x_5, x_4	<1 s	Very low

4.3.4. Summary and Conclusions

Summarily, using data for the oil layer classification based on a well-logging interpretation of the Xiefengqiao Anticline, i.e., the five parameters (true resistivity, acoustictime, porosity, oil saturation, permeability) and an oil test result of 24 oil layers, an explicit nonlinear function (4.17) is constructed by the data mining tool of C-SVM, and the results are consistent with oil test. Again, this function is called *mined knowledge*. In the prediction process, applying this knowledge, i.e., substituting the values of true resistivity, acoustictime, porosity, oil saturation, and permeability in Equations (4.17), respectively, the results are also consistent with the oil test. Therefore, this method can be spread to the oil layer classification based on well-logging interpretation of other areas.

Moreover, it is found that (a) since this case study is a strong nonlinear problem, the major data mining tool can adopt C-SVM but not MRA, R-SVM, or BPNN; and (b) in this case study, MRA established the order of dependence between the oil layer classification and its relative true resistivity, acoustictime, porosity, oil saturation, and permeability under the condition of linearity; thus MRA can be employed as an auxiliary tool.

4.4. DIMENSION-REDUCTION PROCEDURE USING MACHINE LEARNING

4.4.1. Definition and Benefits of Dimension Reduction

The definition of dimension reduction is to reduce the number of dimensions of a data space to as small as possible but to leave the results of the studied problem unchanged. For instance, in Case Study 2 (fractures prediction using conventional well-logging data) of Section 3.3, the studied problem is to solve an 8-D problem ($x_1, x_2, x_3, x_4, x_5, x_6, x_7, y$): $y = f(x_1, x_2, x_3, x_4, x_5, x_6, x_7)$, where x_i ($i=1, 2, ..., 7$) are known variables and y is a prediction variable. If x_1 is eliminated, $y = f(x_2, x_3, x_4, x_5, x_6, x_7)$, that is, a 7-D problem ($x_2, x_3, x_4, x_5, x_6, x_7,$ y); thus the number of dimensions of the data space is deduced from 8-D to 7-D. If $y = f(x_2, x_3,$ $x_4, x_5, x_6, x_7) = y = f(x_1, x_2, x_3, x_4, x_5, x_6, x_7)$, the dimension reduction succeeds, which will be proved below in detail.

The benefits of dimension reduction are to reduce the amount of data to enhance the mining speed, to reduce the attribute (variable) to extend applying ranges, and to reduce the misclassification ratio of prediction samples to enhance mining quality.

4.4.2. Dimension-Reduction Procedure

Taking the "prediction of fractures using conventional well-logging data" as a case study, a dimension-reduction procedure using machine learning (Shi, 2009; Shi and Yang, 2010) was briefly presented, as follows.

4.4.2.1. Using MRA as a Pioneering Dimension-Reduction Tool

From the learning process of MRA, the prediction variable y is shown to depend on the known variables x_i ($i = 1, 2, ..., 7$) in decreasing order: $x_5, x_2, x_4, x_7, x_3, x_6, x_1$ (Table 3.12); their corresponding multiple correlation coefficients are 0.71677, 0.82727, 0.84062, 0.84946, 0.86387, 0.86482, 0.86571. The last two multiple correlation coefficients are very close to the fifth, indicating that the dependences of y on x_6 and x_1 are very low. Hence, x_6 and x_1 could be eliminated, and then the 8-D problem ($x_1, x_2, x_3, x_4, x_5, x_6, x_7, y$) would be reduced to a 6-D problem ($x_2, x_3, x_4, x_5, x_7, y$). However, since MRA is a linear data mining tool, the results of dimension reduction are obtained under linear conditions, and thus a nonlinear data mining tool (C-SVM or BPNN) is required to validate whether the results are feasible.

4.4.2.2. Using C-SVM or BPNN as a Validating Dimension-Reduction Tool

To avoid pointless "do poorly done work over again" scenarios, it is wise to do a successive dimension reduction according to the order indicated by MRA, i.e., to eliminate x_1 at first (to create a 7-D problem) and to have C-SVM or BPNN to validate whether the calculation results are consistent with that of an 8-D problem. If they're not, it indicates that this dimension reduction failed. If they are, eliminate x_6 (create a 6-D problem) and have C-SVM or BPNN validate whether the calculation results are consistent with that of an 8-D problem; if not, only a 7-D problem is allowed; if yes, eliminate x_3 (to create a 5-D problem), and so on.

4.4.3. Case Study

Taking Case Study 2 (fracture prediction using conventional well-logging data) of Section 3.3 as a case study, the aforementioned successive dimension reduction is employed. Before dimension reduction, it is an 8-D problem ($x_1, x_2, x_3, x_4, x_5, x_6, x_7, y$), and its input data, calculation results, and analyses are listed in Tables 3.10, 3.11, and 3.12.

The results of successive dimension reduction are listed in Table 4.7.

We can see from Table 4.7 that the results of a 7-D problem and a 6-D problem are accurate and are consistent with that of the 8-D problem, i.e., not only $\overline{R}_1(\%) = 0$ but also $\overline{R}_2(\%) = 0$.

TABLE 4.7 Successive Dimension-Reduction Results for Fracture Prediction of Wells An1 and An2

						Fracture Identification[a]							
						7-D Program $(x_2, x_3, x_4, x_5, x_6, x_7, y)$ (x_1 Deleted)				6-D Program $(x_2, x_3, x_4, x_5, x_7, y)$ (x_1 and x_6 Deleted)			
						C-SVM		BPNN		C-SVM		BPNN	
Sample Type	Sample No.	Well No.	Depth (m)	y^*	IL	y	$R(\%)$	y	$R(\%)$	y	$R(\%)$	y	$R(\%)$
Learning samples	1	An1	3065.13	1	1	1	0	1	0	1	0	1	0
	2		3089.68	1	1	1	0	1	0	1	0	1	0
	3		3098.21	1	1	1	0	1	0	1	0	1	0
	4		3102.33	1	1	1	0	1	0	1	0	1	0
	5		3173.25	1	1	1	0	1	0	1	0	1	0
	6		3180.37	1	1	1	0	1	0	1	0	1	0
	7		3202.00	2	2	2	0	2	0	2	0	2	0
	8		3265.37	2	2	2	0	2	0	2	0	2	0
	9		3269.87	1	1	1	0	1	0	1	0	1	0
	10		3307.87	1	1	1	0	1	0	1	0	1	0
	11		3357.37	1	1	1	0	1	0	1	0	1	0
	12		3377.03	1	1	1	0	1	0	1	0	1	0
	13		3416.48	2	2	2	0	2	0	2	0	2	0
	14		3445.37	2	2	2	0	2	0	2	0	2	0
	15		3446.12	2	2	2	0	2	0	2	0	2	0
	16		3485.25	2	2	2	0	2	0	2	0	2	0
	17		3575.00	1	1	1	0	1	0	1	0	1	0
	18		3645.00	1	1	1	0	1	0	1	0	1	0
	19		3789.37	1	1	1	0	1	0	1	0	1	0
	20	An2	992.795	2	2	2	0	2	0	2	0	2	0
	21		1525.37	1	1	1	0	1	0	1	0	1	0
	22		1527.25	1	1	1	0	1	0	1	0	1	0
	23		1867.12	1	1	1	0	1	0	1	0	1	0
	24		1880.00	2	2	2	0	2	0	2	0	2	0
	25		2045.87	1	1	1	0	1	0	1	0	1	0
	26		2085.25	2	2	2	0	2	0	2	0	2	0

(*Continued*)

TABLE 4.7   Successive Dimension-Reduction Results for Fracture Prediction of Wells An1 and An2—cont'd

						Fracture Identification[a]						
						7-D Program $(x_2, x_3, x_4, x_5, x_6, x_7, y)$ (x_1 Deleted)				6-D Program $(x_2, x_3, x_4, x_5, x_7, y)$ (x_1 and x_6 Deleted)			
						C-SVM		BPNN		C-SVM		BPNN	
Sample Type	Sample No.	Well No.	Depth (m)	y^* IL	y	$R(\%)$	y	$R(\%)$	y	$R(\%)$	y	$R(\%)$
	27		2112.13	1	1	0	1	0	1	0	1	0
	28		2355.37	1	1	0	1	0	1	0	1	0
	29		2358.00	2	2	0	2	0	2	0	2	0
Prediction samples	30	An1	3164.00	(1)	1	0	1	0	1	0	1	0
	31		3166.50	(1)	1	0	1	0	1	0	1	0
	32	An2	980.485	(1)	1	0	1	0	1	0	1	0
	33		987.018	(1)	1	0	1	0	1	0	1	0

[a] y^* = fracture identification (1—fracture, 2—nonfracture) determined by the imaging log (IL); in y^*, numbers in parenthesis are not input data but are used for calculating R(%). x_1 = acoustictime (Δt), x_2 = compensated neutron density (ρ), x_3 = compensated neutron porosity (φ_N), x_4 = microspherically focused resistivity (R_{xo}), x_5 = deep laterolog resistivity (R_{LLD}), x_6 = shallow laterolog resistivity (R_{LLS}), and x_7 = absolute difference of R_{LLD} and R_{LLS} (R_{DS}).

Therefore, 8-D reduced to 7-D and 7-D reduced to 6-D are successful. Actually, continued dimension reductions are also successful till the problem becomes a 4-D problem (x_2, x_4, x_5, y), i.e., x_1, x_6, x_3, and x_7 are successively eliminated.

4.4.4. Summary and Conclusions

A dimension-reduction procedure using MRA as a pioneering dimension-reduction tool and C-SVM or BPNN as a validating dimension-reduction tool were presented. Taking Case Study 2 (fracture prediction using conventional well-logging data) of Section 3.3 as a case study, this dimension-reduction procedure can reduce the original 8-D problem to a 4-D problem. Though this case study, we conclude the following:

1. MRA is a good pioneering dimension-reduction tool. It can indicate the order of dimension reduction and the possible final number of dimensions. However, since MRA is a linear data mining tool, the results of dimension reduction are obtained under linear conditions, and thus a nonlinear data mining tool (C-SVM or BPNN) is required to validate whether the results are feasible.
2. The action of dimension reduction is to reduce the amount of data so as to speed up the data mining process and especially to extend up the applying ranges. For instance, the previous case study can be reduced to the 4-D problem so that the fracture identification (y) only depends on three well-logging data of deep laterolog resistivity (x_5), compensated neutron density (x_2), and microspherically focused resistivity (x_4), i.e., C-SVM or BPNN can be adopted to the fracture identification without four well-logging data of absolute

difference of R_{LLD} and R_{LLS} (x_7), compensated neutron porosity (x_3), shallow laterolog resistivity (x_6), and acoustictime (x_1).

3. Because of the complexities of geosciences rules, the correlations between different classes of geosciences data are nonlinear in most cases. In general, therefore, it is better to use C-SVM when the nonlinearity is very strong; otherwise, use BPNN. For the sake of unification and safety, it is suggested to simply adopt C-SVM as a validating dimension-reduction tool, since C-SVM is not only applicable to either strongly or weakly nonlinear problems but also runs at speeds as fast as 20 times or more of BPNN. Certainly it is easy to code the BPNN program, whereas it is very complicated to code the C-SVM program, so BPNN is a good validating dimension-reduction tool in case no C-SVM program exists.

EXERCISES

4-1. What are the differences between C-SVM and R-SVM in terms of use?

4-2. What is the advantage of SVM compared with existing methods of statistics?

4-3. To have the comparability of calculation results among various applications in this book, what operation options does SVM take?

4-4. In Case Studies 1 and 2, C-SVM, R-SVM, BPNN, and MRA are adopted, and the results of the four algorithms are compared in Tables 4.3 and 4.6. From the two tables, review the results of each algorithm.

4-5. What are the definitions and actions of dimension reduction?

References

Boser, B.E., Guyon, I., Vapnik, V., 1992. A training algorithm for optimal margin classifiers. In: Proceedings of the Fifth Annual Workshop on Computational Learning Theory. ACM Press, pp. 144–152.

Chang, C., Lin, C., 2001. Training ν-support vector classifiers: Theory and algorithms. Neural Comput. 13 (9), 2119–2147.

Chang, C., Lin, C., 2002. Training ν-support vector regression: Theory and algorithms. Neural Comput. 14 (8), 1959–1977.

Chang, C., Lin, C., 2011. LIBSVM: a library for support vector machines, Version 3.1. Retrieved from www.csie.ntu.edu.tw/~cjlin/libsvm/.

Chen, K., Zhang, S., Ding, X., Su, J., 2006. Gassiness evaluation of gas-bearing layers in tight sandstones. J. Oil Gas Tech. 28 (4), 65–68 (in Chinese with English abstract).

Cortes, C., Vapnik, V., 1995. Support-vector network. Machine Learning 20 (3), 273–297.

Crisp, D.J., Burges, C.J.C., 2000. A geometric interpretation of ν-SVM classifiers. In: Solla, S., Leen, T., Müller, K.R. (Eds.), Advances in Neural Information Processing Systems, vol. 12. MIT Press, Cambridge, MA, USA, pp. 244–250.

Cristianini, N., Shawe-Taylor, J., 2000. An Introduction to Support Vector Machines and other Kernel-Based Learning Methods. Cambridge University Press, Cambridge, UK.

Schölkopf, B., Smola, A.J., Williamson, R.C., Bartlett, P.L., 2000. New support vector algorithms. Neural Comput. 12 (5), 1207–1245.

Shi, G., 2008. Superiorities of support vector machine in fracture prediction and gassiness evaluation. Petro. Expl. Devel. 35 (5), 588–594.

Shi, G., 2009. The use of support vector machine for oil and gas identification in low-porosity and low-permeability reservoirs. Int. J. Math. Model. Numer. Optimisa. 1 (1/2), 75–87.

Shi, G., 2011. Four classifiers used in data mining and knowledge discovery for petroleum exploration and development. Adv. Petro. Expl. Devel. 2 (2), 12–23.

Shi, G., Yang, X., 2010. Optimization and data mining for fracture prediction in geosciences. Procedia Comput. Sci. 1 (1), 1353–1360.

Vapnik, V., 1998. Statistical Learning Theory. Wiley, New York, NY, USA.

Yang, J., 2002. Identification of oil horizons by artificial neural networks in Xiefengqiao structure. Oil Gas Geol. 23 (1), 76–80 (in Chinese with English abstract).

Decision Trees

The *decision tree* (DTR) is a classification approach based on data induction learning. This chapter introduces two DTR algorithms (ID3 and C4.5) as well as their applications in geosciences.

Section 5.1 (methodology) introduces the applying ranges and conditions, basic principles, calculation methods, and calculation flowcharts of ID3 and C4.5 as well as a simple case study. Though this simple case study is small, it reflects the whole process of calculations to benefit readers in understanding and mastering the applied techniques. The simple case study involves the "computer buy" problem, with which we explain how to adopt ID3 and C4.5 for classification. Concretely, we use data for the "computer buy" problem based on four parameters, i.e., the four parameters (age, income, student, credit-rating) and computer buy of 14 customers, a decision tree together with its five classification rules is constructed by the data mining tools of ID3 and C4.5, and the results coincide with practicality. This decision tree, together with its five classification rules, is called *mined knowledge*. In the prediction process, applying this knowledge, i.e., using the four parameters of the other three customers for the "computer buy" problem according to five classification rules, respectively, the prediction results also coincide with practicality. Therefore, this method can be spread to the buy problem for other goods. Moreover, it is found that (a) since this simple case study is a strong nonlinear problem, regression of support vector machines (*R*-SVM; see Chapter 4) or multiple regression analysis (MRA; see Chapter 2) is not applicable; and (b) but the problem is very simple, and ID3, C4.5, classification of support vector machines (*C*-SVM; see Chapter 4) and error back-propagation neural networks (BPNN; see Chapter 3) are applicable. It is noted that *R*-SVM and *C*-SVM are two major algorithms of support vector machines (SVM; see Chapter 4).

Section 5.2 (top coal caving classification [29 learning samples]) introduces Case Study 1 of ID3. Using data for the top coal caving classification based on six parameters, i.e., the six parameters (mining depth, uniaxial compressive strength, dirt band thickness, coal layer thickness, cracking index, and coefficient of fullness at top coal) and coal test result of 29 coal mines in China, a decision tree, together with its 26 classification rules, is constructed by the data mining tool of ID3, and the results are consistent with the coal test. This decision tree, together with its 26 classification rules, is called *mined knowledge*. Therefore, this method can be spread to the top coal caving classification of other coal mines. Moreover, it is found that (a) since this case study is a strong nonlinear problem, MRA is not applicable; (b) since the classification rule is complicated and the number of learning samples is not big enough, neither *C*-SVM nor *R*-SVM is applicable; and (c) since BPNN is adaptive to the structuralist problem, the results of BPNN and ID3 are the same and correct.

Section 5.3 (top coal caving classification [26 learning samples and three prediction samples]) introduces Case Study 2 of ID3. Using data for the top coal caving classification based on six parameters, i.e., the six parameters (mining depth, uniaxial compressive strength, dirt band thickness, coal layer thickness, cracking index, and coefficient of fullness at top coal) and coal test results of 26 coal mines in China, a decision tree, together with its 26 classification rules, is constructed by the data mining tool of ID3, and the results are consistent with coal tests. This decision tree, with its 26 classification rules, is called *mined knowledge*. In the prediction process, applying this knowledge, i.e., using the six parameters of three other coal mines for the top coal caving classification by 26 classification rules, respectively, the results are also consistent with the coal test. Therefore, this method can be spread to the top coal caving classification of other coal mines. Moreover, it is found that (a) since this case study is a strong nonlinear problem, MRA is not applicable; and (b) since the classification rule is complicated and the number of learning samples is not big enough, C-SVM, R-SVM, or BPNN is not applicable.

It is found from the aforementioned Case Study 1 and Case Study 2 that C-SVM, R-SVM, or BPNN may not be applicable in the case of a small number of learning samples; in this case, it is preferable to adopt ID3, which is an advantage over ID3.

5.1. METHODOLOGY

The decision tree (DTR) has been applied gradually in the natural and social sciences since the 1990s and widely applied in this century. The currently popular algorithms are ID3 (Quinlan, 1986) and C4.5 (Quinlan, 1993). In recent years, DTR application to geosciences has occurred to some extent.

The learning samples and prediction samples for ID3 and C4.5 are expressed with Equations (1.1) and (1.3), respectively. The learning process and prediction process of calculation flowcharts are illustrated in Figures 1.2 and 1.3, respectively. In Figure 1.3, if the prediction samples are replaced by the learning samples, Figure 1.3 becomes such a calculation flowchart for learning validation.

The expression of y created using ID3 and C4.5 is a set of rule "expressions" with respect to m parameters $(x_1, x_2, ..., x_m)$:

$$y = DTR(x_1, x_2, ..., x_m) \tag{5.1}$$

where DTR is a nonlinear function ($ID3$ or $C4.5$) calculated by the ID3 or C4.5 algorithm. $ID3$ or $C4.5$ is a set of classification rules expressed in IF—THEN form with respect to $x_1, x_2, ..., x_m$, i.e., a set of logical operations. If Equation (5.1) is regarded as an explicit expression, it is similar to the y expression of MRA, C-SVM, R-SVM, and Bayesian classification (BAC; see Chapter 6); and C-SVM, R-SVM, ID3, C4.5, and BAC are nonlinear algorithms, whereas MRA is a linear algorithm. Equation (5.1) is called the *fitting formula* obtained from the learning process.

The basic idea of DTR is illustrated in Figure 5.1.

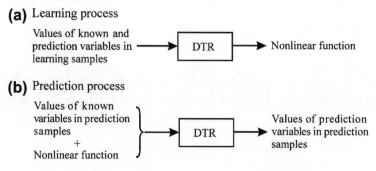

FIGURE 5.1   Learning process (a) and prediction process (b) of DTR (ID3 and C4.5).

5.1.1. Applying Ranges and Conditions

5.1.1.1. Applying Ranges

As Equation (1.1) described, assume that there are n learning samples, each associated with $m + 1$ numbers $(x_1, x_2, ..., x_m, y_i^*)$ and a set of observed values $(x_{1i}, x_{2i}, ..., x_{mi}, y_i^*)$, with $i = 1, 2, ..., n$ for these parameters. In principle, $n > m$, but in actual practice $n >> m$. As Equation (1.3) described, assume that there are k prediction samples, each associated with m parameters $(x_1, x_2, ..., x_m)$ and a set of observed values $(x_{1i}, x_{2i}, ..., x_{mi})$, with $i = n + 1$, $n + 2, ..., n + k$ for these parameters.

5.1.1.2. Applying Conditions

The number of learning samples, n, must be big enough to ensure the accuracy of prediction by DTR. The number of known variables, m, must be greater than one.

5.1.2. Basic Principles

The structure of DTR looks like a tree that is similar to a flowchart (Figure 5.2). (DTR introduced in this section refers to only ID3 and C4.5.) Concretely, the topmost node of the tree is the root node; each internal node denotes a test on a parameter (one of $x_1, x_2, ..., x_m$), each branch represents an output of a test, and each leaf node represents y (a studied parameter). DTR is an induction-learning algorithm based on practical examples and can deduce the classification rule with tree form from a set of unordered and ruleless examples. DTR classifies a practical example in the order from the root node to the leaf nodes. Each node represents a parameter, and each branch represents a possibly selected value on its connected upper node. In classification, several operations are performed: use a top-down recursion, compare parameter values on internal node, discriminate the branch downward from the node

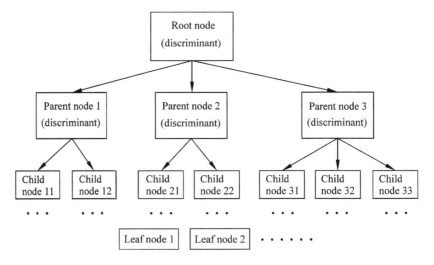

FIGURE 5.2   Sketch map of the structure of DTR.

according to different parameter values, and obtain conclusions on leaf nodes. This procedure is repeated on a subtree with a new node as its root node. Each path from root node to leaf node corresponds to a set of conjunction rules for parameter tests, and thus the whole tree (Figure 5.2) is an expression, Equation (5.1), which represents the conjunction rules of parameter values constrained for the practical example.

The currently popular algorithms are ID3 (Quinlan, 1986) and C4.5 (Quinlan, 1993). They are introduced here.

5.1.3. ID3

The generation of a decision tree is equivalent to the learning process of MRA, BPNN, C-SVM, R-SVM, and BAC, so it is also called the learning process of DTR, based on the same learning samples as the following:

$$x_i = (x_{i1}, x_{i2}, ..., x_{im}, x_{im+1}) \quad (i = 1, 2, ..., n) \tag{5.2}$$

where $x_{i\,m+1} = y_i$; the other symbols were defined in Equations (1.1) and (1.2).

The data expressed by Equation (5.2) cannot be directly used in the tree generation. It is required to divide each parameter into several class values, substituting c for x, and then forming new learning samples:

$$c_i = (c_{i1}, c_{i2}, ..., c_{im}, c_{im+1}) \quad (i = 1, 2, ..., n) \tag{5.3}$$

where c_{ij} ($j = 1, 2, ..., m + 1$) are nonzero integer numbers; in general they can be 1, 2, 3, ...; and $y_i = c_{i\,m+1}$.

In the following tree generation, c_j ($j = 1, 2, ..., m + 1$) expressed by Equation (5.3) are used, but x_j ($j = 1, 2, ..., m + 1$) expressed by Equation (5.2) are not used.

The expected information that is required by classification is

$$Info(D) = -\sum_{k=1}^{K} p_k \log_2 (p_k) \tag{5.4}$$

where D refers to a specific data partition; K is the number of class-values involving D in total; $p_k = $ (The number of class-values related to D)/K, that is, the probability of those class values occurring in K. These three definitions look like nonobjectives in some measure but will be concretely understandable through the following case studies. The expected information, $Info(D)$, is also called *entropy*.

Equation (5.4) can be used to anyone in parameters c_j ($j = 1, 2, ..., m+1$) expressed by Equation (5.3). For c_j ($j = 1, 2, ..., m$), the results of Equation (5.4) are designated as $Info(D)|_{c_j}$; for y (i.e., c_{m+1}), the result of Equation (5.4) is designated as $Info(D)|_y$.

For a parameter c_j ($j = 1, 2, ..., m$), the expected information that is required by classification is

$$E(c_j) = \sum_{k=1}^{K} p_k\, Info(D)\Big|_{c_j} \quad (j = 1, 2, ..., m) \tag{5.5}$$

where all symbols of the right-hand side have been defined in Equation (5.4).

For a parameter c_j ($j = 1, 2, ..., m$), the information gain that is required by classification is

$$Gain(c_j) = Info(D)|_y - E(c_j) \quad (j = 1, 2, ..., m) \tag{5.6}$$

where $E(c_j)$ has been defined by Equation (5.5); and the other symbols have been defined in Equation (5.4).

The advantages of ID3 are the clear basic theory, simple method, and strong learning ability. However, there exist some shortcomings in ID3. For instance, (a) it is sensitive to noises; (b) it is available only when the number of learning samples is small but the tree changes as the number of learning samples increases, and then regenerating a new tree causes big spending; and (c) like BPNN, the local optimal solution but not the global optimal solution occurs.

The procedure of ID3 consists of two steps (Figure 5.3):

1. In the first step, the y values of learning samples are calculated using the classified learning samples.
2. In the second step, the y values of prediction samples are calculated using the classified prediction samples.

5.1.3.1. Learning Process

Learning samples to be used should be those expressed by Equation (5.3).
The learning process consists of the following two steps:

1. *Calculation of expected information required by classification of y in learning samples.* Because y is the aim of the studied problem, the practical value y^* of y is considered at beginning, and y^* will be used as the leaf node. Using the class-value y^* of y in learning samples, and by Equation (5.4), the expected information required by classification of y is obtained:

$$Info(D)|_y$$

2. *Calculation of expected information required by classification of y in learning samples.* c_j ($j = 1, 2, ..., m$) of learning samples will be used as nodes of a decision tree, in which a c_j determined as a node at first is the root node.

At first, using the class values of c_j ($j = 1, 2, ..., m$) in learning samples, and by Equation (5.4), the root node is obtained:

$$Info(D)|_{c_j} \quad (j = 1, 2, ..., m)$$

Then, using Equation (5.5), the expected information that is required by classification of c_j is obtained:

$$E(c_j) = \sum_{k=1}^{K} p_k \, Info(D)\Big|_{c_j} \quad (j = 1, 2, ..., m)$$

Finally, using Equation (5.6), the information gain that is required by classification of c_j is obtained:

$$Gain(c_j) = Info(D)|_y - E(c_j) \quad (j = 1, 2, ..., m)$$

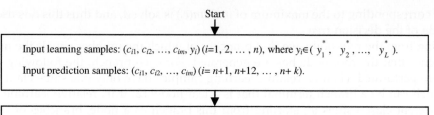

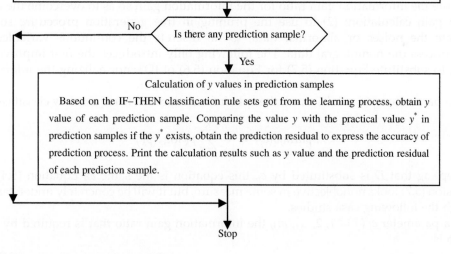

FIGURE 5.3   Calculation flowchart of ID3.

The j corresponding to the maximum of m $Gain(c_j)$ is solved, and thus this c_j is used as the root node of the decision tree.

Starting from the root node, we find branches of which the number equals the number of class values that the root node has are generated. For each branch, the following two processes are performed: (1) if its relative learning samples have only one same class value y^*, this branch can be a branch terminal that is a leaf node; (2) if its relative learning samples have different class values y^*, starting from this branch as a node, branches of which the number equals the number of class values that the branch has are generated; ...; till any branch is in the case of (1), thus a decision tree is obtained. It is required to output the decision tree and to save its corresponding IF—THEN classification rule sets.

Comparing the value y with the practical value y^* in learning samples, the fitting residual $\overline{R}_1(\%)$ is obtained to express the fitness of the learning process. It is required to print the calculation results such as the y value and $\overline{R}_1(\%)$ of each learning sample.

5.1.3.2. Prediction Process

If prediction samples exist, the calculation of the prediction process is conducted.

Using the IF—THEN classification rule sets obtained by the aforementioned learning process, the y value of each learning sample is obtained. Comparing the value y with the practical value y^* if the y^* exists in prediction samples, the prediction residual $\overline{R}_2(\%)$ is obtained to express the accuracy of the prediction process.

5.1.4. C4.5

C4.5 has inherited the advantages of ID3 and made four major improvements: (1) to substitute the information gain ratio for the information gain so as to overcome the deflection in gain calculation; (2) to use the pruning in tree generation procedure so as to eliminate the noises or abnormal data; (3) to discrete the continuous variables; and (4) to process the nonintegral data. The following only introduces the first improvement, that is, to substitute Equation (5.7) for Equation (5.6) of ID3, normalizing the information gain.

For a parameter c_j ($j = 1, 2, ..., m$), the split information that is required by classification is

$$SplitInfo(c_j) = -\sum_{k=1}^{K} p_k \log_2(p_k) \tag{5.7}$$

Excepting that D is substituted by c_j, this equation is the same as Equation (5.4). This expression (5.7) looks nonobjective in some measure, but it will be concretely understandable through the following case studies.

For a parameter c_j ($j = 1, 2, ..., m$), the information gain ratio that is required by classification is

$$GainRatio(c_j) = \frac{Gain(c_j)}{SplitInfo(c_j)} \quad (j = 1, 2, ..., m) \tag{5.8}$$

where $Gain(c_j)$ and $SplitInfo(c_j)$ have been defined by Equations (5.6) and (5.7).

5.1.5. Simple Case Study: Computer Buy

To help readers master the procedure of tree generation, a very simple sample of a computer-buy problem solved by ID3 is introduced here. In Table 5.1, 14 learning samples (customers) are the original data (Han and Kamber, 2006), where $y^* = x_5$; c_j ($j = 1, 2, ..., 5$) can be used by the learning process of ID3 or C4.5, where $y^* = c_5$, and also used by the learning process (customers) of MRA, BPNN, C-SVM, or R-SVM. In Table 5.1, three additional prediction samples are provided and can be used by the prediction process of ID3, C4.5, MRA, BPNN, C-SVM, or R-SVM.

5.1.5.1. *Known Input Parameters*

They are the values of the known variables c_i ($i = 1, 2, 3, 4$) and the prediction variable y^* (i.e., c_5) for 14 learning samples (Table 5.1).

TABLE 5.1 Input Data for a Computer Buy

Sample Type	Sample No.	c_1 (x_1) Age	c_2 (x_2) Income	c_3 (x_3) Student	c_4 (x_4) Credit Rating	y^* c_5 (x_5) Computer Buy[a]
Learning samples	1	1 (Youth)	3 (High)	2 (No)	1 (Fair)	2 (No)
	2	1 (Youth)	3 (High)	2 (No)	2 (Excellent)	2 (No)
	3	2 (Middle-aged)	3 (High)	2 (No)	1 (Fair)	1 (Yes)
	4	3 (Senior)	2 (Medium)	2 (No)	1 (Fair)	1 (Yes)
	5	3 (Senior)	1 (Low)	1 (Yes)	1 (Fair)	1 (Yes)
	6	3 (Senior)	1 (Low)	1 (Yes)	2 (Excellent)	2 (No)
	7	2 (Middle-aged)	1 (Low)	1 (Yes)	2 (Excellent)	1 (Yes)
	8	1 (Youth)	2 (Medium)	2 (No)	1 (Fair)	2 (No)
	9	1 (Youth)	1 (Low)	1 (Yes)	1 (Fair)	1 (Yes)
	10	3 (Senior)	2 (Medium)	1 (Yes)	1 (Fair)	1 (Yes)
	11	1 (Youth)	2 (Medium)	1 (Yes)	2 (Excellent)	1 (Yes)
	12	2 (Middle-aged)	2 (Medium)	2 (No)	2 (Excellent)	1 (Yes)
	13	2 (Middle-aged)	3 (High)	1 (Yes)	1 (Fair)	1 (Yes)
	14	3 (Senior)	2 (Medium)	2 (No)	2 (Excellent)	2 (No)
Prediction samples	15	1 (Youth)	1 (Low)	1 (Yes)	2 (Excellent)	[1 (Yes)]
	16	2 (Middle-aged)	1 (Low)	2 (No)	1 (Fair)	[1 (Yes)]
	17	3 (Senior)	3 (High)	1 (Yes)	1 (Fair)	[1 (Yes)]

[a]y^* = computer-buy (1—buy, 2—no buy). In y^*, numbers in square brackets are not input data but are used for calculating $R(\%)$.

5.1.5.2. Learning Process

The learning process consists of the following three steps:

1. Calculate $Info(D)$. Because a computer buy is the aim of the studied problem, the practical parameter c_5 (i.e., y^*) is considered at the beginning, and $K = 2$ from Table 5.1 refers to two kinds of values of "1(Yes)" and "2(No)". Letting $s1$ be the number of "1(Yes)", $s1 = 9$; letting $s2$ be the number of "2(No)", $s2 = 5$. It is obvious that $p_1 = 9/14$, and $p_2 = 5/14$. Using Equation (5.4), it is calculated out that

$$Info(s1, s2) = -[p_1 Log_2(p_1) + p_2 Log_2(p_2)] = -[9/14 Log_2(9/14) + 5/14 Log_2(5/14)]$$
$$= 0.94$$

which is a concrete expression of $Info(D)$.

2. Calculate $Gain(c_1)$ at first. $Gain(c_1)$ is $Gain(age)$. $Gain(c_1)$ is calculated at first only for explaining a whole procedure of $Gain(c_j)$ calculation as an example. For parameter c_1, $K = 3$ from Table 5.1 refers to three kinds of values: "1(Youth)", "2(Middle-aged)", and "3(Senior)"; combining parameter c_5, $K = 2$ from Table 5.1, as mentioned earlier. Similar to the above definition of $(s1, s2)$, $(s11, s21)$, $(s12, s22)$ and $(s13, s23)$ are defined. Using Equation (5.4), the following three concrete expressions of $Info(D)$ are obtained.

   a. Youth. In c_1, if there are five people, then $p_1 = 5/14$. In c_5, among five people, $s11 = 2$ buy computer, and $s21 = 3$ do not buy computer, then

$$Info(s11, s21) = -[2/5 Log_2(2/5) + 3/5 Log_2(3/5)] = 0.971$$

   b. Middle-aged. In c_1, if there are four people, then $p_2 = 4/14$. In c_5, among four people, $s12 = 4$ buy computer, and $s22 = 0$ do not buy computer, then

$$Info(s12, s22) = -[4/4 Log_2(4/4) + 0/4 Log_2(0/4)] = 0$$

   c. Senior. In c_1, if there are five people, then $p_3 = 5/14$. In c_5, among five people, $s13 = 3$ buy computer, and $s23 = 2$ do not buy computer, then

$$Info(s13, s23) = -[3/5 Log_2(3/5) + 2/5 Log_2(2/5)] = 0.971$$

   Substituting the above three p values and three $Info(D)$ values into Equation (5.5), $E(c_1)$ is obtained:

$$E(age) = 5/14 \times Info(s11, s21) + 4/14 \times Info(s12, s22) + 5/14 \times Info(s13, s23) = 0.694$$

   Finally, using Equation (5.6), $Gain(c_1)$ is obtained:

$$Gain(age) = Info(s1, s2) - E(age) = 0.94 - 0.694 = 0.246$$

3. Similarly calculate $Gain(c_2)$, $Gain(c_3)$, and $Gain(c_4)$. $Gain(c_2)$ is $Gain(income)$, $Gain(c_3)$ is $Gain(student)$, and $Gain(c_4)$ is $Gain(credit-rating)$.

   Similar to the previous calculation of $Gain(age)$, $Gain(income) = 0.029$, $Gain(student) = 0.151$, and $Gain(credit-rating) = 0.048$ are obtained.

Because of

$$Gain(\text{age}) = \max\{Gain(\text{age}), Gain(\text{income}), Gain(\text{student}), Gain(\text{credit-ratimg})\},$$

"Age" is selected as the root node, and three branches are produced from three class values (youth, middle-aged, senior) of "age". (a) For the first branches, "Middle-aged" in c_5 are all "1(Yes)", and the relative samples are sample Nos. 3, 7, 12, and 13. This branch "Middle-aged" is used as a branch terminal, and thus a leaf node is generated. (b) For the second branch, "Youth" in c_5 are "1(Yes)" or "2(No)", so it is required to set "Student" as a node for classification, and two branches are produced. For "Student" that are shown "1(Yes)" in c_3, the relative samples are sample Nos. 9 and 11. For "Student" that are shown "2(No)" in c_3, the relative samples are sample Nos. 1, 2, and 8. Thus two leaf nodes are generated. (c) For the third branch, "Senior" in c_5 are "1(Yes)" or "2(No)", so it is required to set "Credit-rating" as a node for classification, and two branches are produced. For "Credit-rating" that are shown "1(Fair)" in c_4, the relative samples are sample Nos. 4, 5, and 10. For "Credit-rating" that are shown "2(Excellent)" in c_4, the relative samples are sample Nos. 6 and 14. Thus two leaf nodes are also generated. Finally, a decision tree that has five leaf nodes in total is obtained (Figure 5.4).

Figure 5.4 shows that there are five leaf nodes in the tree. Each leaf node expresses the information and its relative sample numbers of the computer-buy problem. It also reflects its classification rule. Corresponding to the five leaf nodes ranging in the order of left to

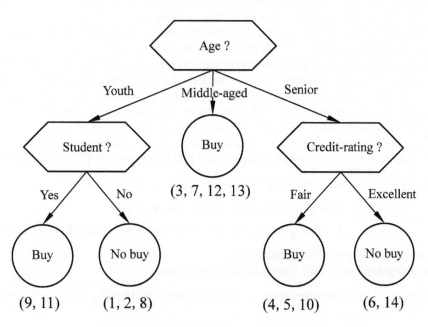

Number in parenthesis denotes sample No.

FIGURE 5.4   A decision tree for a computer buy *(modified from Han and Kamber, 2006).*

right, illustrated in Figure 5.4, five classification rules expressed with IF–THEN are as follows:

(1) IF [(Age = Youth)·and·(= Student)] THEN Computer-buy = Yes (relative to sample Nos. 9, 11);
(2) IF [(Age = Youth)·and·(≠ Student)] THEN Computer–buy = No (relative to sample Nos. 1, 2, 8);
(3) IF (Age = Middle-aged) THEN Computer-buy = Yes (relative to sample Nos. 3, 7, 12, 13);
(4) IF [(Age = Senior)·and·(Credit-rating = Fair)] THEN Computer-buy = Yes (relative to sample Nos. 4, 5, 10);
(5) IF [(Age = Senior)·and·(Credit-rating = Excellent)] THEN Computer-buy = No (relative to sample Nos. 6, 14).

Thus, it is believed that an explicit nonlinear function corresponding to DTR formula (5.1) is obtained:

$$y = ID3(c_1, c_2, c_3, c_4) \tag{5.9}$$

5.1.5.3. Prediction Process

As mentioned, using 14 learning samples (Table 5.1) and the ID3 algorithm, a decision tree (Figure 5.4) and its five classification rules were obtained, and the results of computer-buy (y) are also obtained and then filled into Table 5.2. Now, using three prediction samples (Table 5.1) and the five classification rules, the computer-buy (y) of each prediction sample is predicted and then filled into Table 5.2. That is a prediction process of the ID3 algorithm.

Thus, the application of ID3 to the computer buy is accomplished, showing the learning process and prediction process of ID3 algorithms by an example.

Also by example of the computer-buy problem, the use of Equations (5.7) and (5.8) of C4.5 is explained here. Taking "Income" as an example, the previous calculated $Gain(c_2) = Gain$ (Income) $= 0.029$. Now $SplitInfo$(Income) $= SplitInfo(c_2)$ will be calculated by Equation (5.7). Considering parameter c_2, $K = 3$ from Table 5.1, referring to three kinds of values: "1(Low)", "2(Medium)", and "3(High)". Letting $s1$ be the number of "1(Low)", $s1 = 4$; letting $s2$ be the number of "2(Medium)", $s2 = 6$; and letting $s3$ be the number of "3(High)", $s3 = 4$. It is obvious that $p_1 = 4/14$, $p_2 = 6/14$, and $p_3 = 4/14$. Using Equation (5.7), it is calculated that

$$\begin{aligned} SplitInfo(c_2) &= SplitInfo(\text{Income}) = SplitInfo(s_1, s_2, s_3) \\ &= -\left[p_1 Log_2(p_1) + p_2 Log_2(p_2) + p_3 Log_2(p_3)\right] \\ &= -\left[4/14 Log_2(4/14) + 6/14 Log_2(6/14) + 4/14 Log_2(4/14)\right] = 1.557 \end{aligned}$$

which is a concrete expression of $SplitInfo(c_j)$.

Finally, using Equation (5.8), $GainRatio(c_2)$ is obtained:

$$GainRatio(c_2) = GainRatio(\text{Income}) = Gain(\text{Income})/SplitInfo(\text{Income}) = 0.029/1.557$$

$$= 0.019$$

Also, when we apply C4.5 to the aforementioned computer-buy problem, its results are completely the same as that of ID3, since this problem is very simple.

TABLE 5.2   Prediction Results from Computer Buy

Sample Type	Sample No.	y^* Computer Buy[a]	ID3 or C4.5 y	R(%)	C-SVM y	R(%)	R-SVM y	R(%)	BPNN y	R(%)	MRA y	R(%)
Learning samples	1	2 (No)	2 (No)	0	2 (No)	0	2 (No)	0	2 (No)	0	2 (No)	0
	2	2 (No)	2 (No)	0	2 (No)	0	2 (No)	0	2 (No)	0	2 (No)	0
	3	1 (Yes)	1 (Yes)	0	1 (Yes)	0	1 (Yes)	0	1 (Yes)	0	1 (Yes)	0
	4	1 (Yes)	1 (Yes)	0	1 (Yes)	0	1 (Yes)	0	1 (Yes)	0	1 (Yes)	0
	5	1 (Yes)	1 (Yes)	0	1 (Yes)	0	1 (Yes)	0	1 (Yes)	0	1 (Yes)	0
	6	2 (No)	2 (No)	0	2 (No)	0	2 (No)	0	2 (No)	0	1 (Yes)	50
	7	1 (Yes)	1 (Yes)	0	1 (Yes)	0	1 (Yes)	0	1 (Yes)	0	1 (Yes)	0
	8	2 (No)	2 (No)	0	2 (No)	0	2 (No)	0	2 (No)	0	2 (No)	0
	9	1 (Yes)	1 (Yes)	0	1 (Yes)	0	1 (Yes)	0	1 (Yes)	0	1 (Yes)	0
	10	1 (Yes)	1 (Yes)	0	1 (Yes)	0	1 (Yes)	0	1 (Yes)	0	1 (Yes)	0
	11	1 (Yes)	1 (Yes)	0	1 (Yes)	0	1 (Yes)	0	1 (Yes)	0	1 (Yes)	0
	12	1 (Yes)	1 (Yes)	0	1 (Yes)	0	2 (No)	100	1 (Yes)	0	2 (No)	100
	13	1 (Yes)	1 (Yes)	0	1 (Yes)	0	1 (Yes)	0	1 (Yes)	0	1 (Yes)	0
	14	2 (No)	2 (No)	0	2 (No)	0	2 (No)	0	2 (No)	0	2 (No)	0
Prediction samples	15	1 (Yes)	1 (Yes)	0	1 (Yes)	0	1 (Yes)	0	1 (Yes)	0	1 (Yes)	0
	16	1 (Yes)	1 (Yes)	0	1 (Yes)	0	1 (Yes)	0	1 (Yes)	0	2 (No)	100
	17	1 (Yes)	1 (Yes)	0	1 (Yes)	0	1 (Yes)	0	1 (Yes)	0	1 (Yes)	0

[a]y^* = computer-buy (1—buy, 2—no buy).

5.1.5.4. Application Comparisons Among ID3, C4.5, C-SVM, R-SVM, BPNN, and MRA

Similarly, using the 14 learning samples (Table 5.1) and by C-SVM, R-SVM, BPNN, and MRA, the four functions of computer-buy (y) with respect to four parameters (c_1, c_2, c_3, c_4) have been constructed, i.e., Equations (5.10) and (5.11) corresponding to SVM formula (4.1), Equation (5.12) corresponding to BPNN formula (3.1), and Equation (5.13) corresponding to MRA formula (2.14), respectively. Also similarly to the preceding, using the 14 learning samples and three prediction samples (Table 5.1), C-SVM, R-SVM, BPNN, and MRA are employed for predictions, respectively.

Using C-SVM, the result is an explicit nonlinear function:

$$y = C\text{-}SVM(c_1, c_2, c_3, c_4) \tag{5.10}$$

with $C = 512$, $\gamma = 0.125$, 11 free vectors x_i, and the cross-validation accuracy CVA = 71.4286%.

Using R-SVM, the result is an explicit nonlinear function:

$$y = R\text{-}SVM(c_1, c_2, c_3, c_4) \tag{5.11}$$

with $C = 1$, $\gamma = 0.25$, and 13 free vectors x_i.

The BPNN used consists of four input layer nodes, one output layer node, and 11 hidden layer nodes. Setting the control parameter of t maximum $t_{max} = 100000$, the calculated optimal learning time count $t_{opt} = 97495$. The result is an implicit nonlinear function:

$$y = BPNN(c_1, c_2, c_3, c_4) \tag{5.12}$$

with the root mean square error $RMSE(\%) = 0.1068 \times 10^{-1}$.

Using MRA, the result is an explicit linear function:

$$y = 0.55780 - 0.066474c_1 - 0.034682c_2 + 0.43931c_3 + 0.23988c_4 \tag{5.13}$$

Equation (5.13) yields a residual variance of 0.72113 and a multiple correlation coefficient of 0.52808. From the regression process, computer-buy (y) is shown to depend on the 14 parameters in decreasing order: student (c_3), credit-rating (c_4), age (c_1), and income (c_2).

Substituting the values of c_1, c_2, c_3, c_4 given by the three prediction samples (Table 5.1) in Equations (5.10), (5.11), (5.12), and (5.13), respectively, the computer-buy (y) of each prediction sample is obtained.

Table 5.2 shows the results of the learning and prediction processes by C-SVM, R-SVM, BPNN, and MRA.

As shown by $\overline{R}^*(\%) = 22.02$ of MRA (Table 5.3), the nonlinearity of the relationship between the predicted value y and its relative parameters (c_1, c_2, c_3, c_4) is strong from Table 1.2.

TABLE 5.3   Comparison Among the Applications of ID3, C4.5, C-SVM, R-SVM, BPNN, and MRA to the Computer Buy

Algorithm	Fitting Formula	Mean Absolute Relative Residual			Dependence of the Predicted Value (y) on Parameters (c_1, c_2, c_3, c_4), in Decreasing Order	Time Consumed on PC (Intel Core 2)	Solution Accuracy
		$\overline{R}1(\%)$	$\overline{R}2(\%)$	$\overline{R}^*(\%)$			
ID3	Nonlinear, explicit	0	0	0	N/A	1 s	Very high
C4.5	Nonlinear, explicit	0	0	0	N/A	1 s	Very high
C-SVM	Nonlinear, explicit	0	0	0	N/A	1 s	Very high
R-SVM	Nonlinear, explicit	7.14	0	3.57	N/A	1 s	High
BPNN	Nonlinear, implicit	0	0	0	N/A	20 s	Very high
MRA	Linear, explicit	10.71	33.33	22.02	c_3, c_4, c_1, c_2	<1 s	Very low

Since this case study is very simple, the results of ID3, C4.5, C-SVM, and BPNN are the same, i.e., not only the fitting residual $\overline{R}_1(\%) = 0$ but also the prediction residual $\overline{R}_2(\%) = 0$, and thus the total mean absolute relative residual $\overline{R}^*(\%) = 0$, which coincide with practicality; for R-SVM, though, $\overline{R}_2(\%) = 0$, $\overline{R}_1(\%) = 7.14$, which could not ensure correctly predicting other new prediction samples. As for MRA, the solution accuracy is low, and $\overline{R}_2(\%) > \overline{R}_1(\%)$.

5.1.5.5. Summary and Conclusions

In summary, using data for the computer-buy problem based on four parameters, i.e., the four parameters (age, income, student, credit-rating) and the computer buy of 14 customers, a decision tree (Figure 5.4), together with its five classification rules, is constructed by the data mining tool of ID3 and C4.5, and the results coincide with practicality. This decision tree with its five classification rules is called *mined knowledge*. In the prediction process, applying this knowledge, i.e., using the four parameters of three other customers for the computer-buy problem by five classification rules, respectively, the prediction results also coincide with practicality. Therefore, this method can be spread to the problem of buying other goods.

Moreover, it is found that (a) since this simple case study is a strong nonlinear problem, R-SVM or MRA is not applicable; and (b) but the problem is very simple, and ID3, C4.5, C-SVM, and BPNN are applicable.

5.2. CASE STUDY 1: TOP COAL CAVING CLASSIFICATION (TWENTY-NINE LEARNING SAMPLES)

5.2.1. Studied Problem

The objective of this case study is to conduct a classification of top coal caving (TCC) in coal mines using mining depth, uniaxial compressive strength, dirt band thickness, coal layer thickness, cracking index, and coefficient of fullness at top coal, which has practical value when coal-test data are less limited.

Data from 29 coal mines in China were taken as 29 learning samples, respectively. Each sample contains six parameters (x_1 = mining depth H, x_2 = uniaxial compressive strength Rc, x_3 = dirt band thickness Mt, x_4 = coal layer thickness M, x_5 = cracking index DN, x_6 = coefficient of fullness at top coal K) and a coal test result (x_7 = top coal caving classification R). Dividing each parameter into five classes expressed with 1, 2, 3, 4, and 5, thus (x_1, x_2, x_3, x_4, x_5, x_6, x_7) are transformed to (c_1, c_2, c_3, c_4, c_5, c_6, c_7), listed in Table 5.4. The ID3 algorithm was applied to these data for the TCC classification (Meng et al., 2004).

5.2.2. Known Input Parameters

The input parameters are the values of the known variables c_i ($i = 1, 2, 3, 4, 5, 6$) and the prediction variable y^*(i.e., c_7) for 29 learning samples (Table 5.4).

TABLE 5.4   Input Data for Top Coal Caving Classification I of 29 Coal Mines in China

Sample Type	Sample No.	Coal Mine[a]	Relative Parameters for Top Coal Caving Classification[b]						Coal Test[c]
			c_1 H	c_2 Rc	c_3 Mt	c_4 M	c_5 DN	c_6 K	y^* (c_7) R
Learning samples	1	No. 1	3	1	1	1	1	1	1
	2	No. 2	2	3	1	2	3	1	2
	3	No. 3	2	3	3	1	4	1	3
	4	No. 4	3	2	4	4	2	1	3
	5	No. 5	3	2	3	1	1	3	2
	6	No. 6	2	3	4	4	1	1	2
	7	No. 7	2	5	5	2	3	5	4
	8	No. 8	4	5	2	2	4	4	4
	9	No. 9	2	2	2	3	1	1	2
	10	No. 10	5	2	5	2	3	5	5
	11	No. 11	1	2	3	1	2	4	2
	12	No. 12	1	1	2	1	2	5	1
	13	No. 13	3	4	3	4	3	2	3
	14	No. 14	2	5	4	3	5	5	5
	15	No. 15	2	3	3	4	3	3	3
	16	No. 16	2	2	3	2	3	1	2
	17	No. 17	3	3	1	2	2	4	2
	18	No. 18	1	2	1	2	1	2	2
	19	No. 19	3	3	2	1	2	2	2
	20	No. 20	2	2	3	2	4	1	2
	21	No. 21	1	1	3	5	1	2	2
	22	No. 22	1	2	2	3	1	3	2
	23	No. 23	1	1	1	3	1	1	1
	24	No. 24	3	5	2	1	4	3	4
	25	No. 25	2	3	4	1	4	5	3
	26	No. 26	3	1	2	1	1	4	2
	27	No. 27	2	3	4	5	1	5	3
	28	No. 28	4	2	5	2	3	4	4
	29	No. 29	4	1	2	1	1	1	2

[a]No. 1 = Fenxi Shuiyu #10; No. 2 = Tongchuan Cuijiazhai #5; No. 3 = Yangquan Yikuang #15; No. 4 = Hegang Nanshan #18; No. 5 = Gaoping Tang'an #3; No. 6 = Fuxin Dongliang #2−4; No. 7 = Datong Xinzhouyao #11; No. 8 = Jincheng Fenghuangshan #3; No. 9 = Shenyang Puhe; No. 10 = Taiyuan Wangfeng #8−9; No. 11 = Yanzhou Baodian #3; No. 12 = Huainan Xinji; No. 13 = Tiefa Daming; No. 14 = Datong #2; No. 15 = Yanzhou Nantun; No. 16 = Hebi #6; No. 17 = Xuzhou Sanhejian #7; No. 18 = Yanzhou Xinglongzhuang; No. 19 = Shanxi Guzhuang; No. 20 = Lingwu Yangchangwan; No. 21 = Huating Baicaoyu; No. 22 = Tongchuan Yuhua; No. 23 = Xuzhou Qishan; No. 24 = Jincheng Yongan; No. 25 = Xingtai #2; No. 26 = Taiyuan Dalong #5; No. 27 = Fuxin Wulong #214; No. 28 = Gujiao Jialequan #2−4; and No. 29 = Xuangang Jiaojiazhai #5.

[b]c_1 = mining depth H (m), c_2 = uniaxial compressive strength Rc (MPa), c_3 = dirt band thickness Mt (m), c_4 = coal layer thickness M (m), c_5 = cracking index DN, and c_6 = coefficient of fullness at top coal K; and the data of these six parameters are classified into five numbers (1, 2, 3, 4, 5).

[c]y^* = top coal caving classification (1−excellent, 2−good, 3−average, 4−poor, 5−very poor) determined by the coal test.

5.2.3. Learning Process

This learning process is similar to the aforementioned simple case study (computer buy). At first, since the TCC classification R is the aim of this case study, R is defined as leaf nodes of a decision tree (Figure 5.2). Since $Gain(H) = \max\{Gain(H),\ Gain(Rc),\ Gain(Mt),\ Gain(M),\ Gain(DN),\ Gain(K)\}$, H is chosen as the root node, and five branches are produced from five class values (1, 2, 3, 4, 5) of H; …. Finally the decision tree containing 26 leaf nodes is obtained (Figure 5.5). Besides three leaf nodes relative to two samples, another 23 leaf nodes are related to one sample. Each leaf node expresses a TCC classification and its relative sample numbers as well as reflecting its classification rule. Corresponding to the 26 leaf nodes ranging in the order of left to right, as illustrated by Figure 5.5, 26 classification rules expressed with IF—THEN are as follows:

(1) IF $[(H = 1)\cdot \text{and}\cdot(Mt = 1)\cdot \text{and}\cdot(Rc = 1)]$ THEN $R = 1$ (relative to sample No. 23);
(2) IF $[(H = 1)\cdot \text{and}\cdot(Mt = 1)\cdot \text{and}\cdot(Rc = 2)]$ THEN $R = 2$ (relative to sample No. 18);
(3) IF $[(H = 1)\cdot \text{and}\cdot(Mt = 2)\cdot \text{and}\cdot(Rc = 1)]$ THEN $R = 1$ (relative to sample No. 12);
(4) IF $[(H = 1)\cdot \text{and}\cdot(Mt = 2)\cdot \text{and}\cdot(Rc = 2)]$ THEN $R = 2$ (relative to sample No. 22);
(5) IF $[(H = 1)\cdot \text{and}\cdot(Mt = 3)]$ THEN $R = 2$ (relative to sample Nos. 11, 21);
(6) IF $[(H = 2)\cdot \text{and}\cdot(Mt = 1)]$ THEN $R = 2$ (relative to sample No. 2);
(7) IF $[(H = 2)\cdot \text{and}\cdot(Mt = 2)]$ THEN $R = 2$ (relative to sample No. 9);
(8) IF $[(H = 2)\cdot \text{and}\cdot(Mt = 3)\cdot \text{and}\cdot(Rc = 2)]$ THEN $R = 2$ (relative to sample Nos. 16, 20);
(9) IF $[(H = 2)\cdot \text{and}\cdot(Mt = 3)\cdot \text{and}\cdot(Rc = 3)]$ THEN $R = 3$ (relative to sample Nos. 3, 15);

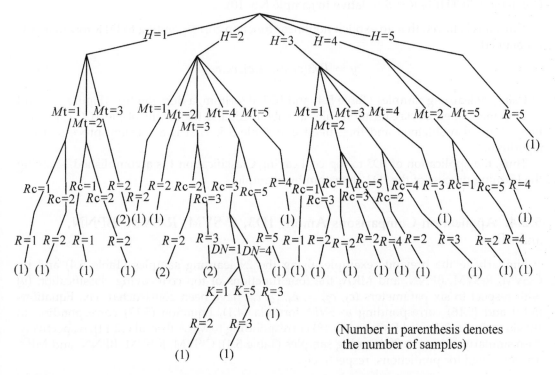

(Number in parenthesis denotes the number of samples)

FIGURE 5.5 A decision tree for Classification I of top coal caving *(from Meng et al., 2004).*

(10) IF $[(H=2)\cdot\text{and}\cdot(Mt=4)\cdot\text{and}\cdot(Rc=3)\cdot\text{and}\cdot(DN=1)\cdot\text{and}\cdot(K=1)]$ THEN $R=2$ (relative to sample No. 6);

(11) IF $[(H=2)\cdot\text{and}\cdot(Mt=4)\cdot\text{and}\cdot(Rc=3)\cdot\text{and}\cdot(DN=1)\cdot\text{and}\cdot(K=5)]$ THEN $R=3$ (relative to sample No. 27);

(12) IF $[(H=2)\cdot\text{and}\cdot(Mt=4)\cdot\text{and}\cdot(Rc=3)\cdot\text{and}\cdot(DN=4)]$ THEN $R=3$ (relative to sample No. 25);

(13) IF $[(H=2)\cdot\text{and}\cdot(Mt=4)\cdot\text{and}\cdot(Rc=5)]$ THEN $R=5$ (relative to sample No. 14);

(14) IF $[(H=2)\cdot\text{and}\cdot(Mt=5)]$ THEN $R=4$ (relative to sample No. 7);

(15) IF $[(H=3)\cdot\text{and}\cdot(Mt=1)\cdot\text{and}\cdot(Rc=1)]$ THEN $R=1$ (relative to sample No. 1);

(16) IF $[(H=3)\cdot\text{and}\cdot(Mt=1)\cdot\text{and}\cdot(Rc=3)]$ THEN $R=2$ (relative to sample No. 17);

(17) IF $[(H=3)\cdot\text{and}\cdot(Mt=2)\cdot\text{and}\cdot(Rc=1)]$ THEN $R=2$ (relative to sample No. 26);

(18) IF $[(H=3)\cdot\text{and}\cdot(Mt=2)\cdot\text{and}\cdot(Rc=3)]$ THEN $R=2$ (relative to sample No. 19);

(19) IF $[(H=3)\cdot\text{and}\cdot(Mt=2)\cdot\text{and}\cdot(Rc=5)]$ THEN $R=4$ (relative to sample No. 24);

(20) IF $[(H=3)\cdot\text{and}\cdot(Mt=3)\cdot\text{and}\cdot(Rc=2)]$ THEN $R=2$ (relative to sample No. 5);

(21) IF $[(H=3)\cdot\text{and}\cdot(Mt=3)\cdot\text{and}\cdot(Rc=4)]$ THEN $R=3$ (relative to sample No. 13);

(22) IF $[(H=3)\cdot\text{and}\cdot(Mt=4)]$ THEN $R=3$ (relative to sample No. 4);

(23) IF $[(H=4)\cdot\text{and}\cdot(Mt=2)\cdot\text{and}\cdot(Rc=1)]$ THEN $R=2$ (relative to sample No. 29);

(24) IF $[(H=4)\cdot\text{and}\cdot(Mt=2)\cdot\text{and}\cdot(Rc=5)]$ THEN $R=4$ (relative to sample No. 8);

(25) IF $[(H=4)\cdot\text{and}\cdot(Mt=5)]$ THEN $R=4$ (relative to sample No. 28);

(26) IF $(H=5)$ THEN $R=5$ (relative to sample No. 10).

Thus, it is believed that an explicit nonlinear function corresponding to DTR formula (5.1) is obtained:

$$y = ID3(c_1, c_2, c_3, c_4, c_5, c_6) \tag{5.14}$$

Using 29 learning samples (Table 5.4) and ID3 algorithm, a decision tree (Figure 5.5) and its aforementioned 26 classification rules are obtained, and the results of the TCC classification (y) are also obtained and then filled into Table 5.5. That is a learning process of ID3 algorithm.

Thus, the application of ID3 to top coal caving Classification I is accomplished, showing the learning process of the ID3 algorithm by example.

5.2.4. Application Comparisons Among ID3, C-SVM, R-SVM, BPNN, and MRA

Similarly to the previous example, using the 29 learning samples (Table 5.4) and by C-SVM, R-SVM, BPNN, and MRA, the four functions of top coal caving classification (y) with respect to six parameters (c_1, c_2, c_3, c_4, c_5, c_6) have been constructed, i.e., Equations (5.15) and (5.16) corresponding to SVM formula (4.1), Equation (5.17) corresponding to BPNN formula (3.1), and Equation (5.18) corresponding to MRA formula (2.14), respectively. Also similarly, using the 29 learning samples (Table 5.4), C-SVM, R-SVM, BPNN, and MRA are employed for predictions, respectively.

TABLE 5.5 Prediction results from top coal caving Classification I of 29 coal mines in China

Sample Type	Sample No.	y^* Coal Test[a]	Top Coal Caving Classification									
			ID3		C-SVM		R-SVM		BPNN		MRA	
			y	$R(\%)$	y	$R(\%)$	y	$R(\%)$	y	$R(\%)$	y	$R(\%)$
Learning samples	1	1	1	0	2	100	2	100	1	0	1	0
	2	2	2	0	2	0	2	0	2	0	2	0
	3	3	3	0	2	33.33	3	0	3	0	3	0
	4	3	3	0	3	0	3	0	3	0	3	0
	5	2	2	0	2	0	2	0	2	0	2	0
	6	2	2	0	2	0	2	0	2	0	3	50
	7	4	4	0	4	0	4	0	4	0	4	0
	8	4	4	0	4	0	4	0	4	0	4	0
	9	2	2	0	2	0	2	0	2	0	2	0
	10	5	5	0	4	20	4	20	5	0	4	20
	11	2	2	0	2	0	2	0	2	0	2	0
	12	1	1	0	2	100	2	100	1	0	2	100
	13	3	3	0	3	0	3	0	3	0	4	33.33
	14	5	5	0	4	20	4	20	5	0	4	20
	15	3	3	0	3	0	3	0	3	0	3	0
	16	2	2	0	2	0	2	0	2	0	2	0
	17	2	2	0	2	0	2	0	2	0	3	50
	18	2	2	0	2	0	2	0	2	0	1	50
	19	2	2	0	2	0	2	0	2	0	2	0
	20	2	2	0	2	0	2	0	2	0	3	50
	21	2	2	0	2	0	2	0	2	0	2	0
	22	2	2	0	2	0	2	0	2	0	2	0
	23	1	1	0	2	100	2	100	1	0	1	0
	24	4	4	0	4	0	4	0	4	0	3	25
	25	3	3	0	4	33.33	3	0	3	0	3	0
	26	2	2	0	2	0	2	0	2	0	2	0
	27	3	3	0	3	0	3	0	3	0	3	0
	28	4	4	0	4	0	4	0	4	0	4	0
	29	2	2	0	2	0	2	0	2	0	2	0

[a]y^* = top coal caving classification (1—excellent, 2—good, 3—average, 4—poor, 5—very poor) determined by the coal test.

Using C-SVM, the result is an explicit nonlinear function:

$$y = C\text{-}SVM(c_1, c_2, c_3, c_4, c_5, c_6) \tag{5.15}$$

with $C = 2048$, $\gamma = 0.0001225$, 28 free vectors c_i, and CVA $= 65.5172\%$.

Using R-SVM, the result is an explicit nonlinear function:

$$y = R\text{-}SVM(c_1, c_2, c_3, c_4, c_5, c_6) \tag{5.16}$$

with $C = 1$, $\gamma = 0.166667$, and 27 free vectors x_i.

The BPNN used consists of six input layer nodes, one output layer node, and 13 hidden layer nodes. Setting $t_{max} = 100000$, the calculated $t_{opt} = 43860$. The result is an implicit nonlinear function:

$$y = BPNN(c_1, c_2, c_3, c_4, c_5, c_6) \tag{5.17}$$

with $RMSE(\%) = 0.4165 \times 10^{-2}$.

Using MRA, the result is an explicit linear function:

$$y = -0.82665 + 0.34578c_1 + 0.23852c_2 + 0.23742c_3 + 0.14853c_4 + 0.25898c_5 + 0.14145c_6 \tag{5.18}$$

Equation (5.18) yields a residual variance of 0.13605 and a multiple correlation coefficient of 0.92949. From the regression process, TCC classification (y) is shown to depend on the six parameters in decreasing order: uniaxial compressive strength (c_2), dirt band thickness (c_3), mining depth (c_1), cracking index (c_5), coefficient of fullness at top coal (c_6), and coal layer thickness (c_4).

Table 5.5 shows the results of learning process by C-SVM, R-SVM, BPNN, and MRA.

$\overline{R}^*(\%) = 13.74$ of MRA (Table 5.6) shows that the nonlinearity of the relationship between the predicted value y and its relative parameters (c_1, c_2, c_3, c_4, c_5, c_6) is strong from Table 1.2. The results of ID3 and BPNN are the same, i.e., $\overline{R}_1(\%) = 0$, and thus $\overline{R}^*(\%) = 0$, which coincide with practicality; as for C-SVM, R-SVM, and MRA, the solution accuracy is low.

TABLE 5.6 Comparison Among the Applications of ID3, C-SVM, R-SVM, BPNN, and MRA to top coal caving Classification I in China

Algorithm	Fitting Formula	Mean Absolute Relative Residual $\overline{R}1(\%)$	$\overline{R}^*(\%)$	Dependence of the Predicted Value (y) on Parameters (c_1, c_2, c_3, c_4, c_5, c_6), in Decreasing Order	Time Consumed on PC (Intel Core 2)	Solution Accuracy
ID3	Nonlinear, explicit	0	0	N/A	2 s	Very high
C-SVM	Nonlinear, explicit	0	14.02	N/A	2 s	Low
R-SVM	Nonlinear, explicit	7.14	11.72	N/A	2 s	Low
BPNN	Nonlinear, implicit	0	0	N/A	35 s	Very high
MRA	Linear, explicit	10.71	13.74	c_2, c_3, c_1, c_5, c_6, c_4	<1 s	Low

5.2.5. Summary and Conclusions

In summary, using data for the top coal caving classification based on six parameters, i.e., the six parameters (mining depth, uniaxial compressive strength, dirt band thickness, coal layer thickness, cracking index, coefficient of fullness at top coal) and coal test results of 29 coal mines in China, a decision tree (Figure 5.5), together with its 26 classification rules, is constructed by the data mining tool of ID3, and the results are consistent with coal tests. This decision tree, together with its 26 classification rules, is called *mined knowledge*. Therefore, this method can be spread to the TCC classification of other coal mines.

Moreover, it is found that (a) since this case study is a strong nonlinear problem, MRA is not applicable; (b) since the classification rule is complicated and the number of learning samples is not big enough, neither C-SVM nor R-SVM is applicable; and (c) since BPNN is adaptive to the structuralist problem, the results of BPNN and ID3 are same and correct.

5.3. CASE STUDY 2: TOP COAL CAVING CLASSIFICATION (TWENTY-SIX LEARNING SAMPLES AND THREE PREDICTION SAMPLES)

5.3.1. Studied Problem

The objective of this case study is to conduct a classification of TCC in coal mines using mining depth, uniaxial compressive strength, dirt band thickness, coal layer thickness, cracking index, and coefficient of fullness at top coal, which has practical value when coal-test data are less limited.

Data from 29 coal mines in China were taken as 29 learning samples, respectively. Each sample contains six parameters (x_1 = mining depth H, x_2 = uniaxial compressive strength Rc, x_3 = dirt band thickness Mt, x_4 = coal layer thickness M, x_5 = cracking index DN, x_6 = coefficient of fullness at top coal K) and a coal test result (x_7 = top coal caving classification R). Dividing each parameter into five classes expressed with 1, 2, 3, 4, and 5, thus (x_1, x_2, x_3, x_4, x_5, x_6, x_7) are transformed to (c_1, c_2, c_3, c_4, c_5, c_6, c_7), listed in Table 5.4. ID3 algorithm was applied to these data for the TCC classification (Meng et al., 2004). Its results and analyses have been described in Section 5.2.

We especially found from the obtained decision tree (Figure 5.5) that (a) this tree contains 26 leaf nodes, in which three leaf nodes are relative to two samples and other 23 leaf nodes are related to one sample; and (b) these three leaf nodes are the fifth leaf node relating to the 11th and 21st samples, the eighth leaf node relating to the 16th and 20th samples, and the ninth leaf node relating to the 3rd and 20th samples, in the order of left to right, corresponding to the fifth, eighth, and ninth items in the aforementioned 26 classification rules. It can be foreseen that if the 21st, 20th, and 20th samples are deleted, the decision tree will be changeless when rerunning ID3 using the remaining 26 learning samples. Therefore, this section uses these 26 samples as learning samples, and the deleted three samples are used as prediction samples (Table 5.7). In this way, the case study is self-contained, and the application comparisons among ID3, C-SVM, R-SVM, BPNN, and MRA are in this self-contained condition. For the sake of the comparison analysis with Section 5.2, sample numbers and corresponding coal mine numbers keep consistent with Table 5.4.

TABLE 5.7 Input Data for Top Coal Caving Classification II of 29 Coal Mines in China

Sample Type	Sample No.	Coal Mine[a]	Relative Parameters for Top Coal Caving Classification[b]						Coal Test[c]
			c_1 H	c_2 Rc	c_3 Mt	c_4 M	c_5 DN	c_6 K	y^* (c_7) R
Learning samples	1	No. 1	3	1	1	1	1	1	1
	2	No. 2	2	3	1	2	3	1	2
	3	No. 3	2	3	3	1	4	1	3
	4	No. 4	3	2	4	4	2	1	3
	5	No. 5	3	2	3	1	1	3	2
	6	No. 6	2	3	4	4	1	1	2
	7	No. 7	2	5	5	2	3	5	4
	8	No. 8	4	5	2	2	4	4	4
	9	No. 9	2	2	2	3	1	1	2
	10	No. 10	5	2	5	2	3	5	5
	11	No. 11	1	2	3	1	2	4	2
	12	No. 12	1	1	2	1	2	5	1
	13	No. 13	3	4	3	4	3	2	3
	14	No. 14	2	5	4	3	5	5	5
	16	No. 16	2	2	3	2	3	1	2
	17	No. 17	3	3	1	2	2	4	2
	18	No. 18	1	2	1	2	1	2	2
	19	No. 19	3	3	2	1	2	2	2
	22	No. 22	1	2	2	3	1	3	2
	23	No. 23	1	1	1	3	1	1	1
	24	No. 24	3	5	2	1	4	3	4
	25	No. 25	2	3	4	1	4	5	3
	26	No. 26	3	1	2	1	1	4	2
	27	No. 27	2	3	4	5	1	5	3
	28	No. 28	4	2	5	2	3	4	4
	29	No. 29	4	1	2	1	1	1	2
Prediction samples	15	No. 15	2	3	3	4	3	3	(3)
	20	No. 20	2	2	3	2	4	1	(2)
	21	No. 21	1	1	3	5	1	2	(2)

[a]See Table 5.4.
[b]See Table 5.4.
[c]y^* = top coal caving classification (1—excellent, 2—good, 3—average, 4—poor, 5—very poor) determined by the coal test. Numbers in parentheses are not input data but are used for calculating R(%).

5.3.2. Known Input Parameters

They are the values of the known variables c_i ($i = 1, 2, 3, 4, 5, 6$) and the prediction variable y^*(i.e., c_7) for 26 learning samples and three prediction samples (Table 5.7).

5.3.3. Learning Process

This learning process is similar to the aforementioned simple case study (computer buy). At first, since the TCC classification R is the aim of this case study, R is defined as leaf nodes of a decision tree (Figure 5.2). Since $Gain(H) = \max\{Gain(H), Gain(Rc), Gain(Mt), Gain(M), Gain(DN), Gain(K)\}$, H is chosen as the root node, and five branches are produced from five class values (1, 2, 3, 4, 5) of H; .... Finally the decision tree containing 26 leaf nodes is obtained (Figure 5.6). Each leaf node is only relative to one sample, which is the only difference between this tree and the tree (Figure 5.5) obtained in Section 5.2. Each leaf node expresses a TCC classification and its relative sample numbers and also reflects its classification rule. Corresponding to the 26 leaf nodes ranging in the order of left

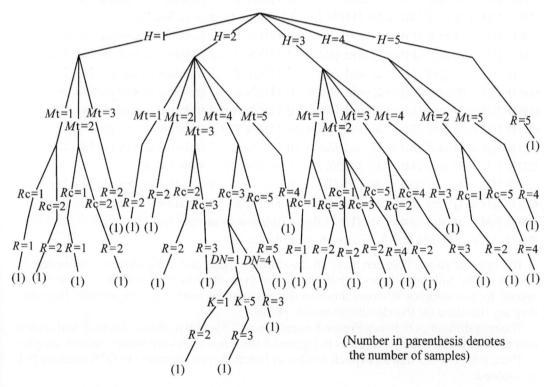

FIGURE 5.6   A decision tree for Classification II of top coal caving (*modified from Meng et al., 2004*).

to right, illustrated by Figure 5.6, 26 classification rules expressed with IF–THEN are as follows:

(1) IF $[(H = 1) \cdot \text{and} \cdot (Mt = 1) \cdot \text{and} \cdot (Rc = 1)]$ THEN $R = 1$ (relative to sample No. 23);
(2) IF $[(H = 1) \cdot \text{and} \cdot (Mt = 1) \cdot \text{and} \cdot (Rc = 2)]$ THEN $R = 2$ (relative to sample No. 18);
(3) IF $[(H = 1) \cdot \text{and} \cdot (Mt = 2) \cdot \text{and} \cdot (Rc = 1)]$ THEN $R = 1$ (relative to sample No. 12);
(4) IF $[(H = 1) \cdot \text{and} \cdot (Mt = 2) \cdot \text{and} \cdot (Rc = 2)]$ THEN $R = 2$ (relative to sample No. 22);
(5) IF $[(H = 1) \cdot \text{and} \cdot (Mt = 3)]$ THEN $R = 2$ (relative to sample No. 11);
(6) IF $[(H = 2) \cdot \text{and} \cdot (Mt = 1)]$ THEN $R = 2$ (relative to sample No. 2);
(7) IF $[(H = 2) \cdot \text{and} \cdot (Mt = 2)]$ THEN $R = 2$ (relative to sample No. 9);
(8) IF $[(H = 2) \cdot \text{and} \cdot (Mt = 3) \cdot \text{and} \cdot (Rc = 2)]$ THEN $R = 2$ (relative to sample No. 16);
(9) IF $[(H = 2) \cdot \text{and} \cdot (Mt = 3) \cdot \text{and} \cdot (Rc = 3)]$ THEN $R = 3$ (relative to sample No. 3);
(10) IF $[(H = 2) \cdot \text{and} \cdot (Mt = 4) \cdot \text{and} \cdot (Rc = 3) \cdot \text{and} \cdot (DN = 1) \cdot \text{and} \cdot (K = 1)]$ THEN $R = 2$ (relative to sample No. 6);
(11) IF $[(H = 2) \cdot \text{and} \cdot (Mt = 4) \cdot \text{and} \cdot (Rc = 3) \cdot \text{and} \cdot (DN = 1) \cdot \text{and} \cdot (K = 5)]$ THEN $R = 3$ (relative to sample No. 27);
(12) IF $[(H = 2) \cdot \text{and} \cdot (Mt = 4) \cdot \text{and} \cdot (Rc = 3) \cdot \text{and} \cdot (DN = 4)]$ THEN $R = 3$ (relative to sample No. 25);
(13) IF $[(H = 2) \cdot \text{and} \cdot (Mt = 4) \cdot \text{and} \cdot (Rc = 5)]$ THEN $R = 5$ (relative to sample No. 14);
(14) IF $[(H = 2) \cdot \text{and} \cdot (Mt = 5)]$ THEN $R = 4$ (relative to sample No. 7);
(15) IF $[(H = 3) \cdot \text{and} \cdot (Mt = 1) \cdot \text{and} \cdot (Rc = 1)]$ THEN $R = 1$ (relative to sample No. 1);
(16) IF $[(H = 3) \cdot \text{and} \cdot (Mt = 1) \cdot \text{and} \cdot (Rc = 3)]$ THEN $R = 2$ (relative to sample No. 17);
(17) IF $[(H = 3) \cdot \text{and} \cdot (Mt = 2) \cdot \text{and} \cdot (Rc = 1)]$ THEN $R = 2$ (relative to sample No. 26);
(18) IF $[(H = 3) \cdot \text{and} \cdot (Mt = 2) \cdot \text{and} \cdot (Rc = 3)]$ THEN $R = 2$ (relative to sample No. 19);
(19) IF $[(H = 3) \cdot \text{and} \cdot (Mt = 2) \cdot \text{and} \cdot (Rc = 5)]$ THEN $R = 4$ (relative to sample No. 24);
(20) IF $[(H = 3) \cdot \text{and} \cdot (Mt = 3) \cdot \text{and} \cdot (Rc = 2)]$ THEN $R = 2$ (relative to sample No. 5);
(21) IF $[(H = 3) \cdot \text{and} \cdot (Mt = 3) \cdot \text{and} \cdot (Rc = 4)]$ THEN $R = 3$ (relative to sample No. 13);
(22) IF $[(H = 3) \cdot \text{and} \cdot (Mt = 4)]$ THEN $R = 3$ (relative to sample No. 4);
(23) IF $[(H = 4) \cdot \text{and} \cdot (Mt = 2) \cdot \text{and} \cdot (Rc = 1)]$ THEN $R = 2$ (relative to sample No. 29);
(24) IF $[(H = 4) \cdot \text{and} \cdot (Mt = 2) \cdot \text{and} \cdot (Rc = 5)]$ THEN $R = 4$ (relative to sample No. 8);
(25) IF $[(H = 4) \cdot \text{and} \cdot (Mt = 5)]$ THEN $R = 4$ (relative to sample No. 28);
(26) IF $(H = 5)$ THEN $R = 5$ (relative to sample No. 10).

The only difference between these 26 classification rules and those obtained in Section 5.2 reflects on the fifth, eighth, and ninth items; the three classification rules in Section 5.2 are related to two samples whereas those in this section are related to one sample. However, they are the same for the classification rule per se.

The only difference between Figure 5.6 and Figure 5.5 is that in Figure 5.6, each leaf node is only relative to one sample, whereas in Figure 5.5 three leaf nodes are relative to two samples.

Thus, it is believed that an explicit nonlinear function corresponding to DTR formula (5.1) is obtained:

$$y = ID3(c_1, c_2, c_3, c_4, c_5, c_6) \tag{5.19}$$

which is the same as Equation (5.14) in Section 5.2.

5.3.4. Prediction Process

As mentioned, using 26 learning samples (Table 5.7) and ID3 algorithm, a decision tree (Figure 5.6) and its 26 classification rules were obtained, and the results of the TCC classification (y) are also obtained and then filled into Table 5.8. Now using three prediction samples (Table 5.7) and the 26 classification rules, the TCC classification (y) of each prediction sample is predicted and then filled into Table 5.8. That is the prediction process of the ID3 algorithm.

Thus, the application of ID3 to top coal caving Classification II is accomplished, showing the learning process and prediction process of the ID3 algorithm by example.

5.3.5. Application Comparisons Among ID3, C-SVM, R-SVM, BPNN, and MRA

Similarly, using the 26 learning samples (Table 5.7) and by C-SVM, R-SVM, BPNN, and MRA, the four functions of the TCC classification (y) with respect to six parameters (c_1, c_2, c_3, c_4, c_5, c_6) have been constructed, i.e., Equations (5.20) and (5.21) corresponding to SVM formula (4.1), Equation (5.22) corresponding to BPNN formula (3.1), and Equation (5.23) corresponding to MRA formula (2.14), respectively. Also similarly, using the 26 learning samples and three prediction samples (Table 5.7), C-SVM, R-SVM, BPNN, and MRA are employed for predictions, respectively.

Using C-SVM, the result is an explicit nonlinear function:

$$y = C\text{-}SVM(c_1, c_2, c_3, c_4, c_5, c_6) \tag{5.20}$$

with $C = 2$, $\gamma = 0.125$, 26 free vectors c_i, and CVA = 61.5385%.

Equation (5.20) differs from Equation (5.15) in Section 5.2.

Using R-SVM, the result is an explicit nonlinear function:

$$y = R\text{-}SVM(c_1, c_2, c_3, c_4, c_5, c_6) \tag{5.21}$$

with $C = 1$, $\gamma = 0.166667$, and 25 free vectors x_i.

Equation (5.21) differs from Equation (5.16) in Section 5.2.

The BPNN used here consists of six input layer nodes, one output layer node, and 13 hidden layer nodes. Setting $t_{max} = 50000$, the calculated $t_{opt} = 49888$. The result is an implicit nonlinear function:

$$y = BPNN(c_1, c_2, c_3, c_4, c_5, c_6) \tag{5.22}$$

with $RMSE(\%) = 0.4400 \times 10^{-2}$.

Equation (5.22) differs from Equation (5.17) in Section 5.2.

Using MRA, the result is an explicit linear function:

$$y = -0.79690 + 0.33807c_1 + 0.20374c_2 + 0.24046c_3 + 0.14493c_4 + 0.32614c_5 + 0.12085c_6 \tag{5.23}$$

which differs from Equation (5.18) in Section 5.2.

Equation (5.23) yields a residual variance of 0.12568 and a multiple correlation coefficient of 0.93505. From the regression process, a TCC classification (y) is shown to depend on the six parameters in decreasing order: cracking index (c_5), dirt band thickness (c_3), mining depth (c_1), uniaxial compressive strength (c_2), coefficient of fullness at top coal (c_6), and coal layer

TABLE 5.8 Prediction Results from Top Coal Caving Classification II of 29 Coal Mines in China

Sample Type	Sample No.	y^* Coal Test[a]	ID3		C-SVM		R-SVM		BPNN		MRA	
			y	$R(\%)$	y	$R(\%)$	y	$R(\%)$	y	$R(\%)$	y	$R(\%)$
Learning samples	1	1	1	0	2	100	2	100	1	0	1	0
	2	2	2	0	2	0	2	0	2	0	2	0
	3	3	3	0	2	33.33	3	0	3	0	3	0
	4	3	3	0	3	0	3	0	3	0	3	0
	5	2	2	0	2	0	2	0	2	0	2	0
	6	2	2	0	3	50	2	0	2	0	2	0
	7	4	4	0	4	0	4	0	4	0	4	0
	8	4	4	0	4	0	4	0	4	0	4	0
	9	2	2	0	2	0	2	0	2	0	2	0
	10	5	5	0	4	20	4	20	5	0	4	20
	11	2	2	0	2	0	2	0	2	0	2	0
	12	1	1	0	2	100	2	100	1	0	2	100
	13	3	3	0	3	0	3	0	3	0	4	33.33
	14	5	5	0	4	20	4	20	5	0	5	0
	16	2	2	0	2	0	2	0	2	0	2	0
	17	2	2	0	2	0	2	0	2	0	2	0
	18	2	2	0	2	0	2	0	2	0	1	50
	19	2	2	0	2	0	2	0	2	0	2	0
	22	2	2	0	2	0	2	0	2	0	2	0
	23	1	1	0	2	100	2	100	1	0	1	0
	24	4	4	0	4	0	4	0	4	0	4	0
	25	3	3	0	4	33.33	3	0	3	0	4	33.33
	26	2	2	0	2	0	2	0	2	0	2	0
	27	3	3	0	3	0	3	0	3	0	3	0
	28	4	4	0	4	0	4	0	4	0	4	0
	29	2	2	0	2	0	2	0	2	0	2	0
Prediction samples	15	3	3	0	3	0	3	0	3	0	3	0
	20	2	2	0	2	0	3	50	1	50	3	50
	21	2	2	0	2	0	2	0	1	50	2	0

[a]y^* = top coal caving classification (1—excellent, 2—good, 3—average, 4—poor, 5—very poor) determined by the coal test.

thickness (c_4). The only difference between this order and the order in Section 5.2 is that the sorting position of c_5 and c_2 is exchanged. It is easy to know that this order is less accurate than the order in Section 5.2 due to the fact that Section 5.2 uses 29 learning samples, whereas this section uses 26 learning samples.

Substituting the values of $c_1, c_2, c_3, c_4, c_5, c_6$ given by the three prediction samples (Table 5.7) in Equations (5.20), (5.21), (5.22), and (5.23), respectively, the TCC classification (y) of each prediction sample is obtained.

Table 5.8 shows the results of the learning and prediction processes by C-SVM, R-SVM, BPNN, and MRA.

From $\overline{R}^*(\%) = 12.89$ of MRA (Table 5.9), we can see that the nonlinearity of the relationship between the predicted value y and its relative parameters ($c_1, c_2, c_3, c_4, c_5, c_6$) is strong from Table 1.2. Only ID3 is applicable, i.e., not only $\overline{R}_1(\%) = 0$, but also $\overline{R}_2(\%) = 0$, and thus $\overline{R}^*(\%) = 0$, which coincide with practicality; comparing C-SVM of Case Study 1 in Section 5.2, $\overline{R}_1(\%)$ is not 0 but $\overline{R}_1(\%) = 17.56$ because only 26 learning samples exist, whereas there are 29 learning samples in Case Study 1. Thus, though $\overline{R}_2(\%) = 0$, $\overline{R}_1(\%) = 17.56$, which could not ensure correct prediction of other new prediction samples. Comparing BPNN of Case Study 1 in Section 5.2, $\overline{R}_2(\%)$ is not 0 but $\overline{R}_2(\%) = 33.33$, because only 26 learning samples exist. As for R-SVM and MRA, the solution accuracy is low, and $\overline{R}_2(\%) > \overline{R}_1(\%)$.

5.3.6. Summary and Conclusions

In summary, using data for the TCC classification based on six parameters, i.e., the six parameters (mining depth, uniaxial compressive strength, dirt band thickness, coal layer thickness, cracking index, and coefficient of fullness at top coal) and coal test results of 26 coal mines in China, a decision tree (Figure 5.6), together with its 26 classification rules, is

TABLE 5.9 Comparison Among the Applications of ID3, C-SVM, R-SVM, BPNN, and MRA to Top Coal Caving Classification **II** in China

Algorithm	Fitting Formula	Mean Absolute Relative Residual			Dependence of the Predicted Value (y) on Parameters ($c_1, c_2, c_3, c_4, c_5, c_6$), in decreasing order	Time Consumed on PC (Intel Core 2)	Solution Accuracy
		$\overline{R}1(\%)$	$\overline{R}2(\%)$	$\overline{R}^*(\%)$			
ID3	Nonlinear, explicit	0	0	0	N/A	2 s	Very high
C-SVM	Nonlinear, explicit	17.56	0	8.78	N/A	2 s	Moderate
R-SVM	Nonlinear, explicit	13.08	16.67	14.88	N/A	2 s	Low
BPNN	Nonlinear, implicit	0	33.33	16.67	N/A	40 s	Low
MRA	Linear, explicit	9.10	16.67	12.89	$c_5, c_3, c_1, c_2, c_6, c_4$	<1 s	Low

constructed by the data mining tool of ID3, and the results are consistent with coal tests. This decision tree, together with its 26 classification rules, is called *mined knowledge*. In the prediction process, applying this knowledge, i.e., using the six parameters of other three coal mines for the TCC classification by 26 classification rules, respectively, the results are also consistent with coal tests. Therefore, this method can be spread to the TCC classification of other coal mines.

Moreover, it is found that (a) since this case study is a strong nonlinear problem, MRA is not applicable; and (b) since the classification rule is complicated and the number of learning samples is not big enough, C-SVM, R-SVM, or BPNN is not applicable.

EXERCISES

5-1. For the classification algorithm, the dependent variable y^* in learning samples must be an integer number, but the independent variables $(x_1, x_2, ..., x_m)$ may be real numbers or integer numbers, e.g., for the two classification algorithms of C-SVM in Chapter 4 and BAC in Chapter 6. For the classification algorithm DTR in this chapter, can the independent variables $(x_1, x_2, ..., x_m)$ be real numbers?

5-2. What are the advantages and shortcomings of ID3?

5-3. C4.5 has inherited the advantages of ID3 and made four major improvements. What are the four improvements?

5-4. From Tables 5.3, 5.6, and 5.9, $\overline{R}^*(\%) = 0$ when DTR and C-SVM are applied to the simple case study in Section 5.1 and Case Study 1 in Section 5.2; $\overline{R}^*(\%) = 0$ and $\overline{R}^*(\%) = 8.78$ when DTR and C-SVM are applied to Case Study 2 in Section 5.3, respectively. Does that mean DTR is superior to C-SVM?

References

Han, J.W., Kamber, M., 2006. Data Mining: Concepts and Techniques, second ed. Morgan Kaufmann, San Francisco, CA, USA.

Meng, X., Lan, H., He, X., 2004. Evaluation arithmetic of difficulty degrees of roof coal caving based on data mining technology. J. Anhui University of Sci. Tech. (Natural Science) 24 (1), 20–24 (in Chinese with English abstract).

Quinlan, J.R., 1986. Induction of decision trees. Machine Learning 1 (1), 81–106.

Quinlan, J.R., 1993. C4.5: Programs for Machine Learning. Morgan Kaufmann, San Mateo, CA, USA.

Bayesian Classification

Bayesian classification (BAC) is a classification approach based on statistics. This chapter introduces naïve Bayesian (NBAY), Bayesian discrimination (BAYD), and Bayesian successive discrimination (BAYSD) of BAC as well as their applications in geosciences.

Section 6.1 (methodology) introduces the applying ranges and conditions, basic principles, calculation methods, and calculation flowcharts of NBAY, BAYD, and BAYSD as well as a simple case study. In this sample, calculation results and analyses are given. Though this simple case study is small, it reflects the whole process of calculations to benefit readers in understanding and mastering the techniques. This case study is about a defaulted borrower; it explains how to adopt NBAY for prediction. Concretely, using data for the defaulted borrower, i.e., the three parameters (homeowner, marital status, annual income) and defaulted borrower of 10 buyers, a calculable NBAY classification formula is constructed by the data mining tool of NBAY, and the results coincide with practicality. This formula is called *mined knowledge*.

In the prediction process, applying this knowledge, i.e., using the three parameters of two other buyers for defaulted borrowers, respectively, the results also coincide with practicality. Therefore, this method can be spread to the defaulted borrower of other cases. Moreover, it is found that since this case study is a strong nonlinear problem, NBAY, error back-propagation neural network (BPNN, see Chapter 3), and classification of support vector machine (C-SVM, see Chapter 4) are applicable, but BAYSD, regression of support vector machine (R-SVM, see Chapter 4), or multiple regression analysis (MRA, see Chapter 2) is not applicable.

Section 6.2 (reservoir classification in the Fuxin Uplift) introduces Case Study 1 of BAC. We used data for the reservoir classification in the Fuxin Uplift based on five parameters, i.e., the five parameters (sandstone thickness, porosity, permeability, carbonate content, mud content) and a well-test result of 20 samples, a set of discriminate functions is constructed by the data mining tool of BAYSD, and the results are consistent with a well test. This set of discriminate functions is called *mined knowledge*. In the prediction process, applying this knowledge, i.e., using the five parameters of three other samples for reservoir classification, respectively, the results are also consistent with a well test. Therefore, this method can be spread to the reservoir classification of other oilfields. Moreover, it is found that (a) since this case study is a strong nonlinear problem, neither R-SVM nor MRA is applicable; and (b) since the number of learning samples is not big enough, neither C-SVM nor BPNN is applicable, but it is preferable to adopt BAYSD, which is an advantage of BAYSD.

Section 6.3 (reservoir classification in the Baibao Oilfield) introduces Case Study 2 of BAC. Using data for the reservoir classification in the Baibao Oilfield based on four parameters, i.e., the four parameters (mud content, porosity, permeability, permeability variation coefficient) and a well-test result of 25 samples, a set of discriminate functions is constructed by the data mining tool of BAYSD, and the results are consistent with a well test except for one sample. This set of discriminate functions is called *mined knowledge*. In the prediction process, applying this knowledge, i.e., using the four parameters of three other samples for reservoir classification, respectively, the results are consistent with a well test. Therefore, this method can be spread to the reservoir classification of other oilfields. Moreover, it is found that (a) since this case study is a very weak nonlinear problem, BAYSD, C-SVM, and MRA are applicable; and (b) since the number of learning samples is not big enough, neither R-SVM nor BPNN is applicable.

Section 6.4 (oil layer classification based on well-logging interpretation) introduces Case Study 3 of BAC. Using data for the oil layer classification in the Xiefengqiao Anticline based on five parameters, i.e., the five parameters (true resistivity, acoustictime, porosity, oil saturation, permeability) and an oil test result of 24 oil layers, a set of discriminate functions is constructed by the data mining tool of BAYSD, but the results are not eligible. This set of discriminate functions is called *mined unqualified knowledge*. In the prediction process, applying this unqualified knowledge, i.e., using the five parameters of three other samples for oil layer classification, respectively, the results are not eligible either. Therefore, this method cannot be spread to the oil layer classification of other oilfields. Moreover, it is found that since this case study is a strong nonlinear problem, only C-SVM is applicable, but BAYSD, BPNN, MRA, or R-SVM is not applicable.

Section 6.5 (integrated evaluation of oil and gas trap quality) introduces Case Study 4 of BAC. Using data for the trap evaluation of the Northern Kuqa Depression, i.e., the 14 parameters (unit structure, trap type, petroliferous formation, trap depth, trap relief, trap closed area, formation HC identifier, data reliability, trap coefficient, source rock coefficient, reservoir coefficient, preservation coefficient, configuration coefficient, resource quantity) and trap quality of 27 traps, a set of discriminate functions is constructed by the data mining tool of BAYSD, and the results coincide with practicality. This set of discriminate functions is called *mined knowledge*. In the prediction process, applying this knowledge, i.e., using the 14 parameters of two other traps for trap evaluation, respectively, the results coincide with the real exploration discovery after this prediction. Therefore, this method can be spread to the trap evaluation of other exploration areas. Moreover, it is found that (a) for this case study, the results of BAYSD, C-SVM, R-SVM, and BPNN are the same and correct; and (b) BAYSD, C-SVM, and R-SVM run much faster than BPNN; however, it is easy to code the BAYSD program, whereas it is very complicated to code the C-SVM or R-SVM program, so BAYSD is a good software for this case study if neither the C-SVM nor R-SVM program exists.

Section 6.6 (coal-gas-outburst classification) introduces Case Study 5 of BAC. We used data for the coal-gas-outburst classification based on five parameters, i.e., the five parameters (initial speed of methane diffusion, coefficient of coal consistence, gas pressure, destructive style of coal, mining depth) and coal test results of 16 coal mines in China, two sets of discriminate functions obtained by two data mining tools of BAYD and BAYSD, respectively, and the results are consistent with coal tests. These two sets of discriminate functions are called *mined knowledge*. In the prediction process, applying this knowledge, i.e., using the five parameters of five other coal mines for coal-gas-outburst classification, respectively, the results are also consistent with coal tests. Therefore, the two methods can be spread to the coal-gas-outburst classification of other coal mines. Moreover, it is found that (a) since this case study is a strong nonlinear problem, MRA, R-SVM, or MRA is not applicable; and (b) since the number of learning samples is not big enough, BAYD or BAYSD is better than C-SVM, which is an advantage of BAYD and BAYSD.

Section 6.7 (top coal caving classification [26 learning samples and three prediction samples]) introduces Case Study 6 of BAC. Using data for the top coal caving classification based on six parameters, i.e., the six parameters (mining depth, uniaxial compressive strength, dirt band thickness, coal layer thickness, cracking index, coefficient of fullness at top coal) and coal test results of 26 coal mines in China, a set of discriminate functions is constructed by the data mining tool of BAYSD, but the results are not eligible. This set of discriminate functions is called mined unqualified knowledge. In the prediction process, applying this unqualified knowledge, i.e. using the six parameters of other three coal mines in China for top-coal-caving classification respectively, the results are not eligible either. Therefore, this method cannot be spread to the top coal caving classification of other coal mines. Moreover, it is found that since this case study is a strong nonlinear problem, only ID3 is applicable, but BAYSD, C-SVM, R-SVM, BPNN, or MRA is not applicable.

Although no dimension-reduction calculations are conducted in seven case studies from Section 6.1 to Section 6.7, the prediction variable y is shown to depend on the known variables x_i ($i = 1, 2, ..., m$) in decreasing order given by BAYSD and MRA, indicating that BAYSD or MRA can serve as a pioneering dimension-reduction tool. For the dependences in each case study obtained by BAYSD and MRA, some are the same, e.g., in a simple case study, Case Studies 1, 3, and 4; but the others, e.g., Case Studies 2, 5, and 6, are not the same. The different dependence results from the fact that MRA is a linear algorithm, whereas BAYSD is a nonlinear algorithm. Hence, the dependence obtained by BAYSD is more reliable than that obtained by MRA, i.e., it is better to adopt BAYSD as a pioneering dimension-reduction tool than to adopt MRA. However, MRA can give the multiple correlation coefficient for y and introduced x_i so as to determine the reducible dimension (see Section 4.4).

Certainly, whether MRA or BAYSD is adopted for dimension reduction, the results are required to be validated by a nonlinear tool (C-SVM or BPNN). Strictly and concretely, since MRA is a regression algorithm, it can serve as a pioneering dimension-reduction tool for regression problems, and the dimension-reduction results are validated by a regression algorithm BPNN, whereas BAYSD is a classification algorithm that can serve as a pioneering dimension-reduction tool for classification problems, and the dimension-reduction results are validated by classification algorithm C-SVM.

In addition, through the seven case studies, it is found that the solution accuracies of BAYSD and C-SVM are (a) high when the nonlinearity of case studies is very weak or moderate, but (b) different when the nonlinearity of case studies is strong, except for Case Study 5, and only one of the two algorithms is applicable.

6.1. METHODOLOGY

Bayesian classification (BAC) has been gradually applied in the natural and social sciences since the 1990s and widely applied in this century. The currently popular algorithms are NBAY (e.g., Domingos and Pazzani, 1997; Ramoni and Sebastiani, 2001), BAYD, and BAYSD (e.g., Zhao, 1992; Logan and Gupta, 1993; Li and Zhao, 1998; Brown et al., 2001; Denison et al., 2002; Shi, 2011). In recent years, NBAY has been applied to geosciences to some extent, whereas BAYD and BAYSD are applied to geosciences relatively more widely.

The learning samples and prediction samples for NBAY, BAYD, and BAYSD are expressed with Equations (1.1) and (1.3), respectively. The learning process and prediction process of calculation flowcharts are illustrated by Figures 1.2 and 1.3, respectively. In Figure 1.3, if the prediction samples are replaced by the learning samples, Figure 1.3 becomes such a calculation flowchart for learning validation.

The expression of y created using NBAY, BAYD, and BAYSD is a nonlinear explicit expression with respect to m parameters $(x_1, x_2, ..., x_m)$:

$$y = BAC(x_1, x_2, ..., x_m) \qquad (6.1)$$

where BAC (*NBAY, BAYD,* or *BAYSD*) is a nonlinear function and calculated out by NBAY algorithm, BAYD algorithm, or BAYSD algorithm, which can be expressed as a usual mathematical formula and so is an explicit expression. Equation (6.1) is called the *fitting formula* obtained from the learning process.

The basic idea of BAC is illustrated in Figure 6.1.

6.1.1. Applying Ranges and Conditions

6.1.1.1. *Applying Ranges*

As Equation (1.1) described, assume that there are n learning samples, each associated with $m + 1$ numbers $(x_1, x_2, ..., x_m, y_i^*)$ and a set of observed values $(x_{1i}, x_{2i}, ..., x_{mi}, y_i^*)$, with $i = 1, 2, ..., n$ for these parameters. In principle, $n > m$, but in actual practice $n >> m$. As Equation (1.3) described, assume that there are k prediction samples, each associated with m parameters $(x_1, x_2, ..., x_m)$ and a set of observed values $(x_{1i}, x_{2i}, ..., x_{mi})$, with $i = n + 1, n + 2, ..., n + k$ for these parameters.

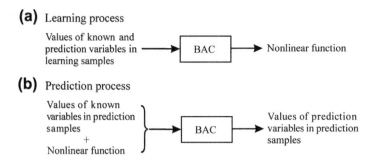

(a) Learning process

Values of known and prediction variables in learning samples → BAC → Nonlinear function

(b) Prediction process

Values of known variables in prediction samples + Nonlinear function → BAC → Values of prediction variables in prediction samples

FIGURE 6.1   Learning process (a) and prediction process (b) of BAC.

6.1.1.2. Applying Conditions

The number of learning samples, n, must be big enough to ensure the accuracy of prediction by BAC. The number of known variables, m, must be greater than one.

BAC introduced in this section refers only to NBAY, BAYD, and BAYSD.

6.1.2. Bayesian Theorem

Bayesian theorem can be expressed by

$$P(y|x_j) = \frac{P(x_j|y)P(y)}{P(x_j)} \quad (j = 1, 2, \ldots, m) \tag{6.2}$$

where (x_j, y) is a pair of random variables; x_j is an independent variable; m is the number of x_j, referring to the relevant definition in Equation (1.1); y is a dependent variable, a class to be solved, referring to a relevant definition in Equation (1.1). $P(x_j)$ is $P(x_j = a)$, the probability of $x_j = a$; here a is any value corresponding to x_j in samples. $P(y)$ is $P(y = b)$, the probability of $y = b$; here b is any value corresponding to y in samples. This probability is called *prior probability*. $P(y \mid x_j)$ is $P(y = b \mid x_j = a)$, the probability of $y = b$ under the condition of $x_j=a$. This probability is called *conditional probability* or *posterior probability*. Finally, $P(x_j \mid y)$ is $P(x_j = a \mid y = b)$, the probability of $x_j = a$ under the condition of $y = b$; this probability is called *conditional probability*.

Relative to conditional probability $P(y \mid x_j)$ and $P(x_j \mid y)$, $P(x_j)$, and $P(y)$ are unconditional probability.

6.1.3. Naïve Bayesian (NBAY)

Assuming the conditional independency between y and x, the posterior probability expression about each b of y is (Tan et al., 2005; Han and Kamber, 2006):

$$P(y|x) = \frac{P(y)\prod_{j=1}^{m}P(x_j|y)}{P(x)} \tag{6.3}$$

where all symbols have been defined in Equations (6.2), (1.1), and (1.2).

Equation (6.3) is the classification formula of NBAY. The b corresponding to the maximum of the calculation results is y value of a given learning sample or prediction sample. Therefore, this important formula can be adopted to calculate the fitness of the learning process as well as the class value y of prediction samples.

In the calculation of the right-hand side of Equation (6.3), since $P(y)$ and $P(x_j)$ are fixed, it is enough to calculate

$$\prod_{j=1}^{m} P(x_j|y) \tag{6.4}$$

in the numerator to get the maximum for classification.

If x_j is a discrete variable (an integer variable), it is easy to calculate $P(x_j \mid y)$ in Equation (6.4) according to the definitions in Equation (6.2); however, if x_j is a continuous variable (a real variable), $P(x_j \mid y)$ is calculated by

$$P(x_j = a|y = b) = \frac{1}{\sigma\sqrt{2\pi}}\exp\left\{\frac{-(a-\mu)^2}{2\sigma^2}\right\} \quad (j = 1, 2, ..., m) \tag{6.5}$$

where μ is the mean of x_j in learning samples when $y = b$; σ is the mean square error of x_j in learning samples when $y = b$; and the other symbols have been defined in Equation (6.2).

If the results of Equation (6.4) are all zeros for each b about y, the classification cannot be performed. This is called the problem of *probability values of zeros*, and a special processing is needed. In the special processing methods, there are an m-estimate (Tan et al., 2005) and Laplacian correction (Han and Kamber, 2006). This m-estimate is explained by an example in the simple case study (defaulted borrower) that follows.

The procedure of NBAY consists of two steps (Figure 6.2). In the first step (learning process), the y values of learning samples are calculated using the classified learning samples. In the second step (prediction process), the y values of prediction samples are calculated using the classified prediction samples.

6.1.3.1. Learning Process

If the y class value in learning samples expressed with Equations (1.1) and (1.2) has not been classified (i.e., y is a continuous variable) or determined, it is required to change y from continuous variable to discrete variable or to assign the discrete class value to y. Assume that y has L class values: $y_1^*, y_2^*, ..., y_L^*$.

The learning process performs in the following three steps:

1. Calculation of unconditional probability of learning samples:

$$P(x_j) \quad (j = 1, 2, ..., m); \; P(y = y_l^*) \quad (l = 1, 2, ..., L)$$

According to the definition of unconditional probability in Equation (6.2),

$$P(x_j) \quad (j = 1, 2, ..., m) \text{ and } P(y = y_l^*) \quad (l = 1, 2, ..., L)$$

are calculated.

Start

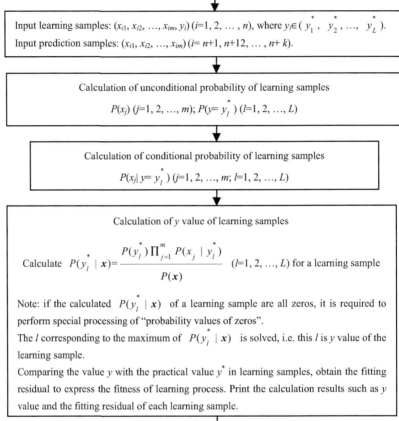

Input learning samples: $(x_{i1}, x_{i2}, ..., x_{im}, y_i)$ $(i=1, 2, ..., n)$, where $y_i \in (y_1^*, y_2^*, ..., y_L^*)$.
Input prediction samples: $(x_{i1}, x_{i2}, ..., x_{im})$ $(i= n+1, n+12, ..., n+k)$.

Calculation of unconditional probability of learning samples

$P(x_j)$ $(j=1, 2, ..., m)$; $P(y= y_l^*)$ $(l=1, 2, ..., L)$

Calculation of conditional probability of learning samples

$P(x_j| y= y_l^*)$ $(j=1, 2, ..., m; l=1, 2, ..., L)$

Calculation of y value of learning samples

Calculate $P(y_l^* \mid x)= \dfrac{P(y_l^*) \prod_{j=1}^{m} P(x_j \mid y_l^*)}{P(x)}$ $(l=1, 2, ..., L)$ for a learning sample

Note: if the calculated $P(y_l^* \mid x)$ of a learning sample are all zeros, it is required to perform special processing of "probability values of zeros".
The l corresponding to the maximum of $P(y_l^* \mid x)$ is solved, i.e. this l is y value of the learning sample.
Comparing the value y with the practical value y^* in learning samples, obtain the fitting residual to express the fitness of learning process. Print the calculation results such as y value and the fitting residual of each learning sample.

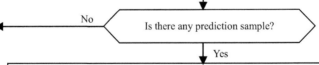

No   Is there any prediction sample?

Yes

Calculation of y value of prediction samples
Similarly to the aforementioned "Calculation of y value of learning samples", obtain y value of each prediction sample. Comparing value y with practical value y^* in prediction samples if the y^* exists, obtain the prediction residual to express the accuracy of prediction process. Print the calculation results such as y value and the prediction residual of each prediction sample.

Stop

FIGURE 6.2   Calculation flowchart of naïve Bayesian (NBAY).

2. Calculation of conditional probability of learning samples:

$$P(x_j|y = y_l^*) \quad (j = 1, 2, ..., m; \ l = 1, 2, ..., L)$$

According to the definition of conditional probability in Equation (6.2),

$$P(x_j|y = y_l^*) \quad (j = 1, 2, ..., m; \ l = 1, 2, ..., L)$$

are calculated.

3. Calculation of y value of learning samples:

$$P(y_l^*|x) = \frac{P(y_l^*)\prod_{j=1}^{m}P(x_j|y_l^*)}{P(x)} \quad (l = 1, 2, ..., L)$$

Using Equation (6.3), the posterior probability of learning samples

$$P(y_l^*|x) = \frac{P(y_l^*)\prod_{j=1}^{m}P(x_j|y_l^*)}{P(x)} \quad (l = 1, 2, ..., L)$$

are calculated.

If the L calculated $P(y_l^*|x)$ of a learning sample are all zeros, it is required to perform a special processing of "probability values of zeros" mentioned earlier.

The l corresponding to the maximum of the L calculated $P(y_l^*|x)$ is the solution, i.e., l is the value of y for a given learning sample. Comparing the y values with the corresponding practical values y^*, the fitting residual $\overline{R}_1(\%)$ is obtained to express the fitness of the learning process.

6.1.3.2. Prediction Process

If prediction samples exist, the calculation of the prediction process is conducted.

$$P(y_l^*|x) = \frac{P(y_l^*)\prod_{j=1}^{m}P(x_j|y_l^*)}{P(x)} \quad (l = 1, 2, ..., L)$$

Using Equation (6.3), the posterior probability of prediction samples

$$P(y_l^*|x) = \frac{P(y_l^*)\prod_{j=1}^{m}P(x_j|y_l^*)}{P(x)} \quad (l = 1, 2, ..., L)$$

are calculated.

If the L calculated $P(y_l^*|x)$ of a prediction sample are all zeros, it is required to perform a special processing of "probability values of zeros" mentioned previously.

The l corresponding to the maximum of the L calculated $P(y_l^*|x)$ is the solution, i.e., l is the value of y for a given prediction sample. Comparing the y values with the corresponding practical values y^* if the y^* exists in prediction samples, the prediction residual $\overline{R}_2(\%)$ is obtained to express the accuracy of prediction process.

The following introduces BAYD and BAYSD. It is assumed for both BAYD and BAYSD that the learning samples of each class have a normal distribution, $N(\boldsymbol{\mu}_l, \boldsymbol{\sigma}_l)$, where $\boldsymbol{\mu}_l$ is a $(m \times 1)$ vector of mathematical expectation for all x_j of Class l, and $\boldsymbol{\sigma}_l$ is a $(m \times m)$ covariance matrix for all x_j of Class l. Moreover, the Mahalanobis distance (de Maesschalck et al., 2000) is used to check whether this assumption is valid for these samples.

6.1.4. Bayesian Discrimination (BAYD)

This algorithm can be called the *discrimination analysis*. Because it uses the Bayesian theorem expressed with Equation (6.2) and normal distribution shown in Equation (6.5) during the deriving process of method, it is also called *Bayesian discrimination* (BAYD).

The procedure of BAYD consists of two steps (Figure 6.3). In the first step (the learning process), using the classified learning samples, a discriminate function is constructed by BAYD. In the second step (the prediction process), this discriminate function is used to classify the prediction samples.

6.1.4.1. Learning Process

If the y class value in learning samples expressed with Equations (1.1) and (1.2) has not been classified (i.e., y is a continuous variable) or determined, it is required to change y from continuous variable to discrete variable or to assign the discrete class value to y. Assume y has L class values: $y_1^*, y_2^*, ..., y_L^*$.

Using the classified learning samples, the discriminate function is constructed in the following four steps (Zhao, 1992; Logan and Gupta, 1993; Li and Zhao, 1998; Brown et al., 2001; Denison et al., 2002; Shi, 2011):

1. Realign learning samples. n learning samples expressed with Equation (1.1) are divided into L classes, and then these samples are realigned in order of class values:

$$x_{il} = (x_{i1l}, x_{i2l}, ..., x_{iml}) \quad (i = 1, 2, ..., n_l; \ l = 1, 2, ..., L) \tag{6.6}$$

where n_l is the number of learning samples with class value y_l. Certainly $n_1 + n_2 + ... + n_L = n$; x_{il} is x vector expressed with Equation (1.1), with class value y_l; x_{ijl} is the value of the jth parameter x_j ($j = 1, 2, ..., m$) in the ith learning sample in the lth class; and the other symbols have been defined in Equations (1.1) and (1.2).

2. Calculate the mean of parameters for each class value

$$\bar{x}_{jl} = \frac{1}{n_l} \sum_{i=1}^{n_l} x_{ijl} \quad (l = 1, 2, ..., L; \ j = 1, 2, ..., m) \tag{6.7}$$

where $\bar{x}_{jl}$ is the mean of x_j in learning samples in the lth class; and the other symbols have been defined in Equation (6.6).

3. Calculate the inverse matrix of the covariance matrix. S^{-1}, the inverse matrix of covariance matrix S, is a $(m \times m)$ matrix:

$$S^{-1} = \left[s_{jj_1}^{-1} \right] \quad (j, j_1 = 1, 2, ..., m) \tag{6.8}$$

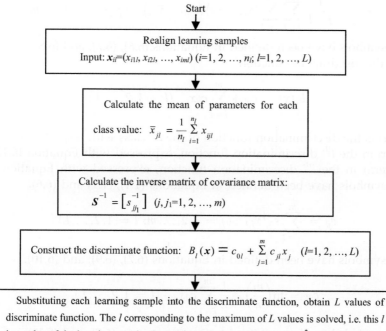

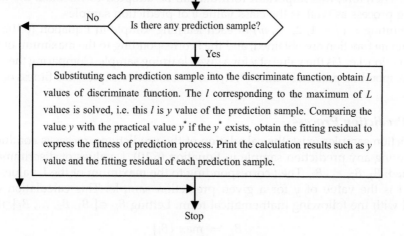

FIGURE 6.3 Calculation flowchart of Bayesian discrimination (BAYD).

where s_{jj_1} is expressed with

$$s_{jj_1} = \frac{1}{n-L} \sum_{l=1}^{L} \sum_{i=1}^{n_l} (x_{ijl} - \bar{x}_{jl})(x_{ij_1l} - \bar{x}_{j_1l}) \quad (j, j_1 = 1, 2, ..., m) \tag{6.9}$$

where all symbols have been defined in Equations (6.6), (6.7), and (6.8).

4. Construct the discriminate function

$$B_l(x) = c_{0l} + \sum_{j=1}^{m} c_{jl}x_j \quad (l = 1, 2, ..., L) \tag{6.10}$$

where $B_l(x)$ is the discrimination function of the lth class with respect to x; c_{jl} is the coefficient of x_j in the lth discrimination function, expressed with Equation (6.11); c_{0l} is the constant term in the lth discrimination function, expressed with Equation (6.12); and the other symbols have been defined in Equations (1.1), (1.2) and (6.6).

$$c_{jl} = \sum_{j_1=1}^{m} s_{jj_1}^{-1} \bar{x}_{j_1l} \quad (j = 1, 2, ..., m; \ l = 1, 2, ..., L) \tag{6.11}$$

where all symbols have been defined in Equations (6.7), (6.9), and (6.10).

$$c_{0l} = \ln\left(\frac{n_l}{n}\right) - \frac{1}{2} \sum_{j=1}^{m} c_{jl}\bar{x}_{jl} \quad (l = 1, 2, ..., L) \tag{6.12}$$

where all symbols have been defined in Equations (6.6), (6.7), and (6.11).

Equation (6.10) is the discriminate function of BAYD. The l corresponding to the maximum of the calculation results is the y value of a given learning sample or prediction sample. Therefore, this important formula can be adopted to calculate the fitness of the learning process as well as the class value y of prediction samples.

Substituting x_j ($j = 1, 2, ..., m$) of each learning sample in Equation (6.10), L values of discriminate function are obtained, and the l corresponding to the maximum of the L values is the solution, i.e., l is the value of y for a given learning sample. Comparing the y values with the corresponding practical values y^*, $\bar{R}_1(\%)$ is obtained to express the fitness of the learning process.

6.1.4.2. Prediction Process

If prediction samples exist, the calculation of the prediction process is conducted.

Substituting any prediction sample in Equation (6.10), L values of discriminate function are obtained: $B_1, B_2, ..., B_L$. The l corresponding to the maximum of the L values is the solution, i.e., l is the value of y for a given prediction sample. This calculation of y can be expressed with the following mathematical form. Letting $B_{l_b} \in [B_1, B_2, ..., B_L]$, if

$$B_{l_b} = \max_{1 \le l \le L} \{B_l\} \tag{6.13}$$

then

$$y = l_b \tag{6.14}$$

The value of y for the prediction sample is l_b, shown in Equation (6.14). Hence, Equation (6.14) can be called $y = BAYD(x_1, x_2, ..., x_m)$, where $BAYD$ is nonlinear function. Comparing the y values with the corresponding practical values y^* if the y^* exists in prediction samples, $\overline{R}_2(\%)$ is obtained to express the accuracy of the prediction process.

6.1.5. Bayesian Successive Discrimination (BAYSD)

This algorithm can be called the *successive discrimination analysis*. Because it uses the Bayesian theorem expressed with Equation (6.2) and normal distribution shown in Equation (6.5) during the deriving process of method, it is also called *Bayesian successive discrimination* (BAYSD).

The procedure of BAYSD consists of two steps (Figure 6.4). In the first step (the learning process), using the classified learning samples, a discriminate function is constructed by BAYSD. In the second step (the prediction process), this discriminate function is used to classify the prediction samples.

6.1.5.1. Learning Process

If the y class value in learning samples expressed with Equations (1.1) and (1.2) has not been classified (i.e., y is a continuous variable) or determined, it is required to change y from continuous variable to discrete variable or to assign the discrete class value to y. Assume that y has L class values: $y_1^*, y_2^*, ..., y_L^*$.

Using the classified learning samples, the construction of a discriminate function consists of the following seven steps (Zhao, 1992; Logan and Gupta, 1993; Li and Zhao, 1998; Brown et al., 2001; Denison et al., 2002; Shi, 2011):

1. Realign learning samples. n learning samples expressed with Equation (1.1) are divided into L classes, and then these samples are realigned in order of class values:

$$x_{il} = (x_{i1l}, x_{i2l}, ..., x_{iml}) \quad (i = 1, 2, ..., n_l; \; l = 1, 2, ..., L) \tag{6.15}$$

where n_l is the number of learning samples with class value y_l, certainly $n_1 + n_2 + ... + n_L = n$; x_{il} is x vector expressed with Equation (1.1), with class value y_l; x_{ijl} is the value of the jth parameter x_j ($j = 1, 2, ..., m$) in the ith learning sample in the lth class; and the other symbols have been defined in Equations (1.1) and (1.2).

2. Calculate the mean of parameters for each class value

$$\overline{x}_{jl} = \frac{1}{n_l} \sum_{i=1}^{n_l} x_{ijl} \quad (l = 1, 2, ..., L; \; j = 1, 2, ..., m) \tag{6.16}$$

where $\overline{x}_{jl}$ is the mean of x_j in learning samples in the lth class, and the other symbols have been defined in Equation (6.15).

3. Calculate the total mean for each parameter

$$\overline{x}_j = \frac{1}{n} \sum_{l=1}^{L} \sum_{i=1}^{n_l} x_{ijl} \quad (j = 1, 2, ..., m) \tag{6.17}$$

where $\overline{x}_j$ is the total mean of x_j in all learning samples, and the other symbols have been defined in Equation (6.15).

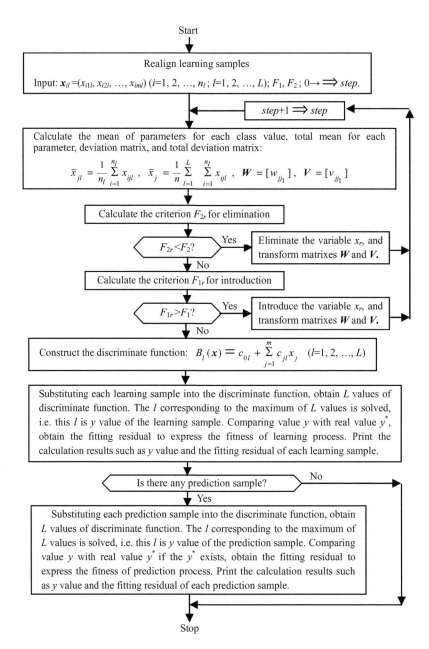

FIGURE 6.4   Calculation flowchart of Bayesian successive discrimination (BAYSD).

4. Calculate the deviation matrix. The deviation matrix W is a $(m \times m)$ matrix:

$$W = [w_{jj_1}] \quad (j, j_1 = 1, 2, ..., m) \tag{6.18}$$

where w_{jj1} is expressed with

$$w_{jj_1} = \sum_{l=1}^{L} \sum_{i=1}^{n_l} \left(x_{ijl} - \overline{x}_{jl} \right) \left(x_{ij_1 l} - \overline{x}_{j_1 l} \right) \quad (j, j_1 = 1, 2, ..., m) \tag{6.19}$$

where all symbols have been defined in Equations (6.15), (6.16), and (6.18).

5. Calculate the total deviation matrix. The total deviation matrix V is a $(m \times m)$ matrix:

$$V = [v_{jj_1}] \quad (j, j_1 = 1, 2, ..., m) \tag{6.20}$$

where v_{jj1} is expressed with

$$v_{jj_1} = \sum_{l=1}^{L} \sum_{i=1}^{n_l} \left(x_{ijl} - \overline{x}_{j} \right) \left(x_{ij_1 l} - \overline{x}_{j_1} \right) \quad (j, j_1 = 1, 2, ..., m) \tag{6.21}$$

where all symbols have been defined in Equations (6.15), (6.17), and (6.20).

The following point discusses the introduction and elimination of parameters. At first, two filtration criteria F_1 and F_2 are assigned, where F_1 is used as a filtration criterion for introduction, whereas F_2 is used as a filtration criterion for elimination, and $F_1 \geq F_2$. For the sake of simplicity, F_1 and F_2 are all assigned to zero, i.e., $F_1 = 0$ and $F_2 = 0$. In general, when $F_1 = 0$ and $F_2 = 0$, the introduced parameter will not be eliminated; when $F_1 \neq 0$ (e.g., $F_1 = 2$) or $F_2 \neq 0$ (e.g., $F_2 = 2$), the introduced parameter will be possibly eliminated.

6. Introduce or eliminate a parameter. If the sth step calculation has been accomplished and p parameters have been introduced, then the calculations on the introduced parameters are conduced:

$$A_j = \frac{v_{jj}^{(s)}}{w_{jj}^{(s)}} \quad (j = 1, 2, ..., p) \tag{6.22}$$

where all symbols of the right-hand side have been defined in Equations (6.19) and (6.21). The criterion for elimination is calculated by

$$F_{2r} = \frac{1 - A_r}{A_r} \cdot \frac{n - p + 1 - L}{L - 1} \tag{6.23}$$

where all symbols excepting A_r of the right-hand side have been defined in Equations (6.15) and (6.22), and A_r is expressed with

$$A_r = \max_{j = 1, 2, ..., p} \{A_j\} \tag{6.24}$$

If $F_{2r} < F_2$, the introduced parameter x_r is eliminated in the discriminate function, and two matrixes W and V are transformed by Equations (6.25) and (6.26).

$$w_{jj_1}^{(s)} = \begin{cases} w_{jj_1}^{(s-1)}/w_{rr}^{(s-1)} & (j = r, j_1 \neq r) \\ w_{jj_1}^{(s-1)} - w_{jr}^{(s-1)} \cdot w_{rj_1}^{(s-1)}/w_{rr}^{(s-1)} & (j \neq r, j_1 \neq r) \\ -w_{jr}^{(s-1)}/w_{rr}^{(s-1)} & (j \neq r, j_1 = r) \\ 1/w_{rr}^{(s-1)} & (j = r, j_1 = r) \end{cases} \quad (j, j_1 = 1, 2, ..., m) \quad (6.25)$$

$$v_{jj_1}^{(s)} = \begin{cases} v_{jj_1}^{(s-1)}/v_{rr}^{(s-1)} & (j = r, j_1 \neq r) \\ v_{jj_1}^{(s-1)} - v_{jr}^{(s-1)} \cdot v_{rj_1}^{(s-1)}/v_{rr}^{(s-1)} & (j \neq r, j_1 \neq r) \\ -v_{jr}^{(s-1)}/v_{rr}^{(s-1)} & (j \neq r, j_1 = r) \\ 1/v_{rr}^{(s-1)} & (j = r, j_1 = r) \end{cases} \quad (j, j_1 = 1, 2, ..., m) \quad (6.26)$$

If $F_{2r} > F_2$, the introduced parameters cannot be eliminated in the discriminate function, but the calculations on the parameters that have not been introduced are conduced:

$$A_j = \frac{w_{jj}^{(s)}}{v_{jj}^{(s)}} \quad (j = 1, 2, ..., m - p) \quad (6.27)$$

where all symbols of the right-hand side have been defined in Equations (6.19) and (6.21). The criterion for introduction is calculated by

$$F_{1r} = \frac{1 - A_r}{A_r} \cdot \frac{n - p - L}{L - 1} \quad (6.28)$$

where all symbols except A_r of the right-hand side have been defined in Equations (6.15) and (6.22), and A_r is expressed with

$$A_r = \min_{j = 1, 2, ..., m-p} \{A_j\} \quad (6.29)$$

If $F_{1r} > F_1$, the parameter x_r is introduced into the discriminate function, and two matrixes W and V are transformed by Equations (6.25) and (6.26).

Thus, the $(s + 1)^{th}$ step calculation is accomplished. Similarly, the $(s + 2)^{th}$ step calculation, the $(s + 3)^{th}$ step calculation, ..., are repeated till we encounter the case that no parameter can be introduced or eliminated. At that time, the whole successive discrimination procedure ends.

7. Construct the discriminate function. When the whole successive discrimination procedure ends, the discriminate function is

$$B_l(x) = c_{0l} + \sum_{j=1}^{m} c_{jl} x_j \quad (l = 1, 2, ..., L) \quad (6.30)$$

where $B_l(x)$ is the discrimination function of the lth class with respect to x; c_{jl} is the coefficient of x_j in the lth discrimination function, expressed with Equation (6.31); c_{0l} is the constant term in the lth discrimination function, expressed with Equation (6.32); and the other symbols have been defined in Equations (1.1), (1.2), and (6.15).

$$c_{jl} = (n - l) \sum_{j_1 = 1}^{m} w_{jj_1}^{-1} \bar{x}_{j_1 l} \quad (j = 1, 2, ..., m; \ l = 1, 2, ..., L) \tag{6.31}$$

where all symbols have been defined in Equations (6.15), (6.16), and (6.19). Note that if some parameters x_j are eliminated, their corresponding c_{jl} will be assigned to zero; thus the calculations of Equation (6.31) for them are not needed.

$$c_{0l} = \ln\left(\frac{n_l}{n}\right) - \frac{1}{2} \sum_{j=1}^{m} c_{jl} \bar{x}_{jl} \quad (l = 1, 2, ..., L) \tag{6.32}$$

where all symbols have been defined in Equations (6.15), (6.16), and (6.31).

As shown here, the learning process of BAYSD is similar to that of MRA, described in Section 2.2.4. The only difference is that BAYSD transforms two matrixes, W and V.

Equation (6.30) is the discriminate function of BAYSD. The l corresponding to the maximum of the calculation results is y value of a given learning sample or prediction sample. Therefore, this important formula can be adopted to calculate the fitness of the learning process as well as the class value y of the prediction samples.

Substituting x_j ($j = 1, 2, ..., m$) of each learning sample in Equation (6.30), L values of discriminate function are obtained, and the l corresponding to the maximum of the L values is the solution, i.e. l is the value of y for a given learning sample. Comparing the y values with the corresponding practical values y^*, $\bar{R}_1(\%)$ is obtained to express the fitness of the learning process.

6.1.5.2. Prediction Process

If prediction samples exist, the calculation of the prediction process is conducted.

Substituting any prediction sample in Equation (6.30), L values of the discriminate function are obtained: $B_1, B_2, ..., B_L$. The l corresponding to the maximum of the L values is the solution, i.e., l is the value of y for a given prediction sample. This calculation of y can be expressed with the following mathematical form. Letting $B_{l_b} \in [B_1, B_2, ..., B_L]$, if

$$B_{l_b} = \max_{1 \leq l \leq L} \{B_l\} \tag{6.33}$$

then

$$y = l_b \tag{6.34}$$

The value of y for the prediction sample is l_b, shown in Equation (6.34). Hence, Equation (6.34) can be called $y = BAYSD(x_1, x_2, ..., x_m)$, where $BAYSD$ is a nonlinear function. Comparing the y values with the corresponding practical values y^* if the y^* exists in prediction samples, $\bar{R}_2(\%)$ is obtained to express the accuracy of the prediction process.

6.1.6. Simple Case Study: Defaulted Borrower by NBAY

To help readers master the aforementioned NBAY and BAYSD, a very simple sample about a problem of a defaulted borrower solved by NBAY and BAYSD is introduced here. In Table 6.1, $x_j (j = 1, 2, 3)$ and y^* of 10 learning samples (customers) are the original data (Tan et al., 2005), which can be used by the learning process of NBAY or BAYSD as well as by the learning process of MRA, BPNN, C-SVM, or R-SVM. In Table 6.1, two prediction samples (customers) are provided; these are additional and can be used by the prediction process of NBAY, BAYSD, MRA, BPNN, C-SVM, or R-SVM.

6.1.6.1. Known Input Parameters

They are the values of the known variables $x_i (i = 1, 2, 3)$ for 10 learning samples and two prediction samples and the value of the prediction variable y^* for 10 learning samples (Table 6.1).

6.1.6.2. Learning Process of NBAY

The following two calculations are conducted in the range of learning samples listed: in Table 6.1.

1. Calculate the unconditional probability $P(x_j)$ and $P(y)$. According to the definitions of $P(x_j)$ and $P(y)$ in Equation (6.2), from Table 6.1 it is obtained that

$$P(x_1 = 1) = 3/10, \ P(x_1 = 2) = 7/10;$$
$$P(x_2 = 1) = 4/10, \ P(x_2 = 2) = 4/10; \ P(x_2 = 3) = 2/10;$$
$$P(y = 1) = 3/10, \ P(y = 2) = 7/10.$$

TABLE 6.1 Input Data for the Defaulted Borrower

Sample Type	Sample No.	x_1 Homeowner	x_2 Marital Status	x_3 Annual Income (US$K)	y^* Defaulted Borrower[a]
Learning samples	1	1 (Yes)	1 (Single)	125	2 (No)
	2	2 (No)	2 (Married)	100	2 (No)
	3	2 (No)	1 (Single)	70	2 (No)
	4	1 (Yes)	2 (Married)	120	2 (No)
	5	2 (No)	3 (Divorced)	95	1 (Yes)
	6	2 (No)	2 (Married)	60	2 (No)
	7	1 (Yes)	3 (Divorced)	220	2 (No)
	8	2 (No)	1 (Single)	85	1 (Yes)
	9	2 (No)	2 (Married)	75	2 (No)
	10	2 (No)	1 (Single)	90	1 (Yes)
Prediction samples	11	2 (No)	2 (Married)	120	[2 (No)]
	12	1 (Yes)	3 (Divorced)	120	[2 (No)]

[a]y^* = defaulted borrower (1—defaulted, 2—no defaulted). In y^*, numbers in square brackets are not input data but are used for calculating R(%).

2. Calculate the conditional probability $P(x_j|y)$. For x_1 (homeowner), it is a discrete variable (an integer variable), and then according to the definition of $P(x_j \mid y)$ in Equation (6.2), from Table 6.1 it is obtained that

$$P(x_1 = 1|y = 1) = 0/3 = 0, \; P(x_1 = 2|y = 1) = 3/3 = 1;$$
$$P(x_1 = 1|y = 2) = 3/7, \; P(x_1 = 2|y = 2) = 4/7.$$

For x_2 (Marital Status), it is a discrete variable (integer variable), and then according to the definition of $P(x_j|y)$ in Equation (6.2), from Table 6.1 it is obtained that

$$P(x_2 = 1|y = 1) = 2/3 = 0, \; P(x_2 = 2|y = 1) = 0/3 = 0, \; P(x_2 = 3|y = 1) = 1/3;$$
$$P(x_2 = 1|y = 2) = 2/7, \; P(x_2 = 2|y = 2) = 4/7, \; P(x_2 = 3|y = 2) = 1/7.$$

For x_3 (Annual Income), it is a continuous variable (real variable), and then μ and σ shown in Equation (6.5) are required to be calculated. From Table 6.1, these calculations will be performed under two conditions below.

a. Under the condition of $y = 1$ (the number of relative samples (NRS) is 3)

$\mu = (95 + 85 + 90)/3 = 90$

$\sigma^2 = \left[(95 - 90)^2 + (85 - 90)^2 + (90 - 90)^2\right]/2$ (*Note* : The denominator is NRS minus one, so it is 2)

$\sigma = \sqrt{25} = 5$

Substituting the preceding two values of μ and σ as well as $j = 3$ in Equation (6.5), it is obtained that

$$P(x_3 = a|y = 1) = \frac{1}{5\sqrt{2\pi}}\exp\left\{-\frac{(a - 90)^2}{2 \times 25}\right\} \qquad (6.35)$$

where all symbols have been defined in Equations (6.2), (6.5), (1.1), and (1.2).

Therefore, under the condition of $y = 1$, as long as we substitute the value of x_3 (Annual Income) in the learning sample for a in Equation (6.35), the conditional probability of that value is obtained. For the requirement of the prediction process, substituting 120 (US$K) of sample Nos. 11 and 12 in Table 6.1 for a in Equation (6.35), it is obtained that

$$P(x_3 = 120|y = 1) = \frac{1}{5\sqrt{2\pi}}\exp\left\{-\frac{(120 - 90)^2}{2 \times 25}\right\} = 1.2 \times 10^{-9}$$

b. Under the condition of $y = 2$ (NRS is 7),

$\mu = (125 + 100 + 70 + 120 + 60 + 220 + 75)/7 = 110$

$\sigma^2 = \left[(125 - 110)^2 + (100 - 110)^2 + (70 - 110)^2 + (120 - 110)^2 + (60 - 110)^2 + (220 - 110)^2 \right.$
$\left. + (75 - 110)^2\right]/6 = 2975$ (*Note* : The denominator is NRS minus one, so it is 6)

$\sigma = \sqrt{2975} = 54.54$

Substituting the previous two values of μ and σ as well as $j = 3$ in Equation (6.5), it is obtained that

$$P(x_3 = a|y = 2) = \frac{1}{54.54\sqrt{2\pi}} \exp\left\{-\frac{(a-110)^2}{2 \times 2975}\right\} \tag{6.36}$$

where all symbols have been defined in Equations (6.2), (6.5), (1.1), and (1.2).

Therefore, under the condition of $y = 2$, as long as we substitute the value of x_3 (Annual Income) in a learning sample for a in Equation (6.36), the conditional probability of that value is obtained. For the requirement of the prediction process that follows, substituting 120 (US\$K) of sample Nos. 11 and 12 in Table 6.1 for a in Equation (6.36), it is obtained that

$$P(x_3 = 120|y = 2) = \frac{1}{54.54\sqrt{2\pi}} \exp\left\{-\frac{(120-110)^2}{2 \times 2975}\right\} = 0.0072$$

Thus, it is believed that an explicit nonlinear function corresponding to BAC formula (6.1) is obtained:

$$y = BAC(x_1, x_2, x_3) \tag{6.37}$$

6.1.6.3. Prediction Process of NBAY

In the preceding, using the 10 learning samples (Table 6.1) and by the learning process of NBAY, we have constructed a calculable Bayesian classification formula (6.3). Substituting the data (x_1, x_2, x_3) of each learning sample into Equation (6.3), respectively, the result y (Defaulted Borrower) of each learning sample is obtained and filled in Table 6.2. Similarly, the results y (Defaulted Borrower) of two prediction samples are obtained and filled in Table 6.2. That is the prediction process of NBAY.

The following two calculations describe the prediction process in detail.

1. Calculate the first prediction sample $(x_1 = 2, x_2 = 2, x_3 = 120)$. Calculations perform under the following two conditions.

   a. Under the condition of $y = 1$. According to Equation (6.4) and the previously related results,

   $$P(x_1 = 2|y = 1) \times P(x_2 = 2|y = 1) \times P(x_3 = 120|y = 1) = 1 \times 0 \times 1.2 \times 10^{-9} = 0$$

   And according to Equation (6.3),

   $$P(y = 1|x) = P(y = 1) \times 0/P(x) = 3/10 \times 0/P(x) = 0/P(x) = 0$$

   b. Under the condition of $y = 2$. According to Equation (6.4) and the previously related results,

   $$P(x_1 = 2|y = 2) \times P(x_2 = 2|y = 2) \times P(x_3 = 120|y = 2) = 4/7 \times 4/7 \times 0.0072 = 0.0024$$

   And according to Equation (6.3),

   $$P(y = 2|x) = P(y = 2) \times 0.0024/P(x) = 7/10 \times 0.0024/P(x) = 0.00168/P(x)$$

TABLE 6.2 Prediction Results from a Defaulted Borrower

Sample Type	Sample No.	y^* Defaulted Borrower[a]	NBAY		BAYSD		C-SVM		R-SVM		BPNN		MRA	
			y	$R(\%)$	y	$R(\%)$	y	$R(\%)$	y	$R(\%)$	y	$R(\%)$	y	$R(\%)$
Learning samples	1	2 (No)	2 (No)	0	2 (No)	0	2 (No)	0	2 (No)	0	2 (No)	0	2 (No)	0
	2	2 (No)	2 (No)	0	2 (No)	0	2 (No)	0	2 (No)	0	2 (No)	0	2 (No)	0
	3	2 (No)	2 (No)	0	2 (No)	0	2 (No)	0	2 (No)	0	2 (No)	0	2 (No)	0
	4	2 (No)	2 (No)	0	2 (No)	0	2 (No)	0	2 (No)	0	2 (No)	0	2 (No)	0
	5	1 (Yes)	1 (Yes)	0	2 (No)	100	1 (Yes)	0	2 (No)	100	1 (Yes)	0	2 (No)	100
	6	2 (No)	2 (No)	0	2 (No)	0	2 (No)	0	2 (No)	0	2 (No)	0	2 (No)	0
	7	2 (No)	2 (No)	0	2 (No)	0	2 (No)	0	2 (No)	0	2 (No)	0	2 (No)	0
	8	1 (Yes)	1 (Yes)	0	1 (Yes)	0	1 (Yes)	0	2 (No)	100	1 (Yes)	0	1 (Yes)	0
	9	2 (No)	2 (No)	0	2 (No)	0	2 (No)	0	2 (No)	0	2 (No)	0	2 (No)	0
	10	1 (Yes)	1 (Yes)	0	1 (Yes)	0	1 (Yes)	0	2 (No)	100	1 (Yes)	0	1 (Yes)	0
Prediction samples	11	2 (No)	2 (No)	0	1 (Yes)	50	2 (No)	0	2 (No)	0	2 (No)	0	1 (Yes)	50
	12	2 (No)	2 (No)	0	2 (No)	0	2 (No)	0	2 (No)	0	2 (No)	0	2 (No)	0

[a] y^* = defaulted borrower (1—defaulted, 2—no defaulted).

It is known from the preceding that $P(y = 2 \mid x) > P(y = 1 \mid x)$, so $y = 2$ (No defaulted borrower). Therefore, $y = 2$ for the first prediction sample ($x_1 = 2$, $x_2 = 2$, $x_3 = 120$).

2. Calculate the second prediction sample ($x_1 = 1$, $x_2 = 3$, $x_3 = 120$). Calculations perform under the following two conditions:

    a. Under the condition of $y = 1$. According to Equation (6.4) and the previously related results,

$$P(x_1 = 1|y = 1) \times P(x_2 = 3|y = 1) \times P(x_3 = 120|y = 1) = 0 \times 1/3 \times 1.2 \times 10^{-9} = 0$$

    And according to Equation (6.3),

$$P(y = 1|x) \times P(y = 1) \times 0/P(x) = 3/10 \times 0/P(x) = 0/P(x) = 0$$

    b. Under the condition of $y = 2$. According to Equation (6.4) and the previously related results,

$$P(x_1 = 1|y = 2) \times P(x_2 = 3|y = 2) \times P(x_3 = 120|y = 2) = 3/7 \times 1/7 \times 0.0072 = 0.00044$$

    And according to Equation (6.3),

$$P(y = 2|x) \times P(y = 2) \times 0.00044/P(x) = 7/10 \times 0.00044/P(x) = 0.00031/P(x)$$

    It is known from the preceding that $P(y = 2 \mid x) > P(y = 1 \mid x)$, so $y = 2$ (No defaulted borrower). Therefore, $y = 2$ for the second prediction sample ($x_1 = 1$, $x_2 = 3$, $x_3 = 120$).

    Thus, the application of NBAY to the defaulted borrower problem is accomplished, showing the learning and prediction processes of the NBAY algorithm by an example.

6.1.6.4. *Processing of "Probability Values of Zeros" in the Prediction Process of NBAY*

In the aforementioned calculation of conditional probability $P(x_j \mid y)$, there are two "probability values of zeros": $P(x_1 = 1 \mid y = 1) = 0 / 3 = 0$, and $P(x_2 = 2 \mid y = 1) = 0 / 3 = 0$. As mentioned, "probability values of zeros" would result in failure of the classification to perform. For example, in the calculation for a prediction sample x, since one of two "probability values of zeros" could result in that $P(y = 1 \mid x) = 0$ and $P(y = 2 \mid x) = 0$, thus the y value of this prediction sample x would not be determined.

The general formula of m-estimate (Tan et al., 2005) is

$$P(x_j = a|y = b) = \frac{n_c + m \cdot P(y = b)}{n_d + m} \tag{6.38}$$

where n_c is the number of $x_j = a$ in learning samples when $y = b$; n_d is the number of $y = b$ in learning samples; and the other symbols have been defined in Equations (1.1), (1.2), and (6.2).

Now, by example of the second "probability values of zeros" $[P(x_2 = 2 \mid y = 1) = 0/3 = 0]$, the estimation of $P(x_2 = 2 \mid y = 1)$ is conducted by Equation (6.38). Since $n_c = 0$, $n_d = 3$, $m = 3$, and $P(y = 1) = 3/10$,

$$P(x_2 = 2|y = 1) = [n_c + m \cdot P(y = 1)]/(n_d + m)[0 + 3 \times 3/10]/(3 + 3) = 0.15$$

6.1.6.5. *Application Comparisons Among NBAY, BAYSD, C-SVM, R-SVM, BPNN, and MRA*

Similar to the above, using the 10 learning samples (Table 6.1) and by BAYSD, C-SVM, R-SVM, BPNN, and MRA, the five functions of defaulted borrower (y) with respect to three parameters (x_1, x_2, x_3) have been constructed, i.e., Equation (6.40) corresponding to BAYSD formula (6.1), Equations (6.41) and (6.42) corresponding to SVM formula (4.1), Equation (6.43) corresponding to BPNN formula (3.1), and Equation (6.44) corresponding to MRA formula (2.14), respectively. Also similar to the above, using the two prediction samples (Table 6.1), BAYSD, C-SVM, R-SVM, BPNN, and MRA are employed for predictions, respectively.

Using BAYSD, the result is a discriminate function corresponding to Equation (6.30):

$$\left.\begin{array}{l} B_1(x) = \ln(0.3) - 45.874 + 34.121x_1 - 3.493x_2 + 0.326x_3 \\ B_2(x) = \ln(0.7) - 37.728 + 30.325x_1 - 2.860x_2 + 0.301x_3 \end{array}\right\} \tag{6.39}$$

Thus, it is believed that an explicit nonlinear function corresponding to BAC formula (6.1) is obtained:

$$y = BAC(x_1, x_2, x_3) \tag{6.40}$$

From the discriminate process of BAYSD, defaulted borrower (y) is shown to depend on the three parameters in decreasing order: homeowner (x_1), annual income (x_3), and marital status (x_2).

Substituting x_j ($j = 1, 2, 3$) of each learning sample in Equation (6.39), two values of the discriminate function are obtained, and the l corresponding to the maximum of the two values is the solution, i.e., l is the value of y for a given learning sample. Comparing the y values with the corresponding practical values y^*, $\bar{R}_1(\%)$ is obtained to express the fitness of the learning process. For instance, for the first learning sample (Table 6.2), $l = 2$, thus $y = 2$; and $y^* = 2$, so $\bar{R}_1(\%) = 0$.

Using C-SVM, the result is an explicit nonlinear function:

$$y = C\text{-}SVM(x_1, x_2, x_3) \tag{6.41}$$

with $C = 32$, $\gamma = 2$, seven free vectors x_i, and the cross-validation accuracy CVA $= 80\%$.

Using R-SVM, the result is an explicit nonlinear function:

$$y = R\text{-}SVM(x_1, x_2, x_3) \tag{6.42}$$

with $C = 1$, $\gamma = 0.333333$, and eight free vectors x_i.

The BPNN consists of three input layer nodes, one output layer node, and seven hidden layer nodes. Setting the control parameter of t maximum $t_{max} = 100000$, the calculated optimal learning time count $t_{opt} = 99994$. The result is an implicit nonlinear function:

$$y = BPNN(x_1, x_2, x_3) \tag{6.43}$$

with the root mean square error $RMSE(\%) = 0.1376 \times 10^{-1}$.

Using MRA, the result is an explicit linear function:

$$y = 3.2602 - 0.75015x_1 + 0.12505x_2 - 0.0049042x_3 \tag{6.44}$$

Equation (6.44) yields a residual variance of 0.75277 and a multiple correlation coefficient of 0.49722. From the regression process, defaulted borrower (y) is shown to depend on the three parameters in decreasing order: homeowner (x_1), annual income (x_3), and marital status (x_2). This order is consistent with the aforementioned one by BAYSD due to the fact that though MRA is a linear algorithm whereas BAYSD is a nonlinear algorithm, the nonlinearity of the studied problem is strong whereas the nonlinearity ability of BAYSD is weak.

Substituting the values of x_1, x_2, x_3 given by the two prediction samples (Table 6.1) in Equations (6.40), (6.41), (6.42), (6.43), and (6.44), respectively, the defaulted borrower (y) of each prediction sample is obtained.

Table 6.2 shows the results of the learning and prediction processes by BAYSD, C-SVM, R-SVM, BPNN, and MRA.

$\overline{R}^*(\%) = 17.5$ of MRA (Table 6.3) shows that the nonlinearity of the relationship between the predicted value y and its relative parameters (x_1, x_2, x_3) is strong from Table 1.2. The results of NBAY, BPNN, and C-SVM are the same, i.e., not only $\overline{R}_1(\%) = 0$, but also $\overline{R}_2(\%) = 0$, and thus the total mean absolute relative residual $\overline{R}^*(\%) = 0$, which coincide with practicality. For R-SVM, though $\overline{R}_2(\%) = 0$, $\overline{R}_1(\%) = 30$, which cannot ensure correct prediction of other new prediction samples. As for BAYSD and MRA, the solution accuracy is low, and $\overline{R}_2(\%) > \overline{R}_1(\%)$.

TABLE 6.3   Comparison Among the Applications of NBAY, BAYSD, C-SVM, R-SVM, BPNN, and MRA to a Defaulted Borrower

Algorithm	Fitting Formula	Mean Absolute Relative Residual			Dependence of the Predicted Value (y) on Parameters (x_1, x_2, x_3), in Decreasing Order	Time Consumed on PC (Intel Core 2)	Solution Accuracy
		$\overline{R}1(\%)$	$\overline{R}2(\%)$	$\overline{R}^*(\%)$			
NBAY	Nonlinear, explicit	0	0	0	N/A	<1 s	Very high
BAYSD	Nonlinear, explicit	10	25	17.5	x_1, x_3, x_2	<1 s	Low
C-SVM	Nonlinear, explicit	0	0	0	N/A	1 s	Very high
R-SVM	Nonlinear, explicit	30	0	15	N/A	1 s	Low
BPNN	Nonlinear, implicit	0	0	0	N/A	15 s	Very high
MRA	Linear, explicit	10	25	17.5	x_1, x_3, x_2	<1 s	Low

6.1.6.6. *Summary and Conclusions*

In summary, using data for the defaulted borrower, i.e., the three parameters (homeowner, marital status, annual income) and a defaulted borrower of 10 buyers, a calculable NBAY classification formula (6.3) is constructed by the data mining tool of NBAY, and the results coincide with practicality. This formula is called *mined knowledge*. In the prediction process, applying this knowledge, i.e., using the three parameters of two other buyers for defaulted borrowers, respectively, the results also coincide with practicality. Therefore, this method can be spread to the defaulted borrower of other cases.

Moreover, it is found that since this case study is a strong nonlinear problem, NBAY, BPNN, and *C*-SVM are applicable, but BAYSD, *R*-SVM, or MRA is not applicable.

6.2. CASE STUDY 1: RESERVOIR CLASSIFICATION IN THE FUXIN UPLIFT

6.2.1. Studied Problem

The objective of this case study is to conduct a reservoir classification using sandstone thickness, porosity, permeability, carbonate content, and mud content, which has practical value when well-test data are less limited.

We used data of 25 learning samples from the fourth stage of Nanquan in the Fuxin Uplift of the Jilin Oilfield in east China, and each sample contains five parameters (x_1 = sandstone thickness, x_2 = porosity, x_3 = permeability, x_4 = carbonate content, x_5 = mud content) and a well-test result (y^* = reservoir type); Fu et al. (2011) adopted BAYD for the reservoir classification. In this case study, among these 25 samples, 20 are taken as learning samples and 3 as prediction samples (Table 6.4).

6.2.2. Known Input Parameters

Input parameters are the values of the known variables x_i (i=1, 2, 3, 4, 5) for 20 learning samples and 3 prediction samples and the value of the prediction variable y^* for 20 learning samples (Table 6.4).

6.2.3. Learning Process

Using the parameters of learning samples and by BAYSD, a discrimination function corresponding to Equation (6.30) has been constructed:

$$\left.\begin{array}{l} B_1(x) = \ln(0.20) - 95.700 + 7.581x_1 + 4.605x_2 + 9.076x_3 - 1.463x_4 + 4.599x_5 \\ B_2(x) = \ln(0.20) - 67.696 + 5.235x_1 + 5.150x_2 + 3.339x_3 + 0.684x_4 + 3.408x_5 \\ B_3(x) = \ln(0.55) - 76.048 + 5.169x_1 + 4.792x_2 + 4.797x_3 - 0.777x_4 + 4.736x_5 \\ B_4(x) = \ln(0.05) - 152.041 + 8.266x_1 + 5.288x_2 + 8.530x_3 - 2.798x_4 + 7.491x_5 \end{array}\right\} \quad (6.45)$$

TABLE 6.4 Input Data for Reservoir Classification in the Fuxin Uplift

Sample Type	Sample No.	Well No.	Layer No.	x_1 Sandstone Thickness (m)	x_2 Porosity (%)	x_3 Permeability (mD)	x_4 Carbonate Content (%)	x_5 Mud Content (%)	$Y*$ Reservoir Type[a]
Learning samples	1	M-138	7	4	15.68	0.82	1.6	6.68	2
	2	M-138	5	7	14.3	0.379	3.07	16.3	3
	3	M-133	4	4.4	14.53	0.73	4.03	10.88	2
	4	M-133	4	5	16.8	1.68	2.55	7.36	2
	5	M-134	10	5.4	14.65	6	2.47	5.6	1
	6	M-134	11	2.4	15.5	10	2.8	6.2	1
	7	M-134	10	1.74	12.5	1.84	6.98	15.37	3
	8	M-134	10	2.4	13.25	1.6	1.9	13.52	3
	9	M-134-1	8	3	14.3	1.35	2.4	9.55	2
	10	M-134-1	7	1.8	13.47	0.89	2.87	16.25	3
	11	M-4	9	4.8	13.26	6.96	3.9	5	1
	12	M-4	9	1.8	13.47	4.33	5.83	15.97	3
	13	M-112	12	1.8	17	0.89	1.3	11.78	3
	14	M-112	3	3.5	9.26	0.2	2.76	18.59	3
	15	M-112	4	7	8.2	0.11	5.5	29.01	4
	16	M-112	2	2.5	6.35	0.02	8.65	25.92	3
	17	M-112	3	5	16.8	4.07	0.8	10.2	1
	18	M-135	7	3.6	11.02	0.5	1.16	14.94	3
	19	M-135	4	4.8	9.38	0.16	4.25	17	3
	20	M-135	10	5.4	8.73	0.08	2.3	13.42	3
Prediction samples	21	M-134-1	10	3	14.5	2.45	13.6	1.7	(2)
	22	M-4	10	6.1	16.88	6.26	2.05	4.5	(1)
	23	M-22-37-39	6	1.8	8.6	0.28	6.9	32.75	(4)

[a] $y*$ = reservoir classification (1–excellent, 2–good, 3–average, 4–poor) determined by the well test. In $y*$, numbers in parentheses are not input data but are used for calculating R(%).

Thus, it is believed that an explicit nonlinear function corresponding to the BAC formula (6.1) is obtained:

$$y = BAC(x_1, x_2, x_3, x_4, x_5) \tag{6.46}$$

From the discriminate process of BAYSD, reservoir type (y) is shown to depend on the five parameters in decreasing order: mud content (x_5), permeability (x_3), sandstone thickness (x_1), carbonate content (x_4), and porosity (x_2).

Substituting x_j ($j = 1, 2, 3, 4, 5$) of each learning sample in Equation (6.45), four values of discriminate function are obtained, and the l corresponding to the maximum of the four values is the solved, i.e., l is the value of y for a given learning sample. Comparing the y values with the corresponding practical values y^*, $\overline{R}_1(\%)$ is obtained to express the fitness of the learning process. For instance, for the first learning sample (Table 6.4), $l = 2$, thus $y = 2$; and $y^* = 2$, so $\overline{R}_1(\%) = 0$.

Table 6.5 shows the results of the learning process by BAYSD.

6.2.4. Prediction Process

Substituting x_j ($j = 1, 2, 3, 4, 5$) of each prediction sample in Equation (6.45), four values of the discriminate function are obtained, and the l corresponding to the maximum of the four values is the solution, i.e., l is the value of y for a given prediction sample. Comparing the y values with the corresponding practical values y^*, $\overline{R}_2(\%)$ is obtained to express the accuracy of the prediction process.

Table 6.5 shows the results of the prediction process by BAYSD.

6.2.5. Application Comparisons Among BAYSD, C-SVM, R-SVM, BPNN, and MRA

Similarly, using the 20 learning samples (Table 6.4) and by C-SVM, R-SVM, BPNN, and MRA, the four functions of reservoir classification (y) with respect to five parameters (x_1, x_2, x_3, x_4, x_5) have been constructed, i.e., Equations (6.47) and (6.48) corresponding to SVM formula (4.1), Equation (6.49) corresponding to BPNN formula (3.1), and Equation (6.50) corresponding to MRA formula (2.14), respectively. Also similar to the preceding, using the three prediction samples (Table 6.4), C-SVM, R-SVM, BPNN, and MRA are employed for predictions, respectively.

Using C-SVM, the result is an explicit nonlinear function:

$$y = C\text{-}SVM(x_1, x_2, x_3, x_4, x_5) \tag{6.47}$$

with $C = 32$, $\gamma = 0.5$, 16 free vectors x_i, and CVA = 90%.

Using R-SVM, the result is an explicit nonlinear function:

$$y = R\text{-}SVM(x_1, x_2, x_3, x_4, x_5) \tag{6.48}$$

with $C = 1$, $\gamma = 0.2$, and 18 free vectors x_i.

TABLE 6.5   Prediction Results from Reservoir Classification in the Fuxin Uplift

Sample Type	Sample No.	y^* Well Test	BAYSD		C-SVM		R-SVM		BPNN		MRA	
			y	$R(\%)$	y	$R(\%)$	y	$R(\%)$	y	$R(\%)$	y	$R(\%)$
Learning samples	1	2	2	0	2	0	2	0	2	0	2	0
	2	3	3	0	3	0	3	0	3	0	3	0
	3	2	2	0	2	0	2	0	2	0	2	0
	4	2	2	0	2	0	2	0	2	0	2	0
	5	1	1	0	1	0	1	0	1	0	1	0
	6	1	1	0	1	0	1	0	1	0	1	0
	7	3	3	0	3	0	3	0	3	0	3	0
	8	3	3	0	3	0	3	0	3	0	3	0
	9	2	2	0	2	0	2	0	2	0	2	0
	10	3	3	0	3	0	3	0	3	0	3	0
	11	1	1	0	1	0	1	0	1	0	1	0
	12	3	3	0	3	0	3	0	3	0	2	33.33
	13	3	3	0	3	0	3	0	3	0	3	0
	14	3	3	0	3	0	3	0	3	0	3	0
	15	4	4	0	4	0	4	0	4	0	4	0
	16	3	3	0	3	0	3	0	3	0	4	33.33
	17	1	1	0	1	0	2	100	1	0	2	100
	18	3	3	0	3	0	3	0	3	0	3	0
	19	3	3	0	3	0	3	0	3	0	3	0
	20	3	3	0	3	0	3	0	3	0	3	0
Prediction samples	21	2	2	0	3	50	3	50	2	0	1	50
	22	1	1	0	1	0	1	0	1	0	1	0
	23	4	4	0	3	25	3	25	2	50	5	25

[a]y^* = reservoir classification (1—excellent, 2—good, 3—average, 4—poor) determined by the well test.

The BPNN used consists of five input layer nodes, one output layer node, and 11 hidden layer nodes. Setting $t_{max} = 50000$, the calculated $t_{opt} = 25598$. The result is an implicit nonlinear function:

$$y = BPNN(x_1, x_2, x_3, x_4, x_5) \tag{6.49}$$

with $RMSE(\%) = 0.5100 \times 10^{-2}$.

Using MRA, the result is an explicit linear function:

$$y = 1.8950 - 0.090332x_1 + 0.0045923x_2 - 0.12934x_3 - 0.043057x_4 + 0.093862x_5 \quad (6.50)$$

Equation (6.50) yields a residual variance of 0.17037 and a multiple correlation coefficient of 0.91084. From the regression process, reservoir type (y) is shown to depend on the five parameters in decreasing order: mud content (x_5), permeability (x_3), sandstone thickness (x_1), carbonate content (x_4), and porosity (x_2). This order is consistent with the aforementioned one by BAYSD due to the fact that though MRA is a linear algorithm whereas BAYSD is a nonlinear algorithm, the nonlinearity of the studied problem is strong whereas the nonlinearity ability of BAYSD is weak.

Substituting the values of x_1, x_2, x_3, x_4, x_5 given by the three prediction samples (Table 6.4) in Equations (6.47), (6.48), (6.49), and (6.50) respectively, the reservoir type (y) of each prediction sample is obtained.

Table 6.5 shows the results of learning and prediction processes by C-SVM, R-SVM, BPNN, and MRA.

$\overline{R}^*(\%) = 16.67$ of MRA (Table 6.6) demonstrates that the nonlinearity of the relationship between the predicted value y and its relative parameters (x_1, x_2, x_3, x_4, x_5) is strong from Table 1.2. Only BAYSD is applicable, i.e., not only $\overline{R}_1(\%) = 0$, but also $\overline{R}_2(\%) = 0$, and thus $\overline{R}^*(\%) = 0$, which coincide with practicality; for C-SVM and BPNN, though $\overline{R}_1(\%) = 0$, $\overline{R}_2(\%) = 25$ and 16.67, respectively, indicating the number of learning samples is not big enough; as for R-SVM and MRA, the solution accuracy is low, and $\overline{R}_2(\%) > \overline{R}_1(\%)$.

TABLE 6.6   Comparison Among the Applications of BAYSD, C-SVM, R-SVM, BPNN, and MRA to Reservoir Classification in the Fuxin Uplift

Algorithm	Fitting Formula	Mean Absolute Relative Residual			Dependence of the Predicted Value (y) on Parameters (x_1, x_2, x_3, x_4, x_5), in Decreasing Order	Time Consumed on PC (Intel Core 2)	Solution Accuracy
		$\overline{R}1(\%)$	$\overline{R}2(\%)$	$\overline{R}^*(\%)$			
BAYSD	Nonlinear, explicit	0	0	0	x_5, x_3, x_1, x_4, x_2	3 s	Very high
C-SVM	Nonlinear, explicit	0	25	12.50	N/A	3 s	Low
R-SVM	Nonlinear, explicit	5	25	15	N/A	3 s	Low
BPNN	Nonlinear, implicit	0	16.67	8.34	N/A	25 s	Moderate
MRA	Linear, explicit	8.33	25	16.67	x_5, x_3, x_1, x_4, x_2	<1 s	Low

6.2.6. Summary and Conclusions

In summary, using data for the reservoir classification in the Fuxin Uplift based on five parameters, i.e., the five parameters (sandstone thickness, porosity, permeability, carbonate content, mud content) and a well-test result of 20 samples, a set of discriminate functions (6.45) is constructed by the data mining tool of BAYSD, and the results are consistent with a well test. This set of discriminate functions is called *mined knowledge*. In the prediction process, applying this knowledge, i.e., using the five parameters of three other samples for reservoir classification, respectively, the results are also consistent with a well test. Therefore, this method can be spread to the reservoir classification of other oilfields.

Moreover, it is found that (a) since this case study is a strong nonlinear problem, neither R-SVM nor MRA is applicable; and (b) since the number of learning samples is not big enough, neither C-SVM nor BPNN is applicable, but it is preferable to adopt BAYSD, which is an advantage of BAYSD.

6.3. CASE STUDY 2: RESERVOIR CLASSIFICATION IN THE BAIBAO OILFIELD

6.3.1. Studied Problem

The objective of this case study is to conduct a reservoir classification using mud content, porosity, permeability, and permeability variation coefficient, which has practical value when well-test data are less limited.

We used data 29 learning samples from the Chang 3 and Chang 4+5 of the Triassic Formation of the Baibao Oilfield in the Ordos Basin in western China. Each sample contains four parameters (x_1 = mud content, x_2 = porosity, x_3 = permeability, x_4 = permeability variation coefficient) and a well-test result (y^* = reservoir type); Zhang et al. (2008) adopted BAYD for the reservoir classification. In this case study, among these 29 samples, 25 are taken as learning samples and 3 as prediction samples (Table 6.7), and y^* are recalculated by Q-mode cluster analysis (see Section 7.5).

6.3.2. Known Input Parameters

Input parameters are the values of the known variables x_i (i = 1, 2, 3, 4) for 25 learning samples and three prediction samples and the value of the prediction variable y^* for 25 learning samples (Table 6.7).

6.3.3. Learning Process

Using the parameters of learning samples and by BAYSD, a discrimination function corresponding to Equation (6.30) has been constructed:

$$\left.\begin{array}{l} B_1(x) = \ln(0.16) - 55.524 + 0.422x_1 + 4.242x_2 + 0.859x_3 + 3.797x_4 \\ B_2(x) = \ln(0.52) - 32.802 + 0.769x_1 + 3.385x_2 + 0.303x_3 + 4.816x_4 \\ B_3(x) = \ln(0.24) - 27.395 + 1.171x_1 + 2.654x_2 - 0.445x_3 + 7.176x_4 \\ B_4(x) = \ln(0.08) - 20.612 + 0.538x_1 - 0.385x_2 + 0.350x_3 + 15.678x_4 \end{array}\right\} \quad (6.51)$$

TABLE 6.7   Input Data for Reservoir Classification in the Baibao Oilfield

Sample Type	Sample No.	Well No.	Layer No.	x_1 Mud Content (%)	x_2 Porosity (%)	x_3 Permeability (mD)	x_4 Permeability Variation Coefficient	y^* Reservoir Type[a]
Learning samples	1	B102	C3^2	27.2	9.074	0.068	2.69	3
	2	B102	C3^3	10.9	21.66	36.6	0.51	1
	3	B102	C4+5^1	13.74	13.64	15.9	1.01	2
	4	B107	C3^2	18.4	0.18	0.025	2.82	4
	5	B108	C3^3	8.48	19.27	28.62	0.57	1
	6	B108	C4+5^1	13.98	12.44	15.72	0.83	2
	7	B110	C4+5^1	10.58	12.03	14.01	1.07	2
	8	B112	C4+5^1	11.91	9.16	0.77	1.93	3
	9	B115	C4+5^1	15.0	16.7	31.56	0.65	2
	10	B123	C3^3	12.5	15.13	22.12	1.27	2
	11	B130	C3^2	14.3	11.9	15.39	1.21	2
	12	B202	C4+5^1	12.2	1.5	1.28	1.4	4
	13	B205	C3^1	13.4	5.68	0.81	1.35	3
	14	B205	C3^3	20.26	9.45	1.89	1.83	3
	15	B205	C4+5^1	12.78	12.34	12.71	1.14	2
	16	B205	C4+5^2	12.8	14.91	33.42	0.44	1
	17	B206	C3^3	13.26	6.35	0.98	1.58	3
	18	B206	C4+5^1	14.2	13.84	21.34	0.92	2
	19	B206	C4+5^2	11.3	15.8	34.37	0.52	1
	20	B210	C3^1	21.89	8.35	0.58	1.37	3
	21	B210	C4+5^1	14.32	11.57	18.72	1.01	2
	22	B215	C3^3	13.6	12.96	15.61	0.98	2
	23	B215	C4+5^1	14.03	13.78	28.23	1.0	2
	24	B217	C3^3	17.98	11.53	20.73	0.84	2
	25	B217	C4+5^1	13.89	12.24	14.53	1.23	2
Prediction samples	26	B219	C3^3	14.74	12.32	23.79	0.84	(2)
	27	H186	C4+5^2	18.65	5.13	0.48	1.15	(3)
	28	W4	C4+5^2	19.9	1.09	0.003	3.84	(4)

[a]y^* = reservoir classification (1—excellent, 2—good, 3—average, 4—poor) determined by the well test. In y^*, numbers in parentheses are not input data but are used for calculating R(%).

Thus, it is believed that an explicit nonlinear function corresponding to BAC formula (6.1) is obtained:

$$y = BAC(x_1, x_2, x_3, x_4) \tag{6.52}$$

From the discriminate process of BAYSD, reservoir type (y) is shown to depend on the four parameters in decreasing order: permeability (x_3), porosity (x_2), permeability variation coefficient (x_4), and mud content (x_1).

Substituting x_j ($j = 1, 2, 3, 4$) of each learning sample in Equation (6.51), four values of the discriminate function are obtained, and the l corresponding to the maximum of the four values is the solution, i.e., l is the value of y for a given learning sample. Comparing the y values with the corresponding practical values y^*, $\overline{R}_1(\%)$ is obtained to express the fitness of learning process. For instance, for the first learning sample (Table 6.7), $l = 3$, thus $y = 3$; and $y^* = 3$, so $\overline{R}_1(\%) = 0$.

Table 6.8 shows the results of learning process by BAYSD.

6.3.4. Prediction Process

Substituting x_j ($j = 1, 2, 3, 4$) of each prediction sample in Equation (6.51), four values of discriminate function are obtained, and the l corresponding to the maximum of the four values is the solution, i.e., l is the value of y for a given prediction sample. Comparing the y values with the corresponding practical values y^*, $\overline{R}_2(\%)$ is obtained to express the accuracy of the prediction process.

Table 6.8 shows the results of prediction process by BAYSD.

6.3.5. Application Comparisons Among BAYSD, C-SVM, R-SVM, BPNN, and MRA

Similarly, using the 25 learning samples (Table 6.7) and by C-SVM, R-SVM, BPNN, and MRA, the four functions of reservoir classification (y) with respect to four parameters (x_1, x_2, x_3, x_4) have been constructed, i.e., Equations (6.53) and (6.54) corresponding to SVM formula (4.1), Equation (6.55) corresponding to BPNN formula (3.1), and Equation (6.56) corresponding to MRA formula (2.14), respectively.

Using C-SVM, the result is an explicit nonlinear function:

$$y = C\text{-}SVM(x_1, x_2, x_3, x_4) \tag{6.53}$$

with $C = 2048$, $\gamma = 0.03125$, nine free vectors x_i, and CVA $= 92\%$.

Using R-SVM, the result is an explicit nonlinear function:

$$y = R\text{-}SVM(x_1, x_2, x_3, x_4) \tag{6.54}$$

with $C = 1$, $\gamma = 0.25$, and 14 free vectors x_i.

The BPNN used consists of four input layer nodes, one output layer node, and nine hidden layer nodes. Setting $t_{max} = 100000$, the calculated $t_{opt} = 70307$. The result is an implicit nonlinear function:

$$y = BPNN(x_1, x_2, x_3, x_4) \tag{6.55}$$

with $RMSE(\%) = 0.5212 \times 10^{-2}$.

TABLE 6.8 Prediction Results from Reservoir Classification in the Baibao Oilfield

			Reservoir Classification									
			BAYSD		C-SVM		R-SVM		BPNN		MRA	
Sample Type	Sample No.	y* Well Test	y	R(%)	y	R(%)	y	R(%)	y	R(%)	y	R(%)
Learning samples	1	3	3	0	3	0	3	0	3	0	3	0
	2	1	1	0	1	0	1	0	1	0	1	0
	3	2	2	0	2	0	2	0	2	0	2	0
	4	4	4	0	4	0	4	0	4	0	4	0
	5	1	1	0	1	0	1	0	1	0	1	0
	6	2	2	0	2	0	2	0	2	0	2	0
	7	2	2	0	2	0	2	0	2	0	2	0
	8	3	3	0	3	0	3	0	3	0	3	0
	9	2	1	50	2	0	1	50	2	0	1	50
	10	2	2	0	2	0	2	0	2	0	2	0
	11	2	2	0	2	0	2	0	2	0	2	0
	12	4	4	0	4	0	3	25	4	0	4	0
	13	3	3	0	3	0	3	0	3	0	3	0
	14	3	3	0	3	0	3	0	3	0	3	0
	15	2	2	0	2	0	2	0	2	0	2	0
	16	1	1	0	1	0	1	0	1	0	1	0
	17	3	3	0	3	0	3	0	3	0	3	0
	18	2	2	0	2	0	2	0	2	0	2	0
	19	1	1	0	1	0	1	0	1	0	1	0
	20	3	3	0	3	0	3	0	3	0	3	0
	21	2	2	0	2	0	2	0	2	0	2	0
	22	2	2	0	2	0	2	0	2	0	2	0
	23	2	2	0	2	0	2	0	2	0	2	0
	24	2	2	0	2	0	2	0	2	0	2	0
	25	2	2	0	2	0	2	0	2	0	2	0
Prediction samples	26	2	2	0	2	0	2	0	2	0	2	0
	27	3	3	0	3	0	3	0	4	33.33	3	0
	28	4	4	0	4	0	3	25	4	0	4	0

[a] y* = reservoir classification (1—excellent, 2—good, 3—average, 4—poor) determined by the well test.

TABLE 6.9   Comparison Among the Applications of BAYSD, C-SVM, R-SVM, BPNN, and MRA to Reservoir Classification in the Baibao Oilfield

Algorithm	Fitting Formula	Mean Absolute Relative Residual			Dependence of the Predicted value (y) on Parameters (x_1, x_2, x_3, x_4), in Decreasing Order	Time Consumed on PC (Intel Core 2)	Solution Accuracy
		$\overline{R1}(\%)$	$\overline{R2}(\%)$	$\overline{R}^*(\%)$			
BAYSD	Nonlinear, explicit	2	0	1	x_3, x_2, x_4, x_1	3 s	Very high
C-SVM	Nonlinear, explicit	0	0	0	N/A	3 s	Very high
R-SVM	Nonlinear, explicit	3	8.33	5.67	N/A	3 s	Moderate
BPNN	Nonlinear, implicit	0	11.11	5.56	N/A	30 s	Moderate
MRA	Linear, explicit	2	0	1	x_2, x_4, x_3, x_1	<1 s	Very high

Using MRA, the result is an explicit linear function:

$$y = 3.2701 + 0.00535x_1 - 0.10083x_2 - 0.015233x_3 + 0.2511x_4 \tag{6.56}$$

Equation (6.56) yields a residual variance of 0.087777 and a multiple correlation coefficient of 0.9551. From the regression process, reservoir type (y) is shown to depend on the four parameters in decreasing order: porosity (x_2), permeability variation coefficient (x_4), permeability (x_3), and mud content (x_1). This order differs from the aforementioned one by BAYSD in the former three parameters because MRA is a linear algorithm whereas BAYSD is a nonlinear algorithm.

Table 6.8 shows the results of the learning and prediction processes by C-SVM, R-SVM, BPNN, and MRA.

$\overline{R}^*(\%) = 1$ of MRA (Table 6.9) shows that the nonlinearity of the relationship between the predicted value y and its relative parameters (x_1, x_2, x_3, x_4) is very weak from Table 1.2. Only C-SVM is applicable, i.e., not only $\overline{R}_1(\%) = 0$, but also $\overline{R}_2(\%) = 0$, and thus $\overline{R}^*(\%) = 0$, which coincide with practicality; for BAYSD and MRA, though $\overline{R}_2(\%) = 0$, $\overline{R}_1(\%) = 2$, which cannot ensure correct prediction of other new prediction samples. As for R-SVM and BPNN, though the solution accuracy is moderate, $\overline{R}_2(\%) >> \overline{R}_1(\%)$.

6.3.6. Summary and Conclusions

In summary, using data for the reservoir classification in the Baibao Oilfield based on four parameters, i.e., the four parameters (mud content, porosity, permeability, permeability variation coefficient) and a well-test result of 25 samples, a set of discriminate functions (6.51) is constructed by the data mining tool of BAYSD, and the results are consistent with a well test, except for one sample. This set of discriminate functions is called *mined knowledge*. In the prediction process, applying this knowledge, i.e., using the four parameters of three other samples for reservoir classification, respectively, the results are consistent with the well test. Therefore, this method can be spread to the reservoir classification of other oilfields.

Moreover, it is found that (a) since this case study is a very weak nonlinear problem, BAYSD, C-SVM, and MRA are applicable; and (b) since the number of learning samples is not big enough, neither R-SVM nor BPNN is applicable.

6.4. CASE STUDY 3: OIL LAYER CLASSIFICATION BASED ON WELL-LOGGING INTERPRETATION

6.4.1. Studied Problem

In Section 4.3, the data in Table 4.4 was used, and MRA, BPNN, C-SVM, and R-SVM were adopted to construct the function of oil layer classification (y) with respect to five relative parameters (x_1, x_2, x_3, x_4, x_5) in the Xiefengqiao Anticlinal, and the calculation results were filled into Tables 4.5 and 4.6. Now Table 4.4 is used as input data for Case Study 3 of this chapter. For the sake of algorithm comparisons, Table 4.5 is accordingly copied into Table 6.10, and Table 4.6 is accordingly copied into Table 6.11.

6.4.2. Known Input Parameters

Input parameters are the values of the known variables x_i ($i = 1, 2, 3, 4, 5$) for 24 learning samples and three prediction samples and the value of the prediction variable y^* for 24 learning samples (Table 4.4).

6.4.3. Learning Process

Using the parameters of learning samples and by BAYSD, a discrimination function corresponding to Equation (6.30) has been constructed:

$$\left.\begin{array}{l} B_1(x) = \ln(0.333) - 93842.2 - 3.892x_1 + 1039.76x_2 \\ \qquad -4468.595x_3 - 0.712x_4 - 73.876x_5 \\ B_2(x) = \ln(0.333) - 93203.96 - 3.807x_1 + 1036.232x_2 \\ \qquad -4454.111x_3 - 0.701x_4 - 73.528x_5 \\ B_3(x) = \ln(0.333) - 92927.46 - 3.674x_1 + 1034.66x_2 \\ \qquad -4447.812x_3 - 0.729x_4 - 73.279x_5 \end{array}\right\} \qquad (6.57)$$

TABLE 6.10 Prediction Results from Oil Layer Classification of the Xiefengqiao Anticline

Sample Type	Sample No.	Well No.	Layer No.	y^* Oil Test[a]	BAYSD		C-SVM		R-SVM		BPNN		MRA	
					y	$R(\%)$	y	$R(\%)$	y	$R(\%)$	y	$R(\%)$	y	$R(\%)$
Learning samples	1	ES4	5	2	2	0	2	0	2	0	3	50	2	0
	2		6	3	3	0	3	0	3	0	3	0	3	0
	3		7	2	2	0	2	0	2	0	3	50	2	0
	4		8	3	3	0	3	0	3	0	3	0	3	0
	5		9	1	1	0	1	0	1	0	1	0	1	0
	6		10	2	2	0	2	0	2	0	2	0	2	0
	7		11	2	2	0	2	0	2	0	2	0	2	0
	8		12	3	3	0	3	0	3	0	3	0	3	0
	9		13	1	1	0	1	0	1	0	1	0	1	0
	10	ES6	4	2	2	0	2	0	2	0	2	0	2	0
	11		5	3	3	0	3	0	2	33.33	3	0	2	33.33
	12		6	3	2	33.33	3	0	2	33.33	3	0	2	33.33
	13		8	3	3	0	3	0	3	0	3	0	4	33.33
	14	ES8	5_1	2	2	0	2	0	2	0	2	0	2	0
	15		5_2	3	2	33.33	3	0	2	33.33	3	0	3	0
	16		5_3	2	2	0	2	0	2	0	2	0	2	0
	17		6	3	3	0	3	0	3	0	3	0	3	0
	18		11	2	2	0	2	0	2	0	2	0	2	0
	19		12_1	1	1	0	1	0	1	0	1	0	1	0
	20		12_2	1	1	0	1	0	1	0	1	0	1	0
	21		12_3	1	1	0	1	0	1	0	1	0	1	0
	22		12_4	1	2	100	1	0	1	0	1	0	1	0
	23		13_1	1	1	0	1	0	1	0	1	0	1	0
	24		13_2	1	2	100	1	0	2	100	1	0	2	100
Prediction samples	25	ES8	13_4	1	2	100	1	0	1	0	1	0	2	100
	26		13_5	2	2	0	2	0	2	0	3	50	2	0
	27		7_2	3	3	0	3	0	3	0	3	0	3	0

[a]y^* = oil layer classification (1—oil layer, 2—poor oil layer, 3—dry layer) determined by the oil test.

TABLE 6.11   Comparison Among the Applications of BAYSD, C-SVM, R-SVM, BPNN, and MRA to Oil Layer Classification of the Xiefengqiao Anticline

| Algorithm | Fitting Formula | Mean Absolute Relative Residual | | | Dependence of the Predicted Value (y) on Parameters (x_1, x_2, x_3, x_4, x_5), in Decreasing Order | Time Consumed on PC (Intel Core 2) | Solution Accuracy |
		$\overline{R}1(\%)$	$\overline{R}2(\%)$	$\overline{R}^*(\%)$			
BAYSD	Nonlinear, explicit	11.11	33.33	22.22	x_1, x_2, x_3, x_5, x_4	3 s	Very low
C-SVM	Nonlinear, explicit	0	0	0	N/A	3 s	Very high
R-SVM	Nonlinear, explicit	8.33	0	4.17	N/A	3 s	High
BPNN	Nonlinear, implicit	4.17	16.67	10.42	N/A	30 s	Low
MRA	Linear, explicit	8.33	33.33	20.83	x_1, x_2, x_3, x_5, x_4	<1 s	Very low

Thus, it is believed that an explicit nonlinear function corresponding to BAC formula (6.1) is obtained:

$$y = BAC(x_1, x_2, x_3, x_4, x_5) \tag{6.58}$$

From the discriminate process of BAYSD, reservoir type (y) is shown to depend on the five parameters in decreasing order: true resistivity (x_1), acoustictime (x_2), porosity (x_3), permeability (x_5), and oil saturation (x_4). This order is consistent with that by MRA (Table 6.11) due to the fact that though MRA is a linear algorithm whereas BAYSD is a nonlinear algorithm, the nonlinearity of the studied problem is strong (Table 6.11) whereas the nonlinearity ability of BAYSD is weak.

Substituting x_j ($j = 1, 2, 3, 4, 5$) of each learning sample in Equation (6.57), three values of discriminate function are obtained, and the l corresponding to the maximum of the three values is the solution, i.e., l is the value of y for a given learning sample. Comparing the y values with the corresponding practical values y^*, $\overline{R}_1(\%)$ is obtained to express the fitness of learning process. For instance, for the first learning sample (Table 4.4), $l = 2$, thus $y = 2$; and $y^* = 2$, so $\overline{R}_1(\%) = 0$.

Table 6.10 shows the results of the learning process by BAYSD.

6.4.4. Prediction Process

Substituting x_j ($j = 1, 2, 3, 4, 5$) of each prediction sample in Equation (6.57), three values of discriminate functions are obtained, and the l corresponding to the maximum of the three values is the solution, i.e., l is the value of y for a given prediction sample. Comparing the y values with the corresponding practical values y^*, $\overline{R}_2(\%)$ is obtained to express the accuracy of prediction process.

Table 6.10 shows the results of the prediction process by BAYSD.

6.4.5. Application Comparisons Among BAYSD, C-SVM, R-SVM, BPNN, and MRA

$\overline{R}^*(\%) = 20.83$ of MRA (Table 6.11) shows that the nonlinearity of the relationship between the predicted value y and its relative parameters (x_1, x_2, x_3) is strong from Table 1.2. Only C-SVM is applicable, i.e., not only $\overline{R}_1(\%) = 0$, but also $\overline{R}_2(\%) = 0$, and thus $\overline{R}^*(\%) = 0$, which coincide with practicality. For R-SVM, though $\overline{R}_2(\%) = 0$, $\overline{R}_1(\%) = 8.33$, which cannot ensure correct prediction of other new prediction samples. As for BAYSD, BPNN, and MRA, the solution accuracy is very low, low, and very low, respectively, and $\overline{R}_2(\%) >> \overline{R}_1(\%)$.

6.4.6. Summary and Conclusions

In summary, using data for the oil layer classification in the Xiefengqiao Anticline based on five parameters, i.e., the five parameters (true resistivity, acoustictime, porosity, oil saturation, permeability) and an oil test result of 24 oil layers, a set of discriminate functions (6.57) is constructed by the data mining tool of BAYSD, but the results are not eligible. This set of discriminate functions is called *mined unqualified knowledge*. In the prediction process, applying this unqualified knowledge, i.e., using the five parameters of three other samples for oil layer classification, respectively, the results are not eligible either. Therefore, this method cannot be spread to the oil layer classification of other oilfields.

Moreover, it is found that since this case study is a strong nonlinear problem, only C-SVM is applicable, but BAYSD, BPNN, MRA, or R-SVM is not applicable.

6.5. CASE STUDY 4: INTEGRATED EVALUATION OF OIL AND GAS TRAP QUALITY

6.5.1. Studied Problem

In Section 3.2, the data in Table 3.6 was used, and BPNN was adopted to construct the function of trap quality with respect to 14 relative parameters in the Northern Kuqa Depression. Now Table 3.6 is used as a case study of BAYSD.

6.5.2. Known Input Parameters

They are the values of the known variables x_i ($i = 1, 2, ..., 14$) for 27 learning samples and two prediction samples and the value of the prediction variable y^* for 27 learning samples (Table 3.6).

6.5.3. Learning Process

Using the parameters of learning samples and by BAYSD, a discrimination function corresponding to Equation (6.30) has been constructed:

$$
\left.\begin{aligned}
B_1(x) &= \ln(0.37) - 6090.107 - 37.076x_1 + 31.65x_2 - 11.927x_3 \\
&\quad + 0.042x_4 + 0.194x_5 + 1.062x_6 + 8.483x_7 - 15.526x_8 + 1305.707x_9 \\
&\quad + 1175.127x_{10} + 1799.427x_{11} + 1870.111x_{12} + 7090.26x_{13} - 1.129x_{14} \\
B_2(x) &= \ln(0.63) - 5746.595 - 37.817x_1 + 30.562x_2 - 11.974x_3 \\
&\quad + 0.038x_4 + 0.184x_5 + 0.993x_6 + 9.75x_7 + 0.806x_8 + 1212.488x_9 \\
&\quad + 1115.348x_{10} + 1705.045x_{11} + 1752.377x_{12} + 7077.386x_{13} - 1.099x_{14}
\end{aligned}\right\}
\tag{6.59}
$$

Thus, it is believed that an explicit nonlinear function corresponding to BAC formula (6.1) is obtained:

$$
y = BAC(x_1, x_2, ..., x_{14}) \tag{6.60}
$$

From the discriminate process of BAYSD, trap quality (y) is shown to depend on the 14 parameters in decreasing order: preservation coefficient (x_{12}), source rock coefficient (x_{10}), trap coefficient (x_9), trap depth (x_4), trap relief (x_5), reservoir coefficient (x_{11}), data reliability (x_8), trap closed area (x_6), resource quantity (x_{14}), unit structure (x_1), trap type (x_2), formation HC identifier (x_7), configuration coefficient (x_{13}), and petroliferous formation (x_3).

Substituting x_j ($j = 1, 2, ..., 14$) of each learning sample in Equation (6.59), two values of a discriminate function are obtained, and the l corresponding to the maximum of the two values is the solution, i.e., l is the value of y for a given learning sample. Comparing the y values with the corresponding practical values y^*, $\overline{R}_1(\%)$ is obtained to express the fitness of the learning process. For instance, for the first learning sample (Table 3.6), $l = 2$, thus $y = 2$; and $y^* = 2$, so $\overline{R}_1(\%) = 0$.

Table 6.12 shows the results of learning process by BAYSD.

6.5.4. Prediction Process

Substituting x_j ($j = 1, 2, ..., 14$) of each prediction sample in Equation (6.59), two values of a discriminate function are obtained, and the l corresponding to the maximum of the two values is the solution, i.e., l is the value of y for a given prediction sample. Comparing the y values with the corresponding practical values y^*, $\overline{R}_2(\%)$ is obtained to express the accuracy of the prediction process.

Table 6.12 shows the results of prediction process by BAYSD.

6.5.5. Application Comparisons Among BAYSD, C-SVM, R-SVM, BPNN, and MRA

Similarly, using the 27 learning samples (Table 3.6) and by C-SVM, R-SVM, BPNN, and MRA, the four functions of trap quality (y) with respect to 14 parameters (x_1, x_2, ..., x_{14}) have been constructed, i.e., Equations (6.61) and (6.62) corresponding to SVM formula (4.1), Equation (6.63) corresponding to BPNN formula (3.1), and Equation (6.64) corresponding to MRA formula (2.14), respectively. Also similar to the preceding, using the two

TABLE 6.12　Prediction Results from Trap Evaluation of the Northern Kuqa Depression

			Trap Quality[a]								
			BAYSD		C-SVM		R-SVM		BPNN		MRA
Sample Type	Trap No.	y^*	y	$R(\%)$	y	$R(\%)$	y	$R(\%)$	y	$R(\%)$	y	$R(\%)$
Learning samples	1	2	2	0	2	0	2	0	2	0	2	0
	2	1	1	0	1	0	1	0	1	0	1	0
	3	1	1	0	1	0	1	0	1	0	1	0
	4	2	2	0	2	0	2	0	2	0	2	0
	5	1	1	0	1	0	1	0	1	0	1	0
	6	2	2	0	2	0	2	0	2	0	2	0
	7	2	2	0	2	0	2	0	2	0	2	0
	8	2	2	0	2	0	2	0	2	0	2	0
	9	2	2	0	2	0	2	0	2	0	2	0
	10	2	2	0	2	0	2	0	2	0	2	0
	11	2	2	0	2	0	2	0	2	0	2	0
	12	1	1	0	1	0	1	0	1	0	1	0
	13	1	1	0	1	0	1	0	1	0	1	0
	14	1	1	0	1	0	1	0	1	0	1	0
	15	2	2	0	2	0	2	0	2	0	2	0
	16	1	1	0	1	0	1	0	1	0	1	0
	17	1	1	0	1	0	1	0	1	0	1	0
	18	2	2	0	2	0	2	0	2	0	2	0
	19	2	2	0	2	0	2	0	2	0	2	0
	20	2	2	0	2	0	2	0	2	0	2	0
	21	2	2	0	2	0	2	0	2	0	2	0
	22	2	2	0	2	0	2	0	2	0	2	0
	23	1	1	0	1	0	1	0	1	0	1	0
	24	1	1	0	1	0	1	0	1	0	1	0
	25	2	2	0	2	0	2	0	2	0	2	0
	26	2	2	0	2	0	2	0	2	0	2	0
	27	2	2	0	2	0	2	0	2	0	2	0
Prediction samples	29	2	2	0	2	0	2	0	2	0	2	0
	30	2	2	0	2	0	2	0	2	0	3	50

[a]y^* = trap quality (1—high, 2—low) assigned by geologists.

prediction samples (Table 3.6), C-SVM, R-SVM, BPNN, and MRA are employed for predictions, respectively.

Using C-SVM, the result is an explicit nonlinear function (Shi, 2011):

$$y = C\text{-}SVM(x_1, x_2, ..., x_{14}) \tag{6.61}$$

with $C = 32$, $\gamma = 0.03125$, 14 free vectors x_i, and CVA = 92.6%.

Using R-SVM, the result is an explicit nonlinear function:

$$y = R\text{-}SVM(x_1, x_2, ..., x_{14}) \tag{6.62}$$

with $C = 1$, $\gamma = 0.0714286$, and 19 free vectors x_i.

The BPNN consists of 14 input layer nodes, one output layer node, and 29 hidden layer nodes. Setting $t_{max} = 40000$, the calculated $t_{opt} = 10372$. The result is an implicit nonlinear function (Shi et al., 2004; Shi, 2011):

$$y = BPNN(x_1, x_2, ..., x_{14}) \tag{6.63}$$

with $RMSE(\%) = 0.1 \times 10^{-1}$.

Using MRA, the result is an explicit linear function (Shi et al., 2004; Shi, 2011):

$$\begin{aligned} y = &\ 13.766 - 0.026405x_1 - 0.038781x_2 - 0.0016605x_3 - 0.00012343x_4 \\ &- 0.00038344x_5 - 0.002442x_6 + 0.045162x_7 + 0.58229x_8 - 3.3236x_9 \\ &- 2.1313x_{10} - 3.3651x_{11} - 4.1977x_{12} - 0.67296x_{13} + 0.0010516x_{14} \end{aligned} \tag{6.64}$$

Equation (6.64) yields a residual variance of 0.14157 and a multiple correlation coefficient of 0.92652. From the regression process, trap quality (y) is shown to depend on the 14 parameters in decreasing order: preservation coefficient (x_{12}), source rock coefficient (x_{10}), trap coefficient (x_9), trap depth (x_4), trap relief (x_5), reservoir coefficient (x_{11}), data reliability (x_8), trap closed area (x_6), resource quantity (x_{14}), unit structure (x_1), trap type (x_2), formation HC identifier (x_7), configuration coefficient (x_{13}), and petroliferous formation (x_3). This order is consistent with the aforementioned one by BAYSD due to the fact that though MRA is a linear algorithm whereas BAYSD is a nonlinear algorithm, the nonlinearity of the studied problem is weak.

Substituting the values of x_1, x_2, ..., x_{14} given by the two prediction samples (Table 3.6) in Equations (6.61), (6.62), (6.63), and (6.64) respectively, the trap quality (y) of each prediction sample is obtained.

Table 6.12 shows the results of the learning and prediction processes by C-SVM, R-SVM, BPNN, and MRA.

$\overline{R}^*(\%) = 8.34$ of MRA (Table 6.13) demonstrates that the nonlinearity of the relationship between the predicted value y and its relative parameters (x_1, x_2, ..., x_{14}) is moderate from Table 1.2. The results of BAYSD, C-SVM, R-SVM, and BPNN are the same, i.e., not only $\overline{R}_1(\%) = 0$, but also $\overline{R}_2(\%) = 0$, and thus $\overline{R}^*(\%) = 0$, which coincide with practicality; as for MRA, the solution accuracy is low, though $\overline{R}_1(\%) = 0$, $\overline{R}_2(\%) = 25$.

6.5.6. Summary and Conclusions

In summary, using data for the trap evaluation of the Northern Kuqa Depression, i.e., the 14 parameters (unit structure, trap type, petroliferous formation, trap depth, trap

TABLE 6.13   Comparison Among the Applications of BAYSD, C-SVM, R-SVM, BPNN, and MRA
             to Trap Evaluation of the Northern Kuqa Depression

Algorithm	Fitting Formula	Mean Absolute Relative Residual			Dependence of the Predicted Value (y) on Parameters ($x_1, x_2, ..., x_{14}$), in Decreasing Order	Time Consumed on PC (Intel Core 2)	Solution Accuracy
		$\overline{R}1(\%)$	$\overline{R}2(\%)$	$\overline{R}^*(\%)$			
BAYSD	Nonlinear, explicit	0	0	0	$x_{12}, x_{10}, x_9, x_4, x_5,$ $x_{11}, x_8, x_6, x_{14}, x_1, x_2,$ x_7, x_{13}, x_3	3 s	Very high
C-SVM	Nonlinear, explicit	0	0	0	N/A	3 s	Very high
R-SVM	Nonlinear, explicit	0	0	0	N/A	3 s	Very high
BPNN	Nonlinear, implicit	0	0	0	N/A	1 min 20 s	Very high
MRA	Linear, explicit	0	25	12.5	$x_{12}, x_{10}, x_9, x_4,$ $x_5, x_{11}, x_8, x_6, x_{14}, x_1,$ x_2, x_7, x_{13}, x_3	<1 s	Low

relief, trap closed area, formation HC identifier, data reliability, trap coefficient, source rock coefficient, reservoir coefficient, preservation coefficient, configuration coefficient, resource quantity) and trap quality of 27 traps, a set of discriminate functions (6.59) is constructed by the data mining tool of BAYSD, and the results coincide with practicality. This set of discriminate functions is called *mined knowledge*. In the prediction process, applying this knowledge, i.e., using the 14 parameters of other two traps for trap evaluation, respectively, the results coincide with the real exploration discovery after this prediction. Therefore, this method can be spread to the trap evaluation of other exploration areas.

Moreover, it is found that (a) for this case study, the results of BAYSD, C-SVM, R-SVM, and BPNN are the same and correct; and (b) BAYSD, C-SVM, and R-SVM run much faster than BPNN; however, it is easy to code the BAYSD program whereas it is very complicated to code the C-SVM or R-SVM program, so BAYSD is a good software for this case study if neither the C-SVM nor R-SVM program exists.

6.6. CASE STUDY 5: COAL-GAS-OUTBURST CLASSIFICATION

6.6.1. Studied Problem

The objective of this case study is to conduct a classification of coal-gas-outburst (CGO) in coal mines using initial speed of methane diffusion, coefficient of coal consistence, gas pressure, destructive style of coal, and mining depth, which has practical value when coal-test data are less limited.

Data from 21 typical coal mines in China were taken as 21 samples, respectively, of which 16 were taken as learning samples and five as prediction samples. Each sample contains five parameters (x_1 = initial speed of methane diffusion, x_2 = coefficient of coal consistence, x_3 = gas pressure, x_4 = destructive style of coal, x_5 = mining depth) and a coal-test result (y^* = coal-gas-outburst type) listed in Table 6.14. The BAYD algorithm was applied to these data for coal-gas-outburst classification (Wang et al., 2010).

TABLE 6.14   Input Data for Coal-Gas-Outburst Classification in China

Sample Type	Sample No.	Relative Parameters for Coal-Gas-Outburst Classification[a]					Outburst Type[b]
		x_1	x_2	x_3	x_4	x_5	y^*
Learning samples	1	19.00	0.31	2.76	3	620	4
	2	6.00	0.24	0.95	5	445	2
	3	18.00	0.16	1.20	3	462	2
	4	5.00	0.61	1.17	1	395	1
	5	8.00	0.36	1.25	3	745	3
	6	7.00	0.48	2.00	1	460	1
	7	14.00	0.22	3.95	3	543	4
	8	11.00	0.28	2.39	3	515	2
	9	4.800	0.60	1.05	2	477	1
	10	6.00	0.24	0.95	3	455	3
	11	14.00	0.34	2.16	4	510	2
	12	4.00	0.58	1.40	3	428	1
	13	4.00	0.53	1.65	2	438	1
	14	6.00	0.54	3.95	5	543	4
	15	7.40	0.37	0.75	4	740	3
	16	3.00	0.51	1.40	3	400	1
Prediction samples	17	11.0	0.37	2.1	3	412	(2)
	18	11.5	0.28	1.9	3	407	(2)
	19	11.8	0.36	2.3	3	403	(2)
	20	10.8	0.30	2.2	3	396	(2)
	21	12.4	0.38	1.8	3	410	(2)

[a] x_1 = initial speed of methane diffusion (m/s); x_2 = coefficient of coal consistence; x_3 = gas pressure (MPa); x_4 = destructive style of coal (1—non-broken coal, 2—broken coal, 3—strongly broken coal, 4—pulverized coal, 5—completely pulverized coal); and x_5 = mining depth (m).
[b] y^* = coal-gas-outburst type (1—non, 2—small, 3—medium, 4—large) determined by the coal test. In y^*, numbers in parentheses are not input data but are used for calculating R(%).

6.6.2. Known Input Parameters

They are the values of the known variables x_i ($i = 1, 2, 3, 4, 5$) for 16 learning samples and five prediction samples and the value of the prediction variable y^* for 16 learning samples (Table 6.14).

6.6.3. Learning Process

Using the parameters of learning samples and by BAYD, a discrimination function corresponding to Equation (6.10) was constructed (Wang et al., 2010):

$$\left.\begin{array}{l} B_1(x) = -43.8693 + 1.4318x_1 + 113.0406x_2 + 6.4055x_3 + 5.2467x_4 - 0.0096x_5 \\ B_2(x) = -65.0341 + 3.0204x_1 + 3.3712x_2 + 11.1590x_3 + 15.6942x_4 + 0.0258x_5 \\ B_3(x) = -44.6246 + 1.4936x_1 - 3.2735x_2 + 5.7412x_3 + 9.9392x_4 + 0.0673x_5 \\ B_4(x) = -93.3579 + 3.5749x_1 + 1.6379x_2 + 16.6684x_3 + 17.3454x_4 + 0.0271x_5 \end{array}\right\} \quad (6.65)$$

Using the parameters of learning samples and by BAYSD, a discrimination function corresponding to Equation (6.30) has been constructed:

$$\left.\begin{array}{l} B_1(x) = \ln(0.375) - 41.102 + 2.388x_1 + 75.806x_2 + 7.787x_3 + 6.479x_4 + 0.012x_5 \\ B_2(x) = \ln(0.250) - 77.072 + 4.181x_1 + 13.656x_2 + 15.403x_3 + 17.300x_4 + 0.018x_5 \\ B_3(x) = \ln(0.188) - 56.466 + 1.952x_1 - 2.080x_2 + 8.002x_3 + 10.679x_4 + 0.087x_5 \\ B_4(x) = \ln(0.188) - 119.157 + 4.964x_1 + 23.494x_2 + 23.884x_3 + 19.672x_4 + 0.015x_5 \end{array}\right\} \quad (6.66)$$

Thus, it is believed that an explicit nonlinear function corresponding to BAC formula (6.1) is obtained:

$$y = BAC(x_1, x_2, x_3, x_4, x_5) \quad (6.67)$$

From the discriminate process of BAYSD, a coal-gas-outburst type (y) is shown to depend on the five parameters in decreasing order: gas pressure (x_3), coefficient of coal consistence (x_2), mining depth (x_5), destructive style of coal (x_4), and initial speed of methane diffusion (x_1).

Substituting x_j ($j = 1, 2, 3, 4, 5$) of each learning sample in Equation (6.65), four values of discriminate function are obtained, and the l corresponding to the maximum of the four values is the solution, i.e., l is the value of y for a given learning sample. Comparing the y values with the corresponding practical values y^*, $\overline{R}_1(\%)$ is obtained to express the fitness of the learning process. For instance, for the first learning sample (Table 6.14), $l = 4$, thus $y = 4$; and $y^* = 4$, so $\overline{R}_1(\%) = 0$. Substituting x_j ($j = 1, 2, 3, 4, 5$) of each learning sample in Equation (6.66), the results are the same as that of the above, substituting x_j ($j = 1, 2, 3, 4, 5$) of each learning sample in Equation (6.65). This indicates that though the expression of discriminate function (6.65) calculated by BAYD differs from that of discriminate function (6.66) calculated by BAYSD, the classification results are the same.

Table 6.15 shows the results of the learning process by BAYD and BAYSD.

TABLE 6.15   Prediction Results from Coal-Gas-Outburst Classification in China

Sample Type	Sample No.	y^* Outburst Type[a]	Coal-Gas-Outburst Classification											
			BAYD		BAYSD		C-SVM		R-SVM		BPNN		MRA	
			y	$R(\%)$	y	$R(\%)$	y	$R(\%)$	y	$R(\%)$	y	$R(\%)$	y	$R(\%)$
Learning samples	1	4	4	0	4	0	4	0	3	25	4	0	3	25
	2	2	2	0	2	0	2	0	2	0	3	50	2	0
	3	2	2	0	2	0	2	0	2	0	3	50	2	0
	4	1	1	0	1	0	1	0	1	0	1	0	0	100
	5	3	3	0	3	0	3	0	3	0	4	33.33	3	0
	6	1	1	0	1	0	1	0	1	0	1	0	2	100
	7	4	4	0	4	0	4	0	3	25	4	0	4	0
	8	2	2	0	2	0	2	0	2	0	3	50	3	50
	9	1	1	0	1	0	1	0	1	0	1	0	1	0
	10	3	3	0	3	0	2	33.33	2	33.33	3	0	2	33.33
	11	2	2	0	2	0	2	0	2	0	3	50	3	50
	12	1	1	0	1	0	1	0	1	0	1	0	1	0
	13	1	1	0	1	0	1	0	1	0	1	0	1	0
	14	4	4	0	4	0	4	0	3	25	4	0	4	0
	15	3	3	0	3	0	3	0	3	0	4	33.33	3	0
	16	1	1	0	1	0	1	0	1	0	1	0	1	0
Prediction samples	17	2	2	0	2	0	2	0	2	0	1	50	2	0
	18	2	2	0	2	0	2	0	2	0	1	50	2	0
	19	2	2	0	2	0	2	0	2	0	1	50	2	0
	20	2	2	0	2	0	2	0	2	0	1	50	2	0
	21	2	2	0	2	0	2	0	1	50	1	50	2	0

[a]y^* = coal-gas-outburst classification (1—non, 2—small, 3—medium, 4—large) determined by the coal test.

6.6.4. Prediction Process

Substituting x_j ($j = 1, 2, 3, 4, 5$) of each prediction sample in Equation (6.65), four values of the discriminate function are obtained, and the l corresponding to the maximum of the four values is the solution, i.e., l is the value of y for a given prediction sample. Comparing the y values with the corresponding practical values y^*, a $\overline{R}_2(\%)$ is obtained to express the accuracy of the prediction process. Similarly, substituting x_j ($j = 1, 2, 3, 4, 5$) of each prediction sample in Equation (6.66), the results are the same as that of the above, substituting x_j ($j = 1, 2, 3, 4, 5$) of each prediction sample in Equation (6.65). This indicates that though the expression of discriminate function (6.65) calculated by BAYD differs from that of discriminate function (6.66) calculated by BAYSD, the classification results are the same.

Table 6.15 shows the results of the prediction process by BAYD and BAYSD.

6.6.5. Application Comparisons Among BAYD, BAYSD, C-SVM, R-SVM, BPNN, and MRA

Similarly, using the 16 learning samples (Table 6.14) and by C-SVM, R-SVM, BPNN, and MRA, the four functions of coal-gas-outburst classification (y) with respect to five parameters (x_1, x_2, x_3, x_4, x_5) have been constructed, i.e., Equations (6.68) and (6.69) corresponding to SVM formula (4.1), Equation (6.70) corresponding to BPNN formula (3.1), and Equation (6.71) corresponding to MRA formula (2.14), respectively. Also similarly, using the five prediction samples (Table 6.14), C-SVM, R-SVM, BPNN, and MRA are employed for predictions, respectively.

Using C-SVM, the result is an explicit nonlinear function:

$$y = C\text{-}SVM(x_1, x_2, x_3, x_4, x_5) \tag{6.68}$$

with $C = 2048$, $\gamma = 0.000122$, 13 free vectors x_i, and CVA = 87.5%.

Using R-SVM, the result is an explicit nonlinear function:

$$y = R\text{-}SVM(x_1, x_2, x_3, x_4, x_5) \tag{6.69}$$

with $C = 1$, $\gamma = 0.2$, and 13 free vectors x_i.

The BPNN consists of five input layer nodes, one output layer node, and 11 hidden layer nodes. Setting $t_{max} = 100000$, the calculated $t_{opt} = 39854$. The result is an implicit nonlinear function:

$$y = BPNN(x_1, x_2, x_3, x_4, x_5) \tag{6.70}$$

with $RMSE(\%) = 0.7510 \times 10^{-2}$.

Using MRA, the result is an explicit linear function:

$$y = -0.69965 - 0.020735x_1 - 2.7151x_2 + 0.57211x_3 + 0.16274x_4 + 0.00513x_5 \tag{6.71}$$

Equation (6.71) yields a residual variance of 0.18264 and a multiple correlation coefficient of 0.90408. From the regression process, outburst type (y) is shown to depend on the five parameters in decreasing order: mining depth (x_5), gas pressure (x_3), coefficient of coal consistence (x_2), destructive style of coal (x_4), and initial speed of methane diffusion (x_1). This order differs from the aforementioned one by BAYSD in the former three parameters because MRA is a linear algorithm whereas BAYSD is a nonlinear algorithm, and the nonlinearity of the studied problem is strong.

Substituting the values of x_1, x_2, x_3, x_4, x_5 given by the five prediction samples (Table 6.14) in Equations (6.68), (6.69), (6.70), and (6.71), respectively, the outburst type (y) of each prediction sample is obtained.

Table 6.15 shows the results of the learning and prediction processes by C-SVM, R-SVM, BPNN, and MRA.

TABLE 6.16 Comparison Among the Applications of BAYD, BAYSD, C-SVM, R-SVM, BPNN, and MRA to Coal-Gas-Outburst Classification in China

Algorithm	Fitting Formula	Mean Absolute Relative Residual			Dependence of the Predicted Value (y) on Parameters (x_1, x_2, x_3, x_4, x_5), in Decreasing Order	Time Consumed on PC (Intel Core 2)	Solution Accuracy
		$\overline{R}1(\%)$	$\overline{R}2(\%)$	$\overline{R}^*(\%)$			
BAYD	Nonlinear, explicit	0	0	0	N/A	2 s	Very high
BAYSD	Nonlinear, explicit	0	0	0	x_3, x_2, x_5, x_4, x_1	2 s	Very high
C-SVM	Nonlinear, explicit	2.08	0	1.04	N/A	2 s	Very high
R-SVM	Nonlinear, explicit	6.77	10	8.39	N/A	2 s	Moderate
BPNN	Nonlinear, implicit	16.67	50	33.34	N/A	25 s	Very low
MRA	Linear, explicit	22.40	0	11.20	x_5, x_3, x_2, x_4, x_1	<1 s	Low

$\overline{R}^*(\%) = 11.20$ of MRA (Table 6.16) shows that the nonlinearity of the relationship between the predicted value y and its relative parameters (x_1, x_2, x_3) is strong from Table 1.2. The results of BAYD and BAYSD are the same, i.e., not only $\overline{R}_1(\%) = 0$, but also $\overline{R}_2(\%) = 0$, and thus $\overline{R}^*(\%) = 0$, which coincide with practicality. For C-SVM and MRA, though $\overline{R}_2(\%) = 0$, $\overline{R}_1(\%) = 2.08$ and 22.40, respectively, which cannot ensure correctly predicting other new prediction samples. As for R-SVM and BPNN, the solution accuracy is moderate and low, respectively, and $\overline{R}_2(\%) > \overline{R}_1(\%)$.

6.6.6. Summary and Conclusions

In summary, using data for the coal-gas-outburst classification based on five parameters, i.e., the five parameters (initial speed of methane diffusion, coefficient of coal consistence, gas pressure, destructive style of coal, mining depth) and coal-test results of 16 coal mines in China, two sets of discriminate functions (6.65) and (6.66) obtained by two data mining tools of BAYD and BAYSD, respectively, and the results are consistent with coal tests. These two sets of discriminate functions are called *mined knowledge*. In the prediction process, applying this knowledge, i.e., using the five parameters of five other coal mines for coal-gas-outburst classification, respectively, the results are also consistent with coal tests. Therefore, the two methods can be spread to the coal-gas-outburst classification of other coal mines.

Moreover, it is found that (a) since this case study is a strong nonlinear problem, MRA, R-SVM, or MRA is not applicable; and (b) since the number of learning samples is not big enough, BAYD or BAYSD is better than C-SVM, which is an advantage of BAYD and BAYSD.

6.7. CASE STUDY 6: TOP COAL CAVING CLASSIFICATION (TWENTY-SIX LEARNING SAMPLES AND THREE PREDICTION SAMPLES)

6.7.1. Studied Problem

In Section 5.3, the data in Table 5.7 was used, and ID3, C-SVM, R-SVM, BPNN, and MRA were adopted to construct the function of top-coal caving classification (y) with respect to six relative parameters ($c_1, c_2, c_3, c_4, c_5, c_6$) for 29 coal mines in China, and the calculation results were filled into Tables 5.8 and 5.9. Now Table 5.7 is used as input data for Case Study 6 of this chapter. For the sake of algorithm comparisons, Table 5.8 is wholly copied into Table 6.17, and Table 5.9 is wholly copied into Table 6.18.

6.7.2. Known Input Parameters

They are the values of the known variables c_i ($i = 1, 2, 3, 4, 5, 6$) for 26 learning samples and three prediction samples and the value of the prediction variable y^* for 26 learning samples (Table 5.7).

6.7.3. Learning Process

Using the parameters of learning samples and by BAYSD, a discrimination function corresponding to Equation (6.30) has been constructed:

$$
\begin{aligned}
B_1(x) &= \ln(0.115) - 18.029 + 5.462c_1 + 0.157c_2 \\
&\quad +2.055c_3 + 5.139c_4 + 7.119c_5 + 2.572c_6 \\
B_2(x) &= \ln(0.462) - 28.947 + 7.202c_1 + 3.037c_2 \\
&\quad +4.032c_3 + 4.977c_4 + 7.057c_5 + 2.516c_6 \\
B_3(x) &= \ln(0.192) - 59.990 + 9.501c_1 + 3.304c_2 \\
&\quad +6.227c_3 + 7.717c_4 + 11.522c_5 + 3.370c_6 \\
B_4(x) &= \ln(0.154) - 77.530 + 11.370c_1 + 6.294c_2 \\
&\quad +7.200c_3 + 6.606c_4 + 11.143c_5 + 3.899c_6 \\
B_5(x) &= \ln(0.077) - 101.285 + 12.656c_1 + 3.860c_2 \\
&\quad +7.778c_3 + 9.423c_4 + 15.304c_5 + 4.999c_6
\end{aligned}
\right\}
\tag{6.72}
$$

Thus, it is believed that an explicit nonlinear function corresponding to BAC formula (6.1) is obtained:

$$y = BAC(c_1, c_2, c_3, c_4, c_5, c_6) \tag{6.73}$$

From the discriminate process of BAYSD, top coal caving classification (y) is shown to depend on the six parameters in decreasing order: cracking index (c_5), dirt band thickness

TABLE 6.17 Prediction Results from Top Coal Caving Classification II of 29 Coal Mines in China

			Top Coal Caving Classification											
			BAYSD		ID3		C-SVM		R-SVM		BPNN		MRA	
Sample Type	Sample No.	y^* Coal Test[a]	y	$R(\%)$	y	$R(\%)$	y	$R(\%)$	y	$R(\%)$	y	$R(\%)$	y	$R(\%)$
Learning samples	1	1	2	100	1	0	2	100	2	100	1	0	1	0
	2	2	2	0	2	0	2	0	2	0	2	0	2	0
	3	3	3	0	3	0	2	33.33	3	0	3	0	3	0
	4	3	3	0	3	0	3	0	3	0	3	0	3	0
	5	2	2	0	2	0	2	0	2	0	2	0	2	0
	6	2	2	0	2	0	3	50	2	0	2	0	2	0
	7	4	4	0	4	0	4	0	4	0	4	0	4	0
	8	4	4	0	4	0	4	0	4	0	4	0	4	0
	9	2	2	0	2	0	2	0	2	0	2	0	2	0
	10	5	5	0	5	0	4	20	4	20	5	0	4	20
	11	2	2	0	2	0	2	0	2	0	2	0	2	0
	12	1	1	0	1	0	2	100	2	100	1	0	2	100
	13	3	3	0	3	0	3	0	3	0	3	0	4	33.33
	14	5	5	0	5	0	4	20	4	20	5	0	5	0
	16	2	2	0	2	0	2	0	2	0	2	0	2	0
	17	2	2	0	2	0	2	0	2	0	2	0	2	0
	18	2	1	50	2	0	2	0	2	0	2	0	1	50
	19	2	2	0	2	0	2	0	2	0	2	0	2	0
	22	2	2	0	2	0	2	0	2	0	2	0	2	0
	23	1	1	0	1	0	2	100	2	100	1	0	1	0
	24	4	4	0	4	0	4	0	4	0	4	0	4	0
	25	3	3	0	3	0	4	33.33	3	0	3	0	4	33.33
	26	2	2	0	2	0	2	0	2	0	2	0	2	0
	27	3	3	0	3	0	3	0	3	0	3	0	3	0
	28	4	5	25	4	0	4	0	4	0	4	0	4	0
	29	2	2	0	2	0	2	0	2	0	2	0	2	0
Prediction samples	15	3	3	0	3	0	3	0	3	0	3	0	3	0
	20	2	3	50	2	0	2	0	3	50	1	50	3	50
	21	2	2	0	2	0	2	0	2	0	1	50	2	0

[a]y^* = top-coal-caving classification (1—excellent, 2—good, 3—average, 4—poor, 5—very poor) determined by the coal test.

TABLE 6.18 Comparison Among the Applications of BAYSD, ID3, C-SVM, R-SVM, BPNN, and MRA to top coal caving Classification II in China

Algorithm	Fitting Formula	Mean Absolute Relative Residual			Dependence of the Predicted Value (y) on Parameters $(c_1, c_2, c_3, c_4, c_5, c_6)$, in Decreasing Order	Time Consumed on PC (Intel Core 2)	Solution Accuracy
		$\overline{R}1(\%)$	$\overline{R}2(\%)$	$\overline{R}^*(\%)$			
BAYSD	Nonlinear, explicit	6.73	16.67	11.70	$c_5, c_3, c_2, c_1, c_4, c_6$	2 s	Low
ID3	Nonlinear, explicit	0	0	0	N/A	2 s	Very high
C-SVM	Nonlinear, explicit	17.56	0	8.78	N/A	2 s	Moderate
R-SVM	Nonlinear, explicit	13.08	16.67	14.88	N/A	2 s	Low
BPNN	Nonlinear, implicit	0	33.33	16.67	N/A	40 s	Low
MRA	Linear, explicit	9.10	16.67	12.89	$c_5, c_3, c_1, c_2, c_6, c_4$	<1 s	Low

(c_3), uniaxial compressive strength (c_2), mining depth (c_1), coal layer thickness (c_4), and coefficient of fullness at top coal (c_6). This order differs from that by MRA (Table 6.18) somewhat in the latter four parameters due to the fact that though MRA is a linear algorithm whereas BAYSD is a nonlinear algorithm, the nonlinearity of the studied problem is strong whereas the nonlinearity ability of BAYSD is weak.

Substituting c_j ($j = 1, 2, 3, 4, 5, 6$) of each learning sample in Equation (6.72), five values of discriminate function are obtained, and the l corresponding to the maximum of the five values is the solution, i.e., l is the value of y for a given learning sample. Comparing the y values with the corresponding practical values y^*, $\overline{R}_1(\%)$ is obtained to express the fitness of learning process. For instance, for the second learning sample (Table 5.7), $l = 2$, thus $y = 2$; and $y^* = 2$, so $\overline{R}_1(\%) = 0$.

Table 6.17 shows the results of learning process by BAYSD.

6.7.4. Prediction Process

Substituting c_j ($j = 1, 2, 3, 4, 5, 6$) of each learning sample in Equation (6.72), five values of discriminate function are obtained, and the l corresponding to the maximum of the five values is the solution, i.e., l is the value of y for a given prediction sample. Comparing the y values with the corresponding practical values y^*, $\overline{R}_2(\%)$ is obtained to express the accuracy of prediction process.

Table 6.17 shows the results of prediction process by BAYSD.

6.7.5. Application Comparisons Among BAYSD, ID3, C-SVM, R-SVM, BPNN, and MRA

$\overline{R}^*(\%) = 12.89$ of MRA (Table 6.18) shows that the nonlinearity of the relationship between the predicted value y and its relative parameters (c_1, c_2, c_3, c_4, c_5, c_6) is strong from Table 1.2. Only ID3 is applicable, i.e., not only $\overline{R}_1(\%) = 0$, but also $\overline{R}_2(\%) = 0$, and thus $\overline{R}^*(\%) = 0$, which coincide with practicality. For C-SVM, though $\overline{R}_2(\%) = 0$, $\overline{R}_1(\%) = 17.56$, which cannot ensure correct prediction of other new prediction samples. For BPNN, though $\overline{R}_1(\%) = 0$, $\overline{R}_2(\%) = 33.33$, indicating the number of learning samples is not big enough. As for BAYSD, R-SVM, and MRA, the solution accuracy is low, and $\overline{R}_2(\%) >> \overline{R}_1(\%)$.

6.7.6. Summary and Conclusions

In summary, using data for the top coal caving classification based on six parameters, i.e., the six parameters (mining depth, uniaxial compressive strength, dirt band thickness, coal layer thickness, cracking index, coefficient of fullness at top coal) and coal-test result of 26 coal mines in China, a set of discriminate functions (6.72) is constructed by the data mining tool of BAYSD, but the results are not eligible. This set of discriminate functions is called *mined unqualified knowledge*. In the prediction process, applying this unqualified knowledge, i.e., using the six parameters of three other coal mines in China for top coal caving classifications, respectively, the results are not eligible either. Therefore, this method cannot be spread to the top coal caving classification of other coal mines.

Moreover, it is found that since this case study is a strong nonlinear problem, only ID3 is applicable, but BAYSD, C-SVM, R-SVM, BPNN, or MRA is not applicable.

Through Tables 6.3, 6.6, 6.9, 6.11, 6.13, 6.16, and 6.18 in seven case studies from Section 6.1 to Section 6.7, respectively, the solution accuracies of BAYSD and C-SVM are listed in Table 6.19.

TABLE 6.19   Solution Accuracy of BAYSD and C-SVM in Seven Case Studies

Case Study	Nonlinearity	Solution Accuracy	
		BAYSD	C-SVM
Simple case study	Strong	Low	Very high
Case Study 1	Strong	Very high	Low
Case Study 2	Very weak	Very high	Very high
Case Study 3	Strong	Very low	Very high
Case Study 4	Moderate	Very high	Very high
Case Study 5	Strong	Very high	Very high
Case Study 6	Strong	Low	Moderate

It is found from Table 6.19 that the solution accuracies of BAYSD and C-SVM are (a) high when the nonlinearity of case studies is very weak or moderate, but (b) different when the nonlinearity of case studies is strong, except for Case Study 5, and only one of the two algorithms is applicable.

EXERCISES

6-1. This chapter introduces naïve Bayesian (NBAY), Bayesian discrimination (BAYD), and Bayesian successive discrimination (BAYSD). What are the differences in methods and application values of these three algorithms?

6-2. In seven case studies from Section 6.1 to Section 6.7, BAYSD and C-SVM in Chapter 4 are applied. What are the solution accuracies of the two classification algorithms?

6-3. Either BAYSD or MRA in Chapter 4 can serve as a pioneering dimension-reduction tool. What are the advantages and shortcomings of the two dimension-reduction tools?

References

Brown, P.J., Kenward, M.G., Bassett, E.E., 2001. Bayesian discrimination with longitudinal data. Biostatistics 2 (4), 417–432.

de Maesschalck, R., Jouan-Rimb, D., Massart, D.L., 2000. The mahalanobis distance. Chemom. Intel. Lab. Sys. 50 (1), 1–18.

Denison, D.G.T., Holmes, C.C., Mallick, B.K., Smith, A.F.M., 2002. Bayesian Methods for Nonlinear Classification and Regression. Wiley, Chichester, UK.

Domingos, P., Pazzani, M., 1997. On the optimality of the simple bayesian classifier under zero-one loss. Machine Learning 29 (2–3), 103–130.

Fu, D., Xu, J., Wang, G., 2011. Reservoir classification and evaluation based on Q cluster analysis combined with Bayesian discrimination algorithm. Sci. Tech. Rev. 29 (3), 29–33 (in Chinese with English abstract).

Han, J.W., Kamber, M., 2006. Data Mining: Concepts and Techniques, second ed. Morgan Kaufmann, San Francisco, CA, USA.

Li, H., Zhao, Y., 1998. Mathematical Petroleum Geology. Petroleum Industry Press, Beijing, China (in Chinese).

Logan, T.P., Gupta, A.K., 1993. Bayesian discrimination using multiple observations. Commun. Stat. Theory Methods 22 (6), 1735–1754.

Ramoni, M., Sebastiani, P., 2001. Robust Bayes classifiers. Artif. Intell. 125 (1–2), 207–224.

Shi, G., 2011. Four classifiers used in data mining and knowledge discovery for petroleum exploration and development. Adv. Petro. Expl. Devel. 2 (2), 12–23.

Tan, P., Steinbach, M., Kumar, V., 2005. Introduction to Data Mining. Pearson Education, Boston, MA, USA.

Wang, C., Wang, E., Xu, J., Liu, X., Ling, L., 2010. Bayesian discriminant analysis for prediction of coal and gas outbursts and application. Mining Sci. Tech. 20 (4), 520–523, 541.

Zhao, X., 1992. Conspectus of Mathematical Petroleum Geology. Petroleum Industry Press, Beijing, China (in Chinese).

Zhang, W., Li, X., Jia, G., 2008. Quantitative classification and evaluation of low permeability reservoir, with Chang 3 and Chang 4+5 of Triassic Formation in Baibao Oil Field as examples. Sci. Tech. Rev. 26 (21), 61–65 (in Chinese with English abstract).

Data Mining and Knowledge Discovery for Geoscientists
http://dx.doi.org/10.1016/B978-0-12-410437-2.00007-2

Cluster analysis is a classification analysis approach to a data set, which is a multiple linear statistical analysis. This chapter introduces Q-mode and R-mode cluster analysis techniques as well as their applications in geosciences.

Section 7.1 introduces the applying ranges and conditions, basic principles, calculation methods, and calculation flowcharts for Q-mode and R-mode cluster analysis.

In the latter four sections of the chapter, a case study is given in each section. Though the four case studies are small, they reflect the whole process of calculations with which to explain how to adopt Q-mode cluster analysis and R-mode cluster analysis to samples and parameters, respectively, to benefit readers in understanding and mastering the techniques.

Section 7.2 (integrated evaluation on oil and gas traps quality) introduces Case Study 1 for cluster analysis. Using data for the trap evaluation of the Northern Kuqa Depression, i.e., the 14 parameters (unit structure, trap type, petroliferous formation, trap depth, trap relief, trap closed area, formation HC identifier, data reliability, trap coefficient, source rock coefficient, reservoir coefficient, preservation coefficient, configuration coefficient, resource quantity) of 30 traps, two cluster pedigree charts are gained by the data mining tool (Q-mode and R-mode cluster analyses). One chart is for 30 traps, indicating the order of dependence between traps; the classification coincidence rate reaches 77%, and the results relatively coincide with practicality. The other chart is for 14 parameters, indicating the order of dependence between the parameters. These two cluster pedigree charts are called *mined knowledge*. Therefore, this method can be spread to the cluster analysis for integrated evaluation of oil and gas trap quality in other exploration areas. Moreover, it is found that (a) Q-mode cluster analysis can be used for trap sample reduction, indicating that Q-mode cluster analysis can serve as a pioneering sample-reduction tool; and (b) whereas R-mode cluster analysis (using analog coefficient) cannot be used for parameter dimension reduction. The two discoveries are also called *mined knowledge*. As for parameter dimension reduction, either multiple regression analysis (MRA, see Chapter 2 and Chapter 4) or Bayesian successive discrimination (BAYSD, see Chapter 6) can serve as a pioneering dimension-reduction tool.

Section 7.3 (oil layer classification based on well-logging information interpretation) introduces Case Study 2 for cluster analysis. Using data for the oil layer classification of the Xiefengqiao Anticline, i.e., the five parameters (true resistivity, acoustictime, porosity, oil saturation, permeability) of 27 oil layers, two cluster pedigree charts are gained by the data mining tool (Q-mode and R-mode cluster analyses). One chart is for 27 oil layers, indicating the order of dependence between oil layers; the classification coincidence rate reaches 70%, and the results relatively coincide with practicality. The other chart is for five parameters, indicating the order of dependence between the parameters. These two cluster pedigree charts are called *mined knowledge*. Therefore, this method can be spread to the cluster analysis for oil layer classification based on well-logging interpretation in other areas. Moreover, it is found that (a) Q-mode cluster analysis can be used for oil layer sample reduction, indicating that Q-mode cluster analysis can serve as a pioneering sample-reduction tool; whereas (b) R-mode cluster analysis (using an analog coefficient) cannot be used for parameter dimension reduction. The two discoveries are also called *mined knowledge*. As for parameter dimension reduction, either MRA or BAYSD can serve as a pioneering dimension-reduction tool.

Section 7.4 (coal-gas-outburst classification) introduces the Case Study 3 for cluster analysis. Using data for the coal-gas-outburst of the typical mines in China, i.e., the five parameters (initial speed of methane diffusion, coefficient of coal consistence, gas pressure, destructive style of coal, mining depth) of 21 samples, two cluster pedigree charts are gained by the data mining tool (Q-mode and R-mode cluster analyses). One chart is for 21 samples, indicating the order of dependence between samples; the classification coincidence rate reaches 81%, and the results relatively coincide with practicality. The other chart is for five parameters, indicating the order of dependence between the parameters. These two cluster pedigree charts are called *mined knowledge*. Therefore, this method can be spread to the cluster analysis for coal-gas-outburst in other mines. Moreover, it is found that (a) Q-mode cluster analysis can be used for coal-gas-outburst sample reduction, indicating that Q-mode cluster analysis can serve as a pioneering sample-reduction tool; and (b) whereas R-mode cluster analysis (using a correlation coefficient) can be used for parameter dimension reduction, indicating that R-mode cluster analysis can serve as dimension-reduction tool, like MRA and BAYSD. The two discoveries are also called *mined knowledge*.

Section 7.5 (reservoir classification) introduces Case Study 4 for cluster analysis. Using data for the reservoir classification in the C3 and C4+5 low-permeability reservoir in the Baibao Oilfield, i.e., the four parameters (shale content, porosity, permeability, permeability variation coefficient) of 28 layers, two cluster pedigree charts are gained by the data mining tool (Q-mode and R-mode cluster analyses). One chart is for 28 layers, indicating the order of dependence between layers; the classification coincidence rate reaches 100%, and the results coincide with practicality. The results are provided for the five algorithms of BAYSD, the classification (C-SVM, see Chapter 4) and the regression (R-SVM, see Chapter 4) of support vector machine (SVM, see Chapter 4), error back-propagation neural network (BPNN, see Chapter 3), and MRA (see Section 6.3). The other chart is for four parameters, indicating the order of dependence between the parameters. These two cluster pedigree charts are called *mined knowledge*. Therefore, this method can be spread to the cluster analysis for low-permeability reservoir classification in other oilfields. Moreover, it is found that (a) Q-mode cluster analysis can be used for layer sample reduction, indicating that Q-mode cluster analysis can serve as a pioneering sample-reduction tool; whereas (b) R-mode cluster analysis (using an analog coefficient) cannot be used for parameter dimension reduction. The two discoveries are also called *mined knowledge*. As for parameter dimension reduction, either MRA or BAYSD can serve as a pioneering dimension-reduction tool.

The preceding four Case Studies 1, 2, 3, and 4 demonstrate that (a) Q-mode cluster analysis provides a cluster pedigree chart for samples, and the classification coincidence rate is passable; (b) if MRA is passed for validation, Q-mode cluster analysis can also serve as a pioneering sample-reduction tool; (c) R-mode cluster analysis provides a cluster pedigree chart for parameters; and (d) if MRA is passed for validation, R-mode cluster analysis (using a correlation coefficient) can also serve as a pioneering dimension-reduction tool. The so-called *pioneering tool* is whether it succeeds or needs a nonlinear tool (C-SVM, R-SVM, or BPNN) for the second validation so as to determine how many samples and how many parameters can be reduced. Why does it need a second validation? Because of the complexities of geosciences rules; the correlations between various classes of geosciences data are nonlinear in most cases. In general, therefore, C-SVM or R-SVM is adopted when nonlinearity is strong, whereas BPNN, DTR, or BAC is adopted when nonlinearity is not strong. The two linear algorithms of cluster analysis and MRA can serve as auxiliary tools, cooperating with major tools (BPNN, C-SVM, R-SVM, DTR, and BAC) for data mining.

It must be pointed out here that since cluster analysis is performed one by one in the light of "two-in-one" aggregation, so the number of the new class is one for the last, two for the previous, four for that before the previous, ... and so on. Obviously, except for the last aggregation, the number of the new class is an even number. So the number of the new class that is obtained in the sample classification by Q-mode cluster analysis is an even number.

7.1. METHODOLOGY

Cluster analysis has been widely applied in the natural and social sciences since the 1970s (Everitt et al., 2011). Q-mode and R-mode cluster analyses are among the most popular methods in cluster analysis and are still very useful tools in some fields, including geosciences.

Assume that there are n samples, each associated with m parameters $(x_1, x_2, ..., x_m)$ and a set of observed values $(x_{1i}, x_{2i}, ..., x_{mi})$, with $i = 1, 2, ..., n$ for these parameters. In principle, $n > m$, but in actual practice $n >> m$. The n samples associated with m parameters are defined as n vectors:

$$x_i = (x_{i1}, x_{i2}, ..., x_{im}) \quad (i = 1, 2, ..., n) \tag{7.1}$$

Equation (7.1) expresses n sample vectors. The cluster analysis to n sample vectors is conducted; thus this kind of cluster analysis is called *Q-mode cluster analysis*.

It is obvious that the only difference between Equation (7.1) and Equation (1.1) is that Equation (1.1) includes y_i^* but Equation (7.1) does not.

$$x_j = (x_{j1}, x_{j2}, ..., x_{jn}) \quad (j = 1, 2, ..., m) \tag{7.2}$$

Equation (7.2) expresses m parameter vectors. The cluster analysis to m parameter vectors is conducted; thus this kind of cluster analysis is called *R-mode cluster analysis*.

It should be noted that x_i defined by Equation (7.1) and x_j defined by Equation (7.2) are definitely different vectors. x_i is a sample vector, and each vector is associated with m elements, whereas x_j is a parameter vector, and each vector is associated with n elements.

Obviously, the matrix defined by Equation (7.1) and the matrix defined by Equation (7.2) are mutually transposed.

We point out that the data used by both Q-mode cluster analysis and R-mode cluster analysis can be taken from the data used by regression algorithms and classification algorithms. First, the data used in regression algorithms and classification algorithms cover learning samples and prediction samples, whereas the data used by cluster analysis cover only samples without the concept of learning samples and prediction samples. Taking the learning sample in regression algorithms and classification algorithms as an example, the data used are x_i vectors expressed by Equation (1.1) that include y_i^*. But the data used by Q-mode cluster analysis are the sample vector x_i expressed by Equation (7.1) that does not include y_i^*, and the data used by R-mode cluster analysis are the parameter vector x_i expressed by Equation (7.2) that does not include y_i^* either. y_i^* is only used for comparison between the calculation results of cluster analysis and y_i^* if y_i^* exist.

7.1.1. Applying Ranges and Conditions

7.1.1.1. Applying Ranges

If there exists a certain relationship between two or more parameters (assume there are m parameters), and the m parameters have a certain amount of sets of observed values (assume there are n samples), then cluster analysis can be adopted. By adopting Q-mode cluster analysis, we can achieve a cluster pedigree chart about n samples, indicating the order of dependence between samples. Therefore, it possibly serves as a pioneering sample-reduction tool. By adopting R-mode cluster analysis, we can achieve a cluster pedigree chart about m parameters, indicating the order of dependence between parameters; therefore, it possibly serves as a pioneering parameter-reduction (i.e., dimension-reduction) tool.

7.1.1.2. Applying Conditions

The number of samples is large enough to ensure the accuracy of cluster analysis, and the number of parameters is greater than one.

7.1.2. Basic Principles

The cluster analysis introduced in this section only refers to Q-mode cluster analysis. As for R-mode cluster analysis, the method is definitely the same in essence as that of Q-mode cluster analysis. The only difference is that the two have different cluster objects. The object for Q-mode cluster analysis is n sample vectors, expressed by Equation (7.1), whereas the object for R-mode cluster analysis is m parameter vectors, expressed by Equation (7.2). So the specific introduction of Q-mode cluster analysis is like the specific introduction of R-mode cluster analysis.

The common method for cluster analysis at present is the aggregation method. Just as the name implies, aggregation method is to merge more classes into less till into one class under the precondition that every cluster object has its own class. For example, Q-mode cluster analysis merges n classes into one class, whereas R-mode cluster analysis merges m classes into one class.

The calculation of Q-mode cluster analysis is performed in the following four steps:

1. To determine the number of cluster objects, i.e., the total number of classes is n at the beginning of cluster analysis.
2. To select one method such as standard difference standardization, maximum difference standardization, and maximum difference normalization, for conducting data normalization for the original data expressed by Equation (7.1).
3. To select an approach involving a cluster statistic and a class-distance method to conduct calculations to every two classes in n classes by a class-distance method (e.g., the shortest distance, the longest distance, and the weighted mean) of a cluster statistic (e.g., distance coefficient, analog coefficient, and correlation coefficient), achieving the order of dependence between classes, and to merge the closest two classes (e.g., to distance coefficient it is minimum class distance, to analog coefficient or correlation coefficient it is maximum class distance) into a new class that is composed of two sample vectors. Thus the total number of classes turns to $(n - 1)$.
4. To use the preceding selected method to conduct calculations to every two classes in $(n - 1)$ classes, achieving the order of dependence between classes, and to merge the closest two classes into a new class, thus the total number of class turns to $(n - 2)$. Repeat this operation in this way until the total number of classes turns to 1. At this time n classes are merged into one class. Then stop the calculation.

The calculation of R-mode cluster analysis is performed in the following four steps:

1. To determine the number of cluster objects, i.e., the total number of clusters is m at the beginning of cluster analysis.
2. To select one method such as standard difference standardization, maximum difference standardization, and maximum difference normalization, for conducting data normalization for the original data expressed by Equation (7.2).
3. To select an approach involving a cluster statistic and a class-distance method to conduct calculations to every two classes in m classes by a class-distance method (e.g., the shortest distance, the longest distance, and the weighted mean) of a cluster statistic (e.g., distance coefficient, analog coefficient, and correlation coefficient), achieving the order of dependence between classes, and to merge the closest two classes (e.g., to the distance coefficient it is minimum class distance, to the analog coefficient or correlation coefficient it is maximum class distance) into a new class that is composed of two sample vectors. Thus the total number of classes turns to $(m - 1)$.
4. To use the selected method to conduct calculations to every two classes in $(m-1)$ classes, achieving the order of dependence between classes, and to merge the closest two classes into a new class. Thus the total number of classes turns to $(m - 2)$. Repeat this operation in this way till the total number of class turns to 1, at which time m classes are merged into 1 class, and then stop the calculation.

As mentioned, it is further indicated that the calculation steps for Q-mode cluster analysis and R-mode cluster analysis are the same, with the only difference that their original data matrixes are mutually transposed.

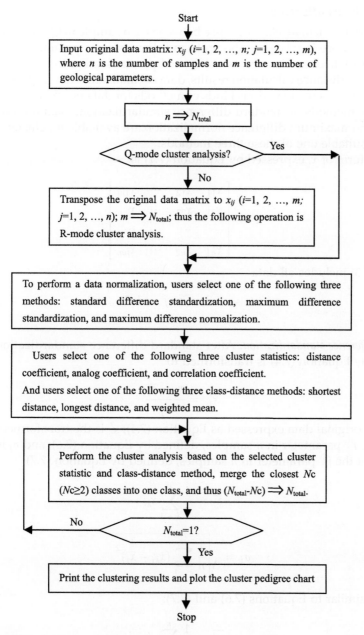

FIGURE 7.1   Calculation flowchart of Q-mode and R-mode cluster analyses.

7.1.3. Q-mode Cluster Analysis

The algorithm for Q-mode cluster analysis consists of data normalization, cluster statistics, and cluster analysis (Zhao, 1992; Li and Zhao, 1998; Everitt et al., 2011).

7.1.3.1. Data Normalization

It is necessary to calculate class distance between each sample in Q-mode cluster analysis, and the class distance is related to the units and dimension of parameters. There are three methods to calculate class distance, as mentioned previously. For the sake of comparability between the class-distance calculation results, data normalization to x_i expressed by Equation (7.1) is carried out. There are various kinds of methods for data normalization. However, the following three methods of standard difference standardization, maximum difference standardization, and maximum difference normalization are available for cluster analysis; users can select the suitable one of these three methods.

The matrix form of x_i expressed by Equation (7.1) is:

$$x = \begin{bmatrix} x_{11} & x_{12} & \cdots & x_{1m} \\ x_{21} & x_{22} & \cdots & x_{2m} \\ \cdots & \cdots & \cdots & \cdots \\ x_{n1} & x_{n2} & \cdots & x_{nm} \end{bmatrix} \tag{7.3}$$

Equation (7.3) includes all data

$$x_{ij} \quad (i = 1, 2, ..., n; \ j = 1, 2, ..., m) \tag{7.4}$$

1. *Standard difference standardization.* After standard difference standardization, the data x_{ij} expressed by Equation (7.4) is transformed to

$$x'_{ij} = \frac{x_{ij} - \overline{x_j}}{\sigma_j} \tag{7.5}$$

where x_{ij} is original data expressed as Equation (7.4); x'_{ij} is the transformed data; $\overline{x_j}$ is the mean of the jth parameter in n samples, defined by Equation (7.6); and σ_j is the standard difference of the jth parameter in n samples, defined by Equation (7.7).

$$\overline{x_j} = \frac{1}{n} \sum_{k=1}^{n} x_{kj} \tag{7.6}$$

$$\sigma_j = \sqrt{\frac{1}{n} \sum_{k=1}^{n} \left(x_{kj} - \overline{x_j}\right)^2} \tag{7.7}$$

For x'_{ij}, it is similar to Equations (7.6) and (7.7):

$$\overline{x'_j} = \frac{1}{n} \sum_{k=1}^{n} x'_{kj} \tag{7.8}$$

$$\sigma'_j = \sqrt{\frac{1}{n} \sum_{k=1}^{n} \left(x'_{kj} - \overline{x'_j}\right)^2} \tag{7.9}$$

It is proved that $\overline{x'_j} = 0$ and $\sigma'_j = 1$, indicating the data x'_{ij} transformed by standard difference standardization are the normalized data.

For instance, an original data shaping is illustrated in Equation (7.3):

$$x = \begin{bmatrix} 0.25 & 0.25 & 0.70 & 0.75 & 0.35 \\ 1.00 & 0.50 & 0.80 & 0.50 & 0.10 \end{bmatrix} \tag{7.10}$$

then the transformed data is

$$x' = \begin{bmatrix} -1 & -1 & -1 & 1 & 1 \\ 1 & 1 & 1 & 1 & -1 \end{bmatrix} \tag{7.11}$$

Obviously, for x', it is $\overline{x'_1} = \overline{x'_2} = \overline{x'_3} = \overline{x'_4} = \overline{x'_5} = 0$ as well as $\sigma'_1 = \sigma'_2 = \sigma'_3 = \sigma'_4 = \sigma'_5 = 1$.

2. *Maximum difference standardization.* After maximum difference standardization, the data x_{ij} expressed by Equation (7.4) is transformed to

$$x'_{ij} = \frac{x_{ij} - \overline{x_j}}{\omega_j} \tag{7.12}$$

where ω_j is the maximum difference of the jth parameter in n samples, which is defined by Equation (7.13): and the other symbols have been defined in Equation (7.5).

$$\omega_j = \max_{1 \le k \le n} \left\{x_{kj}\right\} - \min_{1 \le k \le n} \left\{x_{kj}\right\} \tag{7.13}$$

For x'_{ij}, it is similar to Equation (7.13):

$$\omega'_j = \max_{1 \le k \le n} \left\{x'_{kj}\right\} - \min_{1 \le k \le n} \left\{x'_{kj}\right\} \tag{7.14}$$

It is proved that $\omega'_j = 1$, indicating the data x'_{ij} transformed by maximum difference standardization are the normalized data.

For instance, for the original data expressed by Equation (7.10), the transformed data are

$$x' = \begin{bmatrix} -0.5 & -0.5 & -0.5 & 0.5 & 0.5 \\ 0.5 & 0.5 & 0.5 & 0.5 & -0.5 \end{bmatrix} \tag{7.15}$$

Obviously, for x', it is $\omega'_1 = \omega'_2 = \omega'_3 = \omega'_4 = \omega'_5 = 1$. It must be pointed out that though $\overline{x'_1} = \overline{x'_2} = \overline{x'_3} = \overline{x'_4} = \overline{x'_5} = 0$, this is because of too few and too simple original data expressed by Equation (7.10). This is a very special example. As mentioned, the calculation results of maximum difference standardization show that only $\omega'_j = 1$, without $\overline{x'_j} = 0$.

3. *Maximum difference normalization'.* After maximum difference normalization, the data x_{ij} expressed by Equation (7.4) are transformed to

$$x'_{ij} = \frac{x_{ij} - \min_{1 \le k \le n} \{x_{kj}\}}{\omega_j} \qquad (7.16)$$

where all symbols have been defined in Equations (7.5) and (7.12).

Equation (7.16) is the same as Equation (3.13) of maximum difference normalization of BPNN.

It is proved that $x'_{ij} \in [0, 1]$, indicating the data x'_{ij} transformed by maximum difference normalization are the normalized data. Consequently, maximum difference normalization is a commonly used method for data normalization.

For instance, for the original data expressed by Equation (7.10), the transformed data are

$$x' = \begin{bmatrix} 0 & 0 & 0 & 1 & 1 \\ 1 & 1 & 1 & 0 & 0 \end{bmatrix} \qquad (7.17)$$

Obviously, for x', $x'_{ij} \in [0, 1]$.

7.1.3.2. Cluster Statistic

It is necessary to perform calculation of cluster statistics in Q-mode cluster analysis. There are various kinds of cluster statistic. The following three—distance coefficient, analog coefficient, and correlation coefficient—are available for cluster analysis, and users can select suitable one from these three methods.

It must be noticed that in calculating cluster statistics, the original data x_{ij} expressed by Equation (7.4) will no longer be used, but the transformed data x'_{ij} expressed by Equations (7.5), (7.12), or (7.16) will be used. Therefore, whenever x_{ij} is used in Equations (7.18) − (7.40) is x'_{ij} .

1. *Distance coefficient.* There are various kinds of definitions of distance coefficient. Here only Euclidean distance is used, because Euclidean distance is absolutely consistent with the concept of space distance. It is most popular and most efficient.

Based on the sample vector defined by Equation (7.1), the Euclidean distance between the ith sample x_i and the jth sample x_j is calculated:

$$\sqrt{\sum_{k=1}^{m} (x_{ik} - x_{jk})^2} \quad (i, j = 1, 2, ..., n)$$

To prevent the calculation results using the preceding formula from being too large, the distance coefficient d_{ij} is defined as

$$d_{ij} = \sqrt{\frac{1}{m} \sum_{k=1}^{m} (x_{ik} - x_{jk})^2} \quad (i, j = 1, 2, ..., n) \qquad (7.18)$$

where all symbols of the right-hand side have been defined in Equations (7.1) and (7.3).

$[d_{ij}]_{n \times n}$ is a real symmetrical matrix, and $d_{11} = d_{22} \ldots = d_{nn} = 0$. Obviously, the smaller the distance coefficient d_{ij}, the closer the property of x_i and x_j. This rule is used in the following cluster analysis.

2. *Analog coefficient.* Based on the sample vector defined by Equation (7.1), the analog coefficient ξ_{ij} between the ith sample x_i and the jth sample x_j is calculated:

$$\xi_{ij} = \frac{\sum_{k=1}^{m}(x_{ik}x_{jk})}{\sqrt{\sum_{k=1}^{m}(x_{ik})^2 \cdot \sum_{k=1}^{m}(x_{jk})^2}} \quad (i, j = 1, 2, \ldots, n) \tag{7.19}$$

where all symbols of the right-hand side have been defined in Equation (7.1) or Equation (7.3).

$[\xi_{ij}]_{n \times n}$ is a real symmetrical matrix, and $\xi_{11} = \xi_{22} \ldots = \xi_{nn} = 1$. Obviously, the closer the analog coefficient ξ_{ij} approaches to 1, the closer the property of x_i and x_j. This rule is used in the following cluster analysis.

3. *Correlation coefficient.* Based on the sample vector defined by Equation (7.1), the correlation coefficient r_{ij} between the ith sample x_i and the jth sample x_j is calculated:

$$r_{ij} = \sum_{k=1}^{n} \frac{(x_{ik} - \overline{x_i})(x_{jk} - \overline{x_j})}{\sigma_i \sigma_j} \quad (i, j = 1, 2, \ldots, n) \tag{7.20}$$

where

$$\overline{x_i} = \frac{1}{n}\sum_{k=1}^{n} x_{ik} \qquad \overline{x_j} = \frac{1}{n}\sum_{k=1}^{n} x_{jk} \tag{7.21}$$

$$\sigma_i = \sqrt{\sum_{k=1}^{n}(x_{ik} - \overline{x_i})^2} \qquad \sigma_j = \sqrt{\sum_{k=1}^{n}(x_{jk} - \overline{x_j})^2} \tag{7.22}$$

The other symbols have been defined in Equation (7.1) or Equation (7.3).

The method to create the correlation coefficient is the same as the method to create a single correlation coefficient matrix (i.e., the 0th-step regression matrix) expressed by Equation (2.34) in MRA.

$[r_{ij}]_{n \times n}$ is a real symmetrical matrix, $r_{ij} \in [-1, 1]$, and $r_{11} = r_{22} \ldots = r_{nn} = 1$. Obviously, the closer the correlation coefficient r_{ij} approaches to 1, the closer the property of x_i and x_j. This rule is used in the following cluster analysis.

7.1.3.3. Cluster Analysis

In Section 7.2 (basic principles), taking Q-mode cluster analysis as an example, the calculation steps for cluster analysis are summarized in general. In fact, the algorithm of cluster analysis is relatively complicated, which is mainly reflected by the calculation of cluster statistic such as distance coefficient, analog coefficient, and correlation coefficient. Taking Q-mode cluster analysis as an example, at the beginning when n classes are all sample vectors, calculation is explicit, but when the total number of classes is less than n, a new class will occur after merging, which includes two sample vectors.

Special treatment needs to be given to the calculation of a new class, which is reflected by different definitions of class distance between classes. Three cluster methods have been formed based on three cluster statistics (distance coefficient, analog coefficient, and correlation coefficient), and each cluster method covers three methods of class-distance definition (the shortest distance, the longest distance, and weighted mean). Consequently, nine cluster methods have been formed. Though they are different in basic points, the calculation steps are the same. The following is only the detailed introduction to the group of distance coefficients.

Before clustering, there are n classes, i.e., the total number of classes is n, marked as class $G_1, G_2, ..., G_n$, including sample vectors $x_1, x_2, ..., x_n$ expressed by Equation (7.1), respectively. After the first clustering, there occurs a new class that includes two sample vectors, i.e., the original two classes are merged into a new class, and then the total number of classes is $n - 1$. Continue clustering till the total number of classes is 1, i.e., there is only one class that includes n sample vectors, $x_1, x_2, ..., x_n$.

The following introduces three types of cluster method (distance coefficient, analog coefficient, and correlation coefficient).

Here d_{ij} is designated to express the distance between the sample vectors x_i and x_j, and D_{ij} is designated to express the distance between G_i and G_j.

1. *Aggregation using distance coefficient.* Here are three approaches (shortest distance method, longest distance method, and weighted mean method) to construct the distance coefficient.

   a. *Shortest distance method.* To be brief, the central idea of the shortest distance method is that class distance is defined as the minimum of "sample vector distance" expressed by Equation (7.18), and the two classes to which the minimum in all class-distances corresponds are merged into a new class.

   Letting u and v be the number of any two classes that are different from each other, the distance between classes G_u and G_v is defined to be

$$D_{uv} = \min_{\substack{x_i \in G_u \\ x_j \in G_v}} \{d_{ij}\} \quad (i, j = 1, 2, ..., n) \tag{7.23}$$

   where D_{uv} is the distance between the closest two sample vectors in classes G_u and G_v. *Note:* Here classes G_u and G_v possibly contain one sample vector and also possibly contain two or more sample vectors; the other symbols have been defined in Equation (7.18).

   The shortest distance method is to cluster by using the distance defined by Equation (7.23). It is performed in the following four steps:

   Step 1. To perform calculation of the initial distance coefficient matrix. Let $D^{(0)}$ be the initial distance coefficient matrix:

$$D^{(0)} = \left[D_{ij}^{(0)}\right]_{n \times n} = \left[d_{ij}\right]_{n \times n} \quad (i, j = 1, 2, ..., n) \tag{7.24}$$

   where d_{ij} is defined by Equation (7.18).
   $D^{(0)}$ is the distance coefficient matrix of classes $G_1, G_2, ..., G_n$.

Step 2. To determine which two classes are going to be merged into a new one according to $D^{(0)}$. Because the diagonal elements of the symmetrical matrix $D^{(0)}$ are all 0 (the distance of each class to itself is 0), it is necessary to search for a $D_{ij}^{(0)}$ that is the minimum one in the upper or lower triangular matrix elements of $D^{(0)}$, let it be d_{pq}, to merge class G_p and class G_q into a new class G_r, i.e., $G_r = \{G_p, G_q\}$. This is called *two classes in one class.*

Step 3. To calculate the distance between the new class G_r and the other class G_k,

$$D_{rk} = \min\left\{D_{pk}, D_{qk}\right\} \quad (k = 1, 2, ..., n, \text{ but } k \neq p, q) \tag{7.25}$$

where D_{pk} is the distance between class G_p and the other class G_k, which is defined by Equation (7.26); and D_{qk} is the distance between class G_q and the other class G_k, which is defined by Equation (7.27).

In terms of the all-purpose definition of Equation (7.23), D_{pk} and D_{qk} are defined as

$$D_{pk} = \min_{\substack{x_i \in G_p \\ x_k \in G_k}} \left\{d_{ij}\right\} \quad (k = 1, 2, ..., n \text{ but } k \neq p, q) \tag{7.26}$$

$$D_{qk} = \min_{\substack{x_i \in G_q \\ x_k \in G_k}} \left\{d_{ij}\right\} \quad (k = 1, 2, ..., n \text{ but } k \neq p, q) \tag{7.27}$$

It might be as well to let $p < q$ and keep p as the number of the new class. By eliminating the elements in the qth row and the qth column of $D^{(0)}$, $D^{(0)}$ becomes $(n-1) \times (n-1)$ matrix. The off-diagonal elements in the pth row and the qth column are substituted by $(n-2)$ numbers calculated by Equation (7.25), respectively, forming the distance coefficient matrix resulted from the first clustering:

$$D^{(1)} = \left[D_{ij}^{(1)}\right]_{(n-1)\times(n-1)} \quad (i, j = 1, 2, ..., n-1) \tag{7.28}$$

Step 4. To run the operation of Step 2 and Step 3 alternately. Now $(n-1)$ classes are available. Substitute $D^{(1)}$ for $D^{(0)}$, analogously run the operation of Step 2 so as to make the total number of classes be $n-2$, and analogously run the operation of Step 3 to obtain $D^{(2)}$. Substitute $D^{(2)}$ for $D^{(1)}$, analogously run the operation of Step 2 so as to make the total number of classes be $n-3$, and analogously run the operation of Step 3 to obtain $D^{(3)}$; ...; substitute $D^{(n-3)}$ for $D^{(n-4)}$, analogously run the operation of Step 2 so as to make the total number of classes be 2, and analogously run the operation of Step 3 to obtain $D^{(n-2)}$. Substitute $D^{(n-2)}$ for $D^{(n-3)}$, analogously run the operation of Step 2 so as to make the total number of classes be 1, and then stop calculation because it is not necessary to run Step 3 operation to obtain $D^{(n-1)}$ that is no longer a matrix but a number.

Note: If two or more elements, the minimum value of which is equal, are found in $D^{(s)}$ ($s = 1, 2, ...$) during operation of Step 2, then the operation that "four classes or more will be merged into one" is performed. For this purpose, it is necessary to make a simple modification to the operation of Step 2 to Step 4, which is not difficult.

b. *Longest distance method.* To be brief, the central idea of the longest distance method is that class distance is defined as the maximum value of *sample vector distance* expressed by Equation (7.18), and the two classes to which the minimum in all class distances corresponds are merged into a new class.

The difference from this shortest distance method is only the different class-distance definition. The calculation process is absolutely the same. So only the formulas related to class-distance definition in the longest distance method are listed here, since the other contents are the same as the shortest distance method.

Equation (7.23) is rewritten to

$$D_{uv} = \max_{\substack{x_i \in G_u \\ x_j \in G_v}} \{d_{ij}\} \quad (i, j = 1, 2, ..., n) \tag{7.29}$$

Equation (7.25) is rewritten to

$$D_{rk} = \max \{D_{pk}, D_{qk}\} \quad (k = 1, 2, ..., n \text{ but } k \neq p, q) \tag{7.30}$$

Equation (7.26) is rewritten to

$$D_{pk} = \max_{\substack{x_i \in G_p \\ x_k \in G_k}} \{d_{ij}\} \quad (k = 1, 2, ..., n \text{ but } k \neq p, q) \tag{7.31}$$

Equation (7.27) is rewritten to

$$D_{qk} = \max_{\substack{x_i \in G_q \\ x_k \in G_k}} \{d_{ij}\} \quad (k = 1, 2, ..., n \text{ but } k \neq p, q) \tag{7.32}$$

c. *Weighted mean method.* To be brief, the clou of the weighted mean method is that class distance is defined to be the weighted mean value of *sample vector distance* expressed by Equation (7.18), and the two classes that the minimum in all class distances corresponds to are merged into a new class.

The difference from the previous shortest-distance method and longest-distance method is only the different class-distance definition. The calculation process is absolutely the same. So, only the formulas related to class-distance definition in the weighted mean method are listed here, since the other contents are the same as the shortest-distance method.

Equation (7.23) is rewritten to

$$D_{uv}^2 = \frac{1}{n_u n_v} \sum_{\substack{x_i \in G_u \\ x_j \in G_v}} d_{ij}^2 \quad (i, j = 1, 2, ..., n) \tag{7.33}$$

where n_u is the number of sample vectors of class G_u; n_v is the number of sample vectors of class G_v; and the other symbols have been defined in Equation (7.23).

Equation (7.25) is rewritten to

$$D_{rk} = \frac{n_p}{n_p + n_q} D_{pk} + \frac{n_q}{n_p + n_q} D_{qk} \quad (k = 1, 2, ..., n \text{ but } k \neq p, q) \tag{7.34}$$

where D_{pk} is the distance between class G_p and the other class G_k, which is defined by Equation (7.35). D_{qk} is the distance between class G_q and the other class G_k, which is defined by Equation (7.36). n_p is the number of sample vectors of class G_p. n_q is the number of sample vectors of class G_q. and the other symbols have been defined in Equation (7.25).

Equation (7.26) is rewritten to

$$D_{pk}^2 = \frac{1}{n_p n_k} \sum_{\substack{x_i \in G_p \\ x_j \in G_k}} d_{ij}^2 \quad (i, j = 1, 2, ..., n) \tag{7.35}$$

where n_k is the number of sample vectors of class G_k, and the other symbols have been defined in Equations (7.26) and (7.34).

Equation (7.27) is rewritten to

$$D_{qk}^2 = \frac{1}{n_q n_k} \sum_{\substack{x_i \in G_q \\ x_j \in G_k}} d_{ij}^2 \quad (i, j = 1, 2, ..., n) \tag{7.36}$$

where all symbols have been defined in Equations (7.27), (7.34), and (7.35).

2. *Aggregation using analog coefficient.* In the aforementioned aggregation using distance coefficient, the distance coefficient d_{ij} defined in Equation (7.18) is substituted by the analog coefficient ξ_{ij} defined in Equation (7.19), which introduces three methods for aggregation using analog coefficient of the shortest-distance method, the longest-distance method, and the weighted mean method. Moreover, note that since the *distance coefficient matrix* is substituted by the *analog coefficient matrix*, it is not like aggregation using a distance coefficient to search for the minimum class distance in the upper or lower triangular matrix elements of the distance coefficient matrix to determine which two classes are to be merged into a new class; rather, it is to search for the maximum class distance in the upper or lower triangular matrix of the analog coefficient matrix to determine which two classes are to be merged into a new class.

3. *Aggregation using correlation coefficient.* In the aforementioned aggregation using a distance coefficient, the distance coefficient d_{ij} defined in Equation (7.18) is substituted by the correlation coefficient r_{ij} defined in Equation (7.20), which introduces three methods for aggregation using the correlation coefficient of the shortest-distance method, the longest-distance method, and the weighted mean method. Moreover, note that since the distance coefficient matrix is substituted by the correlation coefficient matrix, it is not like aggregation using the distance coefficient to search for the minimum class distance in the upper or lower triangular matrix elements of the distance coefficient matrix to determine which two classes to be merged into a new class. Rather, it is to search for the maximum

class distance in the upper or lower triangular matrix of the correlation coefficient matrix to determine which two classes should be merged into a new class.

Up to now, nine clustering methods have been introduced: three class-distance methods of the shortest-distance method, the longest-distance method, and the weighted mean method under the definition of each of the three cluster statistics (distance coefficient, analog coefficient, and correlation coefficient). Even if the same kind of data normalization (to select one form of standard difference standardization, maximum difference standardization, and maximum difference normalization) is used to the original data, the calculation results are found to be different with these nine methods. Then which method is the best one? A suitable one will be chosen depending on the practical application and specific data.

7.1.4. R-mode Cluster Analysis

Section 7.1.3 introduces the Q-mode cluster analysis. As for R-mode cluster analysis, the method is essentially the same as Q-mode cluster analysis. The only difference is different objects. The clustering object for Q-mode is n sample vectors expressed by Equation (7.1), whereas the clustering object for R-mode is m parameter vectors expressed by Equation (7.2) (Zhao, 1992; Li and Zhao, 1998; Everitt et al., 2011). Therefore, the only way is to transpose the original data matrix x expressed by Equation (7.3) at the beginning of the software of Q-mode cluster analysis, forming new original data x^T that is the transposed matrix of x:

$$x^T = \begin{bmatrix} x_{11} & x_{12} & \cdots & x_{1n} \\ x_{21} & x_{22} & \cdots & x_{2n} \\ \cdots & \cdots & \cdots & \cdots \\ x_{m1} & x_{m2} & \cdots & x_{mn} \end{bmatrix} \tag{7.37}$$

Q-mode cluster analysis is performed using x^T as original data and exchanging the software control parameters n with m. The obtained calculation results are the calculation results of R-mode cluster analysis.

This chapter provides four case studies coming from Chapter 6. Each case study uses the weighted mean method to perform Q-mode and R-mode cluster analyses. But the original data normalization method and cluster statistics are not the same. The reason is that each case study tries to use nine methods (i.e., three methods for data normalization of standard difference standardization, maximum difference standardization, and maximum difference normalization as well as the three methods for calculating cluster statistics of distance coefficient, analog coefficient, and correlation coefficient) for calculation, and then compare nine calculation results y by Q-mode cluster analysis with y^* of the corresponding case studies in Chapter 6, to select one result that is closest. The method obtaining the closest result is the method that should be used for this case study. For example: Case Study 1 uses standard difference standardization and analog coefficient; Case Study 2 uses maximum difference normalization and analog coefficient; Case Study 3 uses maximum difference standardization and correlation coefficient, and Case Study 4 uses maximum difference standardization and analog coefficient. Therefore, the optimal selection of methods depends on applications, i.e., different applications have different optimal selections.

7.2. CASE STUDY 1: INTEGRATED EVALUATION OF OIL AND GAS TRAP QUALITY

7.2.1. Studied Problem

In Sections 3.2 and 6.5, the data in Table 3.6 were used, and BAYSD, C-SVM, R-SVM, BPNN, and MRA were adopted to construct the function of trap quality (y) with respect to 14 relative parameters (x_1, x_2, ..., x_{14}) in the Northern Kuqa Depression. Now Table 3.6 is used as input data (excepting y^*) for Case Study 1 of this chapter, and it is wholly copied into Tables 7.1 so we can observe whether the results of Q-mode cluster analysis coincide with y^*.

Table 7.1 lists the quality of 30 traps in the Northern Kuqa Depression and 14 relative parameters (unit structure, trap type, petroliferous formation, trap depth, trap relief, trap closed area, formation HC identifier, data reliability, trap coefficient, source rock coefficient, reservoir coefficient, preservation coefficient, configuration coefficient, resource quantity). This table has 30 samples, and each sample contains 14 parameters (x_1, x_2, x_3, x_4, x_5, x_6, x_7, x_8, x_9, x_{10}, x_{11}, x_{12}, x_{13}, x_{14}), i.e., $n = 30$, $m = 14$ calculated out from Equation (7.1). The method of standard difference standardization is used for normalization to the original data, and the weighted mean method is used to conduct Q-mode and R-mode cluster analyses, taking analog coefficients as cluster statistics.

7.2.2. Input Data

7.2.2.1. Input Known Parameters

They are the value of x_i ($i = 1, 2, ..., 14$) of 30 traps in Table 7.1.

7.2.2.2. Input Control Information

The control information is related to the methods to be used: standard difference standardization, analog coefficient taken as cluster statistic, weighted mean method, and Q-mode and R-mode cluster analyses.

7.2.3. Results and Analysis

7.2.3.1. Q-mode Cluster Analysis

The calculation results from Q-mode cluster analysis are shown in Table 7.1 and Figure 7.2. Now the following two understandings have been gained through analysis.

1. *Sample cluster.* Figure 7.2 illustrates a cluster pedigree chart of 30 samples, and each sample contains unit structure, trap type, petroliferous formation, trap depth, trap relief, trap closed area, formation HC identifier, data reliability, trap coefficient, source rock coefficient, reservoir coefficient, preservation coefficient, configuration coefficient, resource quantity. It illustrates the whole process that 30 classes (i.e., 30 samples) are merged step by step into one class. Users can see the order of dependence between samples in the cluster pedigree chart. Two classes can be divided in Figure 7.2 (classes with even number), based on which the class value of each sample is filled in the y column of Table 7.1. It is found from the comparison of y with y^* that the coincidence rate reaches to 23 / 30 = 77%.

TABLE 7.1 Case Study 1 of Cluster Analysis: Integrated Evaluation of Oil and Gas Trap Quality

Sample Type	Trap No.	Relative Parameters for Trap Evaluation[a]														Trap Quality[b] y*	Q-mode Cluster Analysis y
		x_1	x_2	x_3	x_4	x_5	x_6	x_7	x_8	x_9	x_{10}	x_{11}	x_{12}	x_{13}	x_{14}		
Learning samples	1	1	1	2	2362	300	58	2	0.45	0.753	0.960	0.935	0.808	0.900	6.6	(2)	1
	2	1	2	1.5	3150	350	42	2	0.85	1.000	1.000	0.935	0.921	0.900	210.5	(1)	1
	3	1	2	2	3650	350	12	2	0.51	0.975	1.000	0.935	0.763	0.900	8.3	(1)	1
	4	1	2	2	2630	150	17	2	0.51	0.818	0.898	0.935	0.763	0.900	1.9	(2)	1
	5	1	3	2	5950	750	135	2	0.45	0.895	0.940	0.820	0.808	0.900	171.9	(1)	1
	6	1	2	2	3970	300	28	2	0.75	0.950	0.868	0.820	0.763	0.900	5.5	(2)	1
	7	1	2	2	4680	300	27	2	0.75	0.828	0.868	0.820	0.808	0.900	12.6	(2)	1
	8	1	1	2	1450	700	54	1	0.45	0.778	0.898	0.935	0.751	0.900	7.13	(2)	2
	9	1	1	3	1450	1000	74	1	0.45	0.778	0.898	0.935	0.808	0.900	9.8	(2)	2
	10	1	2	3	1200	750	23	1	0.45	0.888	0.970	0.840	0.681	0.900	1.4	(2)	2
	11	1	1	2	1550	1780	34	2	0.45	0.778	0.860	0.820	0.856	0.900	43.2	(2)	1
	12	1	1	2	6700	250	11	2	0.45	0.693	0.930	0.935	0.936	0.900	13.6	(1)	1
	13	1	1	2	5500	500	16	2	0.45	0.693	1.000	0.935	0.936	0.900	20.3	(1)	1
	14	1	1	2	5500	200	11	2	0.45	0.753	1.000	0.935	0.936	0.900	13.6	(1)	1
	15	1	2	1	850	550	50	1	0.45	1.000	0.868	0.820	0.681	0.900	1.82	(2)	2

Sample	x_1	x_2	x_3	x_4	x_5	x_6	x_7	x_8	x_9	x_{10}	x_{11}	x_{12}	x_{13}	x_{14}	y^*	y
16	1	2	3	1510	750	57	1	0.45	1.000	1.000	0.840	0.794	0.930	3.48	(1)	2
17	1	1	3	3510	1150	161	2	0.75	0.865	1.000	0.900	0.794	0.930	179.9	(1)	1
18	1	1	3	2700	300	56	1	0.45	0.888	0.970	0.840	0.714	0.900	3.41	(2)	2
19	2	3	1	4220	460	66	2	0.51	0.808	0.882	0.840	0.756	0.930	9.8	(2)	2
20	2	2	3	5600	300	27	2	0.51	0.828	0.882	0.840	0.756	0.930	29.2	(2)	2
21	2	2	3	8580	300	17	2	0.51	0.780	0.898	0.840	0.748	0.930	18.5	(2)	2
22	2	2	3	4940	260	46	2	0.51	0.808	0.798	0.840	0.756	0.930	3.9	(2)	2
23	3	3	1	1855	1800	42	2	0.85	0.865	0.798	0.840	0.909	0.900	4	(1)	1
24	3	3	2	4755	700	83	2	0.85	0.808	1.000	0.850	0.920	0.900	169.4	(1)	1
25	1	2	3	1000	400	82	1	0.45	0.913	0.970	0.885	0.673	0.930	2.61	(2)	2
26	1	1	4	3670	300	133	2	0.45	0.753	0.970	0.885	0.673	0.900	37.9	(2)	2
27	3	3	1	2750	1100	118	2	0.85	0.955	0.898	0.780	0.763	0.900	10.2	(2)	1
Prediction samples																
28	1	2	3	4450	250	35	2	0.60	0.853	0.898	0.900	0.898	0.900	58.9	(1)	1
29	2	3	3	5660	340	56	2	0.51	0.808	0.860	0.850	0.748	0.930	71.8	(2)	2
30	2	1	3	5850	180	17	2	0.51	0.753	0.758	0.840	0.728	0.930	19.1	(2)	2

[a] x_1 = unit structure (1−linear anticline belt, 2−Yangxia sag, 3−Qiulitake anticline belt); x_2 = trap type (1−faulted nose, 2−anticline, 3−faulted anticline); x_3 = petroliferous formation (1 − E, 1.5 − E + K, 2 − K, 3 − J, 4 − T); x_4 = trap depth (m); x_5 = trap relief (m); x_6 = trap closed area (km²); x_7 = formation HC identifier (1−oil, 2−gas); x_8 = data reliability (0−1); x_9 = trap coefficient (0−1); x_{10} = source rock coefficient (0−1); x_{11} = reservoir coefficient (0−1); x_{12} = preservation coefficient (0−1); x_{13} = configuration coefficient (0−1); and x_{14} = resource quantity (Mt, million-ton oil equivalent).

[b] y^* = trap quality (1−high, 2−low) assigned by geologists. In y^*, numbers in parentheses are not input data but are used for calculating $R(\%)$.

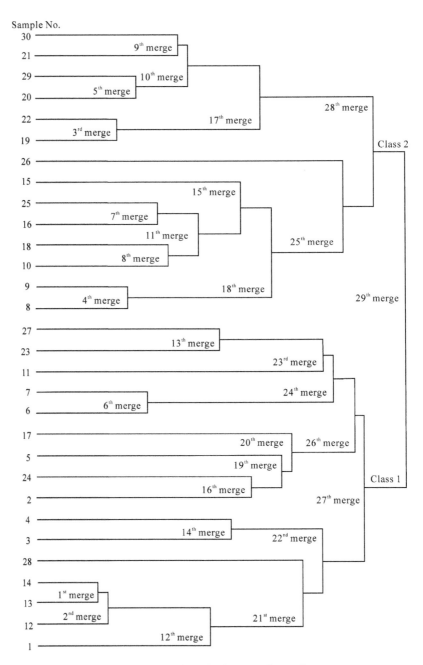

FIGURE 7.2   Case Study 1 of cluster analysis: Q-mode cluster pedigree chart.

2. *Sample reduction.* When the analog coefficient $r \geq 0.96$ (this value can be adjusted by users) between two samples, it is possible to eliminate one of the two samples to make the number of samples minus 1. From Table 7.2, it can be seen that $r = 0.9628 > 0.96$ between sample No. 13 and sample No. 14 in the first merging, which indicates one of the two samples (e.g., sample No. 13) is probably eliminated from the original data when data analysis (e.g., MRA, BPNN, C-SVM, R-SVM, DTR, and BAC) is performed, leading the number of samples minus 1, from 30 to 29. Is this method of sample reduction reasonable or not? Now it will be validated by using MRA calculation.

Using 27 learning samples, the calculation results by MRA are expressed by Equation (3.44) as follows:

$$y = 13.766 - 0.026405x_1 - 0.038781x_2 - 0.0016605x_3 - 0.00012343x_4$$
$$- 0.00038344x_5 - 0.002442x_6 + 0.045162x_7 + 0.58229x_8 - 3.3236x_9$$
$$- 2.1313x_{10} - 3.3651x_{11} - 4.1977x_{12} - 0.67296x_{13} + 0.0010516x_{14}$$

The corresponding residual variance is 0.14157, multiple correlation coefficient is 0.92652, $\overline{R}(\%)$ is 9.93. y is shown to depend on the 14 parameters in decreasing order: $x_{12}, x_{10}, x_9, x_4, x_5, x_{11}, x_8, x_6, x_{14}, x_1, x_2, x_7, x_{13}$, and x_3.

Now using 26 learning samples after sample No. 13 is reduced, the calculation results by MRA are as follows:

$$y = 13.665 - 0.033215x_1 - 0.037455x_2 - 0.0091141x_3 - 0.00012103x_4$$
$$- 0.0003717x_5 - 0.0024783x_6 + 0.040856x_7 + 0.60288x_8 - 3.3958x_9 \qquad (7.38)$$
$$- 1.9872x_{10} - 3.3847x_{11} - 4.2085x_{12} - 0.60149x_{13} + 0.00097989x_{14}$$

The corresponding residual variance is 0.15091, multiple correlation coefficient is 0.92146, $\overline{R}(\%)$ is 10.16. y is shown to depend on the 14 parameters in decreasing order: $x_{12}, x_{10}, x_9, x_4, x_5, x_{11}, x_8, x_6, x_{14}, x_1, x_2, x_7, x_{13}$, and x_3.

From comparison of the two calculation results (mainly the residual variance and $\overline{R}(\%)$) of MRA before and after sample No.13 is reduced, it is found that there is very little difference between the two calculation results, showing that the sample-reduction probably succeeds.

Table 7.2 notes that (a) as mentioned, $p < q$, and keep p as the number of new class; (b) for Class p or q; the number without superscript * contains only one sample, whereas the number with superscript * contains two or more samples; and (c) the same Class p^* possibly occurs two or more times due to its more than one merge, and this Class p^* contains more or more samples.

7.2.3.2. R-mode Cluster Analysis

The calculation results from R-mode cluster analysis are shown in Table 7.3 and Figure 7.3. Now the following two understandings have been gained through analysis:

1. *Parameter cluster.* Figure 7.3 illustrates a cluster pedigree chart of 14 parameters (unit structure, trap type, petroliferous formation, trap depth, trap relief, trap closed area, formation HC identifier, data reliability, trap coefficient, source rock coefficient, reservoir coefficient, preservation coefficient, configuration coefficient, resource quantity). It illustrates the whole process that 14 classes (i.e., 14 parameters) are merged step by step

TABLE 7.2   Case Study 1 of Cluster Analysis: Q-mode Cluster Analysis

| | Merged Classes | | | | | | New Class After Merge | |
| | Class G_p | | Class G_q | | Analog Coefficient Between Class G_p and Class G_q | | Class G_{p^*} | |
Merge No.	Class No. p	Sample Nos. Contained by Class G_p	Class No. q	Sample Nos. Contained by Class G_q		Class No. p^*	Sample Nos. Contained by Class G_{p^*}
1	13	13	14	14	0.9628	13^*	13, 14
2	12	12	13^*	13, 14	0.9415	12^*	12, 13, 14
3	19	19	22	22	0.9197	19^*	19, 22
4	8	8	9	9	0.8725	8^*	8, 9
5	20	20	29	29	0.8381	20^*	20, 29
6	6	6	7	7	0.8282	6^*	6, 7
7	16	16	25	25	0.8115	16^*	16, 25
8	10	10	18	18	0.8015	10^*	10, 18
9	21	21	30	30	0.7861	21^*	21, 30
10	20^*	20, 29	21^*	21, 30	0.7474	20^*	20, 21, 29, 30
11	10^*	10, 18	16^*	16, 25	0.6988	10^*	10, 16, 18, 25
12	1	1	12^*	12, 13, 14	0.6638	1^*	1, 12, 13, 14
13	23	23	27	27	0.6140	23^*	23, 27
14	3	3	4	4	0.5954	3^*	3, 4
15	10^*	10, 16, 18, 25	15	15	0.5626	10^*	10, 15, 16, 18, 25
16	2	2	24	24	0.5332	2^*	2, 24
17	19^*	19, 22	20^*	20, 21, 29, 30	0.4801	19^*	19, 20, 21, 22, 29, 30
18	8^*	8, 9	10^*	10, 15, 16, 18, 25	0.4452	8^*	8, 9, 10, 15, 16, 18, 25
19	2^*	2, 24	5	5	0.4246	2^*	2, 5, 24
20	2^*	2, 5, 24	17	17	0.4246	2^*	2, 5, 17, 24
21	1^*	1, 12, 13, 14	28	28	0.4212	1^*	1, 12, 13, 14, 28
22	1^*	1, 12, 13, 14, 28	3^*	3, 4	0.4212	1^*	1, 3, 4, 12, 13, 14, 28
23	11	11	23^*	23, 27	0.3509	11^*	11, 23, 27
24	6^*	6, 7	11^*	11, 23, 27	0.2128	6^*	6, 7, 11, 23, 27
25	8^*	8, 9, 10, 15, 16, 18, 25	26	26	0.1561	8^*	8, 9, 10, 15, 16, 18, 25, 26
26	2^*	2, 5, 17, 24	6^*	6, 7, 11, 23, 27	0.0917	2^*	2, 5, 6, 7, 11, 17, 23, 24, 27

(Continued)

TABLE 7.2   Case Study 1 of Cluster Analysis: Q-mode Cluster Analysis—cont'd

| | Merged Classes | | | | | New Class After Merge | |
| | Class G_p | | Class G_q | | Analog Coefficient Between Class G_p and Class G_q | Class G_{p^*} | |
Merge No.	Class No. p	Sample Nos. Contained by Class G_p	Class No. q	Sample Nos. Contained by Class G_q		Class No. p^*	Sample Nos. Contained by Class G_{p^*}
27	1^*	1, 3, 4, 12, 13, 14, 28	2^*	2, 5, 6, 7, 11, 17, 23, 24, 27	−0.2044	1^*	1, 2, 3, 4, 5, 6, 7, 11, 12, 13, 14, 17, 23, 24, 27, 28
28	8^*	8, 9, 10, 15, 16, 18, 25, 26	19^*	19, 20, 21, 22, 29, 30	−0.3411	8^*	8, 9, 10, 15, 16, 18, 19, 20, 21, 22, 25, 26, 29, 30
29	1^*	1, 2, 3, 4, 5, 6, 7, 11, 12, 13, 14, 17, 23, 24, 27, 28	8^*	8, 9, 10, 15, 16, 18, 19, 20, 21, 22, 25, 26, 29, 30	−0.6607	1^*	1, 2, 3, 4, 5, 6, 7, 8, 9, 10, 11, 12, 13, 14, 15, 16, 17, 18, 19, 20, 21, 22, 23, 24, 25, 26, 27, 28, 29, 30

into one class. Users can see the order of dependence between parameters in the cluster pedigree chart.

2. *Parameter dimension reduction.* It is impossible to perform the parameter dimension reduction in this case study. The reason is that the variety of analog coefficient listed in Table 7.3 is not consistent with the results of MRA used in this case study [see Equation (3.44)]. Table 7.3 shows that x_9 (trap coefficient) or x_{13} (configuration coefficient) is the first parameter for dimension reduction, whereas Equation (3.44) shows that x_3 (petroliferous formation) is the first parameter for dimension reduction.

It is noted in Table 7.3 that (a) as mentioned, $p < q$, and keep p as the number of the new class; (b) for Class p or q, the number without superscript * contains only one parameter, whereas the number with superscript * contains two or more parameters; and (c) the same Class p^* possibly occurs two or more times due to its more than one merge, and this Class p^* contains more or more parameters.

It is noted in Figure 7.3 that the definition of the geological parameter No is: 1, unit structure; 2, trap type; 3, petroliferous formation; 4, trap depth; 5, trap relief; 6, trap closed area; 7, formation HC identifier; 8, data reliability; 9, trap coefficient; 10, source rock coefficient; 11, reservoir coefficient; 12, preservation coefficient; 13, configuration coefficient; and 14, resource quantity.

7.2.4. Summary and Conclusions

In summary, using data for the trap evaluation of the Northern Kuqa Depression, i.e., the 14 parameters (unit structure, trap type, petroliferous formation, trap depth, trap relief, trap closed area, formation HC identifier, data reliability, trap coefficient, source rock coefficient, reservoir coefficient, preservation coefficient, configuration coefficient, resource quantity) of 30 traps, two cluster pedigree charts are gained by the data mining tool (Q-mode and R-mode cluster analyses). One chart is for 30 traps (Figure 7.2), indicating the order of dependence

TABLE 7.3 Case Study 1 of Cluster Analysis: R-mode Cluster Analysis

Merge No.	Merged Classes						New Class After Merge	
	Class G_p		Class G_q		Analog Coefficient Between Class G_p and Class G_q			Class G_p
	Class No. p	Parameter Nos. Contained by Class G_p	Class No. q	Parameter Nos. Contained by Class G_q		Class No. p^*	Parameter Nos. Contained by Class G_{p^*}	
1	9	9	13	13	0.9960	9^*	9, 13	
2	8	8	9^*	9, 13	0.9924	8^*	8, 9, 13	
3	7	7	8^*	8, 9, 13	0.9762	7^*	7, 8, 9, 13	
4	7^*	7, 8, 9, 13	10	10	0.9603	7^*	7, 8, 9, 10, 13	
5	7^*	7, 8, 9, 10, 13	11	11	0.9366	7^*	7, 8, 9, 10, 11, 13	
6	7^*	7, 8, 9, 10, 11, 13	12	12	0.8289	7^*	7, 8, 9, 10, 11, 12, 13	
7	6	6	7^*	7, 8, 9, 10, 11, 12, 13	0.8213	6^*	6, 7, 8, 9, 10, 11, 12, 13	
8	1	1	3	3	0.8192	1^*	1, 3	
9	1^*	1, 3	2	2	0.5361	1^*	1, 2, 3	
10	4	4	5	5	0.2664	4^*	4, 5	
11	1^*	1, 2, 3	6^*	6, 7, 8, 9, 10, 11, 12, 13	0.0720	1^*	1, 2, 3, 6, 7, 8, 9, 10, 11, 12, 13	
12	4^*	4, 5	14	14	−0.2190	4^*	4, 5, 14	
13	1^*	1, 2, 3, 6, 7, 8, 9, 10, 11, 12, 13	4^*	4, 5, 14	−0.9389	1^*	1, 2, 3, 4, 5, 6, 7, 8, 9, 10, 11, 12, 13, 14	

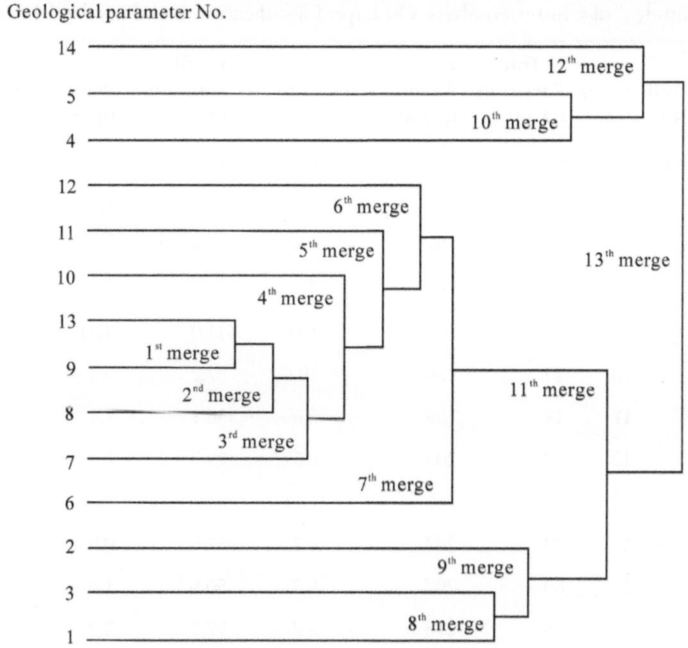

Geological parameter No.

FIGURE 7.3   Case Study 1 of cluster analysis: R-mode cluster pedigree chart.

between traps; the classification coincidence rate reaches 77%, and the results relatively coincide with practicality. The other chart is for 14 parameters (Figure 7.3), indicating the order of dependence between the parameters. These two cluster pedigree charts are called *mined knowledge*. Therefore, this method can be spread to the cluster analysis for integrated evaluation of oil and gas trap quality in other exploration areas. Moreover, it is found that (a) Q-mode cluster analysis can be used for trap sample-reduction, indicating that Q-mode cluster analysis can serve as a pioneering sample-reduction tool; whereas (b) R-mode cluster analysis (using an analog coefficient) cannot be used for parameter dimension reduction. The two discoveries are also called *mined knowledge*. As for parameter-dimension reduction, either MRA or BAYSD can serve as a pioneering dimension-reduction tool.

7.3. CASE STUDY 2: OIL LAYER CLASSIFICATION BASED ON WELL-LOGGING INTERPRETATION

7.3.1. Studied Problem

In Sections 4.3 and 6.4, the data in Table 4.4 was used, and C-SVM, R-SVM, BPNN, MRA, and BAYSD were adopted to construct the function of oil layer classification (y) with respect to five relative parameters (x_1, x_2, x_3, x_4, x_5) in the Xiefengqiao Anticline. Now Table 4.4 is wholly copied into Table 7.4 as input data (excepting y^*) for Case Study 2 of this chapter so as to observe whether the results of Q-mode cluster analysis coincide with y^*.

TABLE 7.4　Case Study 2 of Cluster Analysis: Oil Layer Classification Based on Well-Logging Interpretation

Sample Type	Sample No.	Well No.	Layer No.	x_1 True Resistivity $(\Omega \cdot m)$	x_2 Acoustictime $(\mu s/m)$	x_3 Porosity (%)	x_4 Oil Saturation (%)	x_5 Permeability (mD)	y^* Oil Test[a]	y Q-mode Cluster Analysis
Learning samples	1	ES4	5	64	206	6.0	48.6	1.1	(2)	3
	2		6	140	208	6.5	41.5	1.3	(3)	3
	3		7	63	206	6.0	36.4	1.1	(2)	2
	4		8	116	196	3.8	0.7	0.5	(3)	3
	5		9	17	267	19.6	44.0	32.4	(1)	1
	6		10	49	226	10.6	57.2	5.2	(2)	1
	7		11	44	208	6.6	36.1	1.4	(2)	2
	8		12	90	208	6.5	29.7	1.3	(3)	3
	9		13	69	260	18.1	81.7	16.8	(1)	1
	10	ES6	4	49	207	6.2	67.5	0.9	(2)	3
	11		5	80	207	6.3	50.9	1.0	(3)	3
	12		6	95	218	8.7	77.5	2.2	(3)	3
	13		8	164	212	7.5	67.5	1.5	(3)	3
	14	ES8	5_1	21	202	5.1	22.2	1.0	(2)	2
	15		5_2	56	192	2.9	24.2	0.6	(3)	2
	16		5_3	36	198	4.1	28.8	0.8	(2)	2
	17		6	128	196	3.4	19.2	0.6	(3)	3
	18		11	34	197	3.9	28.4	0.7	(2)	2
	19		12_1	10	208	6.4	42.4	1.7	(1)	2
	20		12_2	6	226	10.4	45.6	5.8	(1)	1
	21		12_3	6	225	10.3	50.4	6.1	(1)	1
	22		12_4	10	206	6.0	44.0	1.7	(1)	2
	23		13_1	7	224	9.9	44.4	5.2	(1)	1
	24		13_2	15	197	3.8	34.2	0.6	(1)	2
Prediction samples	25	ES8	13_4	11	201	4.8	39.3	0.8	(1)	2
	26		13_5	25	197	3.8	16.9	0.6	(2)	2
	27		7_2	109	199	4.4	17.8	0.8	(3)	3

[a]y^* = oil layer classification (1—oil layer, 2—poor oil layer, 3—dry layer) determined by the oil test. In y^*, numbers in parentheses are not input data but are used for calculating R(%).

Table 7.4 lists the oil test results of 27 layers in the Xiefengqiao Anticline and five relative parameters (true resistivity, acoustictime, porosity, oil saturation, permeability). This table has 27 samples, and each sample contains five parameters $(x_1, x_2, x_3, x_4, x_5)$, i.e., $n = 27$, $m = 5$ calculated out from Equation (7.1). The method of maximum difference normalization is used for normalization to the original data, and the weighted mean method is used to conduct Q-mode and R-mode cluster analyses, taking analog coefficients as cluster statistics.

7.3.2. Input Data

7.3.2.1. Known Input Parameters

They are the values of x_i $(i = 1, 2, 3, 4, 5)$ of 27 layers in Table 7.4.

7.3.2.2. Input Control Information

The control information is related to the methods to be used: maximum difference normalization, analog coefficient taken as cluster statistic, weighted mean method, and Q-mode and R-mode cluster analyses.

7.3.3. Results and Analyses

7.3.3.1. Q-mode Cluster Analysis

The calculation results from Q-mode cluster analysis are shown in Table 7.4 and Figure 7.4. Now the following two understandings have been gained through analysis:

1. *Sample cluster.* Figure 7.4 illustrates a cluster pedigree chart of 27 samples, and each sample contains true resistivity, acoustictime, porosity, oil saturation, and permeability. It illustrates the whole process as 27 classes (i.e., 27 samples) are merged step by step into one class. Users can see the order of dependence between samples in the cluster pedigree chart. Three classes can be limpingly divided in Figure 7.4 (classes with odd numbers), based on which the class value of each sample is filled in the y column of Table 7.4. It is found from the comparison of y with y^* that the coincidence rate reaches to $19 / 27 = 70\%$.
2. *Sample reduction.* When the analog coefficient $r \geq 0.999$ (this value can be adjusted by users) between two samples, it is possible to eliminate one of the samples to make the number of samples minus 1. From Table 7.5 it can be seen that $r = 0.9997 > 0.999$ between sample No. 16 and sample No. 18 in the first merging, which indicates that one of the two samples (e.g., sample No. 16) is probably eliminated from the original data when data analysis (e.g., MRA, BPNN, C-SVM, R-SVM, DTR, and BAC) is performed, leading the number of samples to be minus 1, from 27 to 26. Is this method of sample reduction reasonable or not? Now we will validate it by using MRA calculation.

Using 24 learning samples, the calculation results by MRA are expressed by Equation (4.22) as follows:

$$y = 62.641 + 0.0134x_1 - 0.3378x_2 + 1.3781x_3 - 0.0007x_4 + 0.0386x_5$$

The corresponding residual variance is 0.16649, multiple correlation coefficient is 0.91297, $\overline{R}(\%)$ is 8.33 (based on y, which is taken as integer). y is shown to depend on the five parameters in decreasing order: x_1, x_2, x_3, x_5, and x_4.

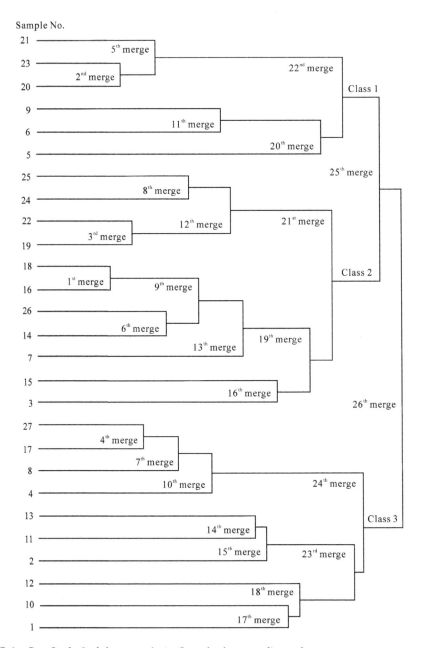

FIGURE 7.4   Case Study 2 of cluster analysis: Q-mode cluster pedigree chart.

TABLE 7.5 Case Study 2 of Cluster Analysis: Q-mode Cluster Analysis

Merge No.	Merged Classes					New Class After Merge	
	Class G_p		Class G_q		Analog Coefficient Between Class G_p and Class G_q	Class G_{p^*}	
	Class No. p	Sample Nos. Contained by Class G_p	Class No. q	Sample Nos. Contained by Class G_q		Class No. p^*	Sample Nos. Contained by Class G_{p^*}
1	16	16	18	18	0.9997	16*	16, 18
2	20	20	23	23	0.9969	20*	20, 23
3	19	19	22	22	0.9952	19*	19, 22
4	17	17	27	27	0.9859	17*	17, 27
5	20*	20, 23	21	21	0.9845	20*	20, 21, 23
6	14	14	26	26	0.9780	14*	14, 26
7	8	8	17*	17, 27	0.9719	8*	8, 17, 27
8	24	24	25	25	0.9654	24*	24, 25
9	14*	14, 26	16*	16, 18	0.9550	14*	14, 16, 18, 26
10	4	4	8*	8, 17, 27	0.9536	4*	4, 8, 17, 27
11	6	6	9	9	0.9218	6*	6, 9
12	19*	19, 22	24*	24, 25	0.9075	19*	19, 22, 24, 25
13	7	7	14*	14, 16, 18, 26	0.9068	7*	7, 14, 16, 18, 26
14	11	11	13	13	0.9029	11*	11, 13

(Continued)

TABLE 7.5 Case Study 2 of Cluster Analysis: Q-mode Cluster Analysis—cont'd

Merge No.	Merged Classes				Analog Coefficient Between Class G_p and Class G_q	New Class After Merge	
	Class G_p		Class G_q				Class G_{p^*}
	Class No. p	Sample Nos. Contained by Class G_p	Class No. q	Sample Nos. Contained by Class G_q		Class No. p^*	Sample Nos. Contained by Class G_{p^*}
15	2	2	11^*	11, 13	0.8822	2^*	2, 11, 13
16	3	3	15	15	0.8312	3^*	3, 15
17	1	1	10	10	0.7799	1^*	1, 10
18	1^*	1, 10	12	12	0.7799	1^*	1, 10, 12
19	3^*	3, 15	7^*	7, 14, 16, 18, 26	0.7681	3^*	3, 7, 14, 15, 16, 18, 26
20	5	5	6^*	6, 9	0.7294	5^*	5, 6, 9
21	3^*	3, 7, 14, 15, 16, 18, 26	19^*	19, 22, 24, 25	0.6047	3^*	3, 7, 14, 15, 16, 18, 19, 22, 24, 25, 26
22	5^*	5, 6, 9	20^*	20, 21, 23	0.5889	5^*	5, 6, 9, 20, 21, 23
23	1^*	1, 10, 12	2^*	2, 11, 13	0.5843	1^*	1, 2, 10, 11, 12, 13
24	1^*	1, 2, 10, 11, 12, 13	4^*	4, 8, 17, 27	−0.0091	1^*	1, 2, 4, 8, 10, 11, 12, 13, 17, 27
25	3^*	3, 7, 14, 15, 16, 18, 19, 22, 24, 25, 26	5^*	5, 6, 9, 20, 21, 23	−0.1525	3^*	3, 5, 6, 7, 9, 14, 15, 16, 18, 19, 20, 21, 22, 23, 24, 25, 26
26	1^*	1, 2, 4, 8, 10, 11, 12, 13, 17, 27	3^*	3, 5, 6, 7, 9, 14, 15, 16, 18, 19, 20, 21, 22, 23, 24, 25, 26	−0.8917	1^*	1, 2, 3, 4, 5, 6, 7, 8, 9, 10, 11, 12, 13, 14, 15, 16, 17, 18, 19, 20, 21, 22, 23, 24, 25, 26, 27

Now for the 23 learning samples: After sample No. 16 is reduced, the calculation results by MRA are as follows:

$$y = 63.363 + 0.0135x_1 - 0.3419x_2 + 1.3993x_3 - 0.0008x_4 + 0.0375x_5 \qquad (7.39)$$

The corresponding residual variance is 0.16585, multiple correlation coefficient is 0.91332, $\overline{R}(\%)$ is 8.7 (based on y, which is taken as integer). y is shown to depend on the five parameters in decreasing order: x_1, x_2, x_3, x_5, and x_4.

From comparison of the two calculation results (mainly the residual variance and $\overline{R}(\%)$) of MRA before and after sample No.16 is reduced, it is found that there is little difference between the two calculation results, showing that the sample reduction probably succeeds.

It is noted in Table 7.5 that (a) as mentioned, $p < q$, and keep p as the number of new class; (b) for Class p or q, the number without superscript * contains only one sample, whereas the number with superscript * contains two or more samples; and (c) the same Class p^* possibly occurs two or more times due to its more than one merge, and this Class p^* contains more or more samples.

7.3.3.2. R-mode Cluster Analysis

The calculation results from R-mode cluster analysis are shown in Table 7.6 and Figure 7.5. Now the following two understandings have been gained through analysis:

1. *Parameter cluster.* Figure 7.5 illustrates a cluster pedigree chart of five parameters (true resistivity, acoustictime, porosity, oil saturation, permeability). It illustrates the whole process that five classes (i.e., five parameters) are merged step by step into one class. Users can see the order of dependence between parameters in the cluster pedigree chart.
2. *Parameter dimension reduction.* It is impossible to perform the parameter-dimension reduction in this case study. The reason is that the variety of analog coefficients listed in Table 7.6 is not consistent with the results of MRA used in this case study [see Equation (4.22)]. Table 7.6 shows that x_3 (porosity) or x_5 (permeability) is the first parameter for dimension reduction, whereas Equation (4.22) shows that x_4 (oil saturation) is the first parameter for dimension reduction.

TABLE 7.6 Case Study 2 of Cluster Analysis: R-Mode Cluster Analysis

Merge No.	Merged Classes						New Class After Merge	
	Class G_p		Class G_q		Analog Coefficient Between Class G_p and Class G_q		Class G_p	
	Class No. p	Parameter Nos. Contained by Class G_p	Class No. q	Parameter Nos. Contained by Class G_q			Class No. p^*	Parameter Nos. Contained by Class G_{p^*}
1	3	3	5	5	0.7346		3^*	3, 5
2	3^*	3, 5	4	4	0.3713		3^*	3, 4, 5
3	1	1	2	2	−0.1492		1^*	1, 2
4	1^*	1, 2	3^*	3, 4, 5	−1.0000		1^*	1, 2, 3, 4, 5

Geological parameter No.

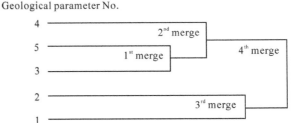

FIGURE 7.5   Case Study 2 of cluster analysis: R-mode cluster pedigree chart.

Table 7.6 shows that (a) as mentioned, $p < q$, and keep p as the number of new class; (b) for Class p or q, the number without superscript * contains only one parameter, whereas the number with superscript * contains two or more parameters; and (c) the same Class p^* possibly occurs two or more times due to its more than one merge, and this Class p^* contains more or more parameters.

It is noted in Figure 7.5 that the definition of the geological parameter number is: 1, true resistivity; 2, acoustictime, 3, porosity; 4, oil saturation; and 5, permeability.

7.3.4. Summary and Conclusions

In summary, using data for the oil layer classification of the Xiefengqiao Anticline, i.e., the five parameters (true resistivity, acoustictime, porosity, oil saturation, permeability) of 27 oil layers, two cluster pedigree charts are gained by the data mining tool (Q-mode and R-mode cluster analyses). One chart is for 27 oil layers (Figure 7.4), indicating the order of dependence between oil layers; the classification coincidence rate reaches 70%, and the results relatively coincide with practicality. The other chart is for five parameters (Figure 7.5), indicating the order of dependence between the parameters. These two cluster pedigree charts are called *mined knowledge*. Therefore, this method can be spread to the cluster analysis for oil layer classification based on well-logging interpretation in other areas. Moreover, it is found that (a) Q-mode cluster analysis can be used for oil layer sample-reduction, indicating that Q-mode cluster analysis can serve as a pioneering sample-reduction tool; and (b) whereas R-mode cluster analysis (using the analog coefficient) cannot be used for parameter dimension reduction. The two discoveries are also called *mined knowledge*. As for parameter dimension reduction, either MRA or BAYSD can serve as a pioneering dimension-reduction tool.

7.4. CASE STUDY 3: COAL-GAS-OUTBURST CLASSIFICATION

7.4.1. Studied Problem

In Section 6.6, the data in Table 6.14 were used, and BAYD, BAYSD, C-SVM, R-SVM, BPNN, and MRA were adopted to construct the function of coal-gas-outburst classification (y) with respect to five relative parameters (x_1, x_2, x_3, x_4, x_5) from the 21 typical coal mines in China. Now Table 6.14 is wholly copied into Table 7.7 as input data (excepting y^*) for Case Study 3 of this chapter so as to observe whether the results of Q-mode cluster analysis coincide with $y.^*$

Table 7.7 lists the outbursts of Chinese typical coal mines and five relative parameters (initial speed of methane diffusion, coefficient of coal consistence, gas pressure, destructive style of coal, mining depth). This table has 21 samples, and each sample contains five

TABLE 7.7　Case Study 3 of Cluster Analysis: Coal-Gas-Outburst Classification in China

Sample Type	Sample No.	Relative Parameters for Coal-Gas-Outburst Classification[a]					Outburst Type[b]	Q-mode Cluster Analysis
		x_1	x_2	x_3	x_4	x_5	y^*	y
Learning samples	1	19.00	0.31	2.76	3	620	(4)	4
	2	6.00	0.24	0.95	5	445	(2)	3
	3	18.00	0.16	1.20	3	462	(2)	4
	4	5.00	0.61	1.17	1	395	(1)	1
	5	8.00	0.36	1.25	3	745	(3)	3
	6	7.00	0.48	2.00	1	460	(1)	1
	7	14.00	0.22	3.95	3	543	(4)	4
	8	11.00	0.28	2.39	3	515	(2)	4
	9	4.800	0.60	1.05	2	477	(1)	1
	10	6.00	0.24	0.95	3	455	(3)	3
	11	14.00	0.34	2.16	4	510	(2)	2
	12	4.00	0.58	1.40	3	428	(1)	1
	13	4.00	0.53	1.65	2	438	(1)	1
	14	6.00	0.54	3.95	5	543	(4)	1
	15	7.40	0.37	0.75	4	740	(3)	3
	16	3.00	0.51	1.40	3	400	(1)	1
Prediction samples	17	11.0	0.37	2.1	3	412	(2)	2
	18	11.5	0.28	1.9	3	407	(2)	2
	19	11.8	0.36	2.3	3	403	(2)	2
	20	10.8	0.30	2.2	3	396	(2)	2
	21	12.4	0.38	1.8	3	410	(2)	2

[a] x_1 = initial speed of methane diffusion (m/s); x_2 = coefficient of coal consistence; x_3 = gas pressure (MPa); x_4 = destructive style of coal (1—nonbroken coal, 2—broken coal, 3—strongly broken coal, 4—pulverized coal, 5—completely pulverized coal); and x_5 = mining depth (m).
[b] y^* = coal-gas-outburst classification (1—non, 2—small, 3—medium, 4—large) determined by the coal test. In y^*, numbers in parentheses are not input data but are used for calculating R(%).

parameters $(x_1, x_2, x_3, x_4, x_5)$, i.e., $n = 21$, $m = 5$ calculated from Equation (7.1). The method of maximum difference standardization is used for normalization to the original data, and the weighted mean method is used to conduct Q-mode and R-mode cluster analyses, taking correlation coefficients as cluster statistics.

7.4.2. Input Data

7.4.2.1. Known Input Parameters

They are the values of x_i $(i = 1, 2, 3, 4, 5)$ of 21 samples in Table 7.7.

7.4.2.2. Input Control Information

The control information is related to the methods to be used: maximum difference standardization, correlation coefficient taken as cluster statistic, weighted mean method, and Q-mode and R-mode cluster analyses.

7.4.3. Results and Analyses

7.4.3.1. Q-mode Cluster Analysis

The calculation results from Q-mode cluster analysis are shown in Table 7.8 and Figure 7.6. Now the following two understandings have been gained through analysis:

1. *Sample cluster.* Figure 7.6 illustrates a cluster pedigree chart of 21 samples, and each sample contains initial speed of methane diffusion, coefficient of coal consistence, gas pressure, destructive style of coal, mining depth. It illustrates the whole process via which 21 classes (i.e., 21 samples) are merged step by step into one class. Users can see the order of dependence between samples in the cluster pedigree chart. Four classes can be divided in Figure 7.6 (classes with even number), based on which the class value of each sample is filled in the y column of Table 7.7. It is found from the comparison of y with y^* that the coincidence rate reaches to $17 / 21 = 81\%$.

2. *Sample reduction.* When the correlation coefficient $r \geq 0.986$ (this value can be adjusted by users) between two samples, it is possible to eliminate one of the samples to make the number of samples minus 1. From Table 7.8 it can be seen that $r = 0.9865 > 0.986$ between sample No. 12 and sample No. 16 in the first merging, which indicates one of the two samples (e.g., sample No. 12) is probably eliminated from the original data when data analysis (e.g., MRA, BPNN, C-SVM, R-SVM, DTR, and BAC) is performed, leading the number of samples minus 1, from 21 to 20. Is this method of sample reduction reasonable or not? Now we will validate it using MRA calculation.

Using 16 learning samples, the calculation results by MRA are expressed by Equation (6.69) as follows:

$$y = -0.69965 - 0.020735x_1 - 2.7151x_2 + 0.57211x_3 + 0.16274x_4 + 0.00513x_5$$

The corresponding residual variance is 0.18264, multiple correlation coefficient is 0.90408, $\overline{R}(\%)$ is 22.40 (based on y, which is taken as an integer). y is shown to depend on the five parameters in decreasing order: x_5, x_3, x_2, x_4, and x_1.

TABLE 7.8   Case Study 3 of Cluster Analysis: Q-mode Cluster Analysis

| | Merged Classes | | | | | New Class After Merge | |
| | Class G_p | | Class G_q | | Analog Coefficient Between Class G_p and Class G_q | Class G_{p*} | |
Merge No.	Class No. p	Sample Nos. Contained by Class G_p	Class No. q	Sample Nos. Contained by Class G_q		Class No. $p*$	Sample Nos. Contained by Class G_{p*}
1	12	12	16	16	0.9865	12*	12, 16
2	17	17	19	19	0.9856	17*	17, 19
3	2	2	10	10	0.9816	2*	2, 10
4	18	18	20	20	0.9623	18*	18, 20
5	17*	17, 19	18*	18, 20	0.9441	17*	17, 18, 19, 20
6	4	4	9	9	0.9415	4*	4, 9
7	4*	4, 9	13	13	0.9415	4*	4, 9, 13
8	5	5	15	15	0.9299	5*	5, 15
9	17*	17, 18, 19, 20	21	21	0.9240	17*	17, 18, 19, 20, 21
10	4*	4, 9, 13	12*	12, 16	0.8773	4*	4, 9, 12, 13, 16
11	7	7	8	8	0.8447	7*	7, 8
12	1	1	3	3	0.8170	1*	1, 3
13	11	11	17*	17, 18, 19, 20, 21	0.7769	11*	11, 17, 18, 19, 20, 21
14	4*	4, 9, 12, 13, 16	6	6	0.7707	4*	4, 6, 9, 12, 13, 16
15	1*	1, 3	7*	7, 8	0.4943	1*	1, 3, 7, 8
16	1*	1, 3, 7, 8	11*	11, 17, 18, 19, 20, 21	0.4402	1*	1, 3, 7, 8, 11, 17, 18, 19, 20, 21
17	4*	4, 6, 9, 12, 13, 16	14	14	0.4013	4*	4, 6, 9, 12, 13, 14, 16
18	2*	2, 10	5*	5, 15	0.2966	2*	2, 5, 10, 15
19	2*	2, 5, 10, 15	4*	4, 6, 9, 12, 13, 14, 16	−0.1018	2*	2, 5, 4, 6, 9, 10, 12, 13, 14, 15, 16
20	1*	1, 3, 7, 8, 11, 17, 18, 19, 20, 21	2*	2, 5, 4, 6, 9, 10, 12, 13, 14, 15, 16	−0.2559	1*	1, 2, 3, 4, 5, 6, 7, 8, 9, 10, 11, 12, 13, 14, 15, 16, 17, 18, 19, 20, 21

Sample No.

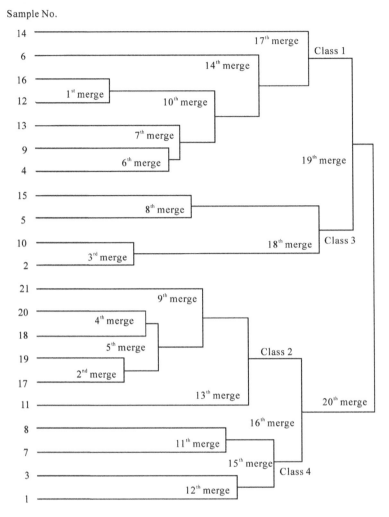

FIGURE 7.6   Case Study 3 of cluster analysis: Q-mode cluster pedigree chart.

Now using 15 learning samples after sample No. 16 is reduced, the calculation results by MRA are as follows:

$$y = -0.73246 - 0.019225x_1 - 2.5827x_2 + 0.5651x_3 + 0.17221x_4 + 0.00506x_5 \qquad (7.40)$$

The corresponding residual variance is 0.19605, multiple correlation coefficient is 0.89663, $\overline{R}(\%)$ is 23.89 (based on y, which is taken as an integer). y is shown to depend on the five parameters in decreasing order: x_5, x_3, x_2, x_4, and x_1.

From comparison of the two calculation results (mainly the residual variance and $\overline{R}(\%)$) of MRA before and after sample No.12 is reduced, it is found that there is little difference between the two calculation results, showing that the sample reduction probably succeeds.

TABLE 7.9   Case Study 3 of Cluster Analysis: R-mode Cluster Analysis

| | | Merged Classes | | | | | New Class After a Merge | |
| | | Class G_p | | Class G_q | | | | Class G_p |
Merge No.	Class No. p	Parameter Nos. Contained by Class G_p	Class No. q	Parameter Nos. Contained by Class G_q	Analog Coefficient Between Class G_p and Class G_q		Class No. p^*	Parameter Nos. Contained by Class G_{p^*}
1	1	1	3	3	0.9982		1^*	1, 3
2	1^*	1, 3	2	2	0.9947		1^*	1, 2, 3
3	1^*	1, 2, 3	5	5	0.0984		1^*	1, 2, 3, 5
4	1^*	1, 2, 3, 5	4	4	−0.4524		1^*	1, 2, 3, 4, 5

It is noted in Table 7.8 that (a) as mentioned, $p < q$, and keep p as the number of new class; (b) for Class p or q, the number without superscript * contains only one sample, whereas the number with superscript * contains two or more samples; and (c) the same Class p^* possibly occurs two or more times due to its more than one merge, and this Class p^* contains more or more samples.

7.4.3.2. R-mode Cluster Analysis

The calculation results from R-mode cluster analysis are shown in Table 7.9 and Figure 7.7. Now the following two understandings have been gained through analysis:

1. *Parameter cluster.* Figure 7.7 illustrates a cluster pedigree chart of five parameters (initial speed of methane diffusion, coefficient of coal consistence, gas pressure, destructive style of coal, mining depth). It illustrates the whole process that five classes (i.e., five parameters) are merged step by step into one class. Users can see the order of dependence between samples in the cluster pedigree chart.
2. *Parameter-dimension reduction.* It is possible to perform the parameter-dimension reduction in this case study. The reason is that the variety of correlation coefficients listed in Table 7.9 is consistent with the results of MRA used in this case study [see Equation (6.69)]. Table 7.9 shows that x_1 (initial speed of methane diffusion) or x_3 (gas pressure) is the first parameter

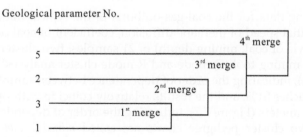

Geological parameter No.

FIGURE 7.7   Case Study 3 of cluster analysis: R-mode cluster pedigree chart.

for dimension reduction, whereas Equation (6.69) shows that x_1(initial speed of methane diffusion) is the first parameter for dimension reduction. If x_1 (initial speed of methane diffusion) is eliminated, leading the number of parameters minus 1, resulting in 5 reduced to 4, that is, the studied 6-D (plus y) problem is reduced to a 5-D (plus y) problem. Is this method of parameter reduction reasonable or not? Now it will be validated by MRA calculation.

Using five parameters (x_1, x_2, x_3, x_4, x_5), the calculation results by MRA have been expressed by Equation (6.69) as follows:

$$y = -0.69965 - 0.020735x_1 - 2.7151x_2 + 0.57211x_3 + 0.16274x_4 + 0.00513x_5$$

The corresponding residual variance is 0.18264, and the multiple correlation coefficient is 0.90408, $\overline{R}(\%)$ is 22.40 (based on y, which is taken as integer). y is shown to depend on the five parameters in decreasing order: x_5, x_3, x_2, x_4, and x_1.

Eliminating x_1, i.e., using four parameters (x_2, x_3, x_4, x_5), the calculation results by MRA are expressed as follows:

$$y = -1.0012 - 2.2113x_2 + 0.53461x_3 + 0.18767x_4 + 0.0049663x_5 \qquad (7.41)$$

The corresponding residual variance is 0.18548, multiple correlation coefficient is 0.90251, and $\overline{R}(\%)$ is 16.15 (based on y which is taken as integer). y is shown to depend on the four parameters in decreasing order: x_5, x_3, x_2, and x_4.

It is found from comparison of the two calculation results [mainly the residual variance and $\overline{R}(\%)$] of MRA before and after x_1 is reduced that the fitting residual is smaller after x_1 is reduced, that is, fitness is improved, so the parameter-dimension reduction probably succeeds.

It is noted in Table 7.9 that (a) as mentioned, $p < q$, and keep p as the number of new class; (b) for Class p or q, the number without superscript * contains only one parameter, whereas the number with superscript * contains two or more parameters; and (c) the same Class p^* possibly occurs two or more times due to its more than one merger, and this Class p^* contains more or more parameters.

Figure 7.7 notes that the definition of the geological parameter number is 1, initial speed of methane diffusion; 2, coefficient of coal consistence; 3, gas pressure; 4, destructive style of coal; and 5, mining depth.

7.4.4. Summary and Conclusions

In summary, using data for the coal-gas-outburst of the typical mines in China, i.e., the five parameters (initial speed of methane diffusion, coefficient of coal consistence, gas pressure, destructive style of coal, mining depth) of 21 samples, two cluster pedigree charts are gained by the data mining tool (Q-mode and R-mode cluster analyses). One chart is for 21 samples (Figure 7.6), indicating the order of dependence between samples; the classification coincidence rate reaches 81%, and the results relatively coincide with practicality. The other chart is for five parameters (Figure 7.7), indicating the order of dependence between the parameters. These two cluster pedigree charts are called *mined knowledge*. Therefore, this method can be spread to the cluster analysis for coal-gas-outburst in other mines. Moreover, it is found that (a) Q-mode cluster analysis can be used for coal-gas-outburst sample-reduction, indicating that Q-mode cluster analysis can serve as a pioneering sample-reduction tool; whereas (b) R-mode cluster analysis (using a correlation coefficient) can be used for parameter-dimension reduction, indicating that R-mode cluster analysis can serve as a dimension-reduction tool like MRA and BAYSD. The two discoveries are also called *mined knowledge*.

7.5. CASE STUDY 4: RESERVOIR CLASSIFICATION IN THE BAIBAO OILFIELD

7.5.1. Studied Problem

In Section 6.3, the data in Table 6.7 was used, and BAYSD, C-SVM, R-SVM, BPNN, and MRA were adopted to construct the function of reservoir classification (y) with respect to four relative parameters (x_1, x_2, x_3, x_4) in the Baibao Oilfield. Now Table 6.7 is wholly copied into Table 7.10 as input data (excepting y^*) for Case Study 4 of this chapter so as to observe whether the results of Q-mode cluster analysis coincide with y^*.

Table 7.10 lists the reservoir classification in the C3 and C4+5 low-permeability reservoir in the Baibao Oilfield and four relative parameters (shale content, porosity, permeability, permeability variation coefficient). This table has 28 samples, and each sample contains four parameters (x_1, x_2, x_3, x_4), i.e., $n = 28$, $m = 4$ calculated out from Equation (7.1). The method of maximum difference standardization is used for normalization to the original data, and the weighted mean method is used to conduct Q-mode and R-mode cluster analyses taking analog coefficients as cluster statistics.

7.5.2. Input Data

7.5.2.1. Known Input Parameters

They are the values of x_i ($i = 1, 2, 3, 4$) of 28 layers in Table 7.10.

7.5.2.2. Input Control Information

The control information is related to the methods to be used: maximum difference standardization, analog coefficient taken as cluster statistic, weighted mean method, and Q-mode and R-mode cluster analyses.

7.5.3. Results and Analyses

7.5.3.1. Q-mode Cluster Analysis

The calculation results from Q-mode cluster analysis are shown in Table 7.10 and Figure 7.8. Now the following two understandings have been gained through analysis:

1. *Sample cluster.* Figure 7.8 illustrates a cluster pedigree chart of 28 samples, and each sample contains shale content, porosity, permeability, and permeability variation coefficient. It illustrates the whole process that 28 classes (i.e., 28 samples) are merged step by step into

TABLE 7.10   Case Study 4 of Cluster Analysis: Reservoir Classification in the Baibao Oilfield

Sample Type	Sample No.	Well No.	Layer No.	x_1 Mud Content (%)	x_2 Porosity (%)	x_3 Permeability (mD)	x_4 Permeability Variation Coefficient	y^* Reservoir Type[a]	y Q-mode Cluster Analysis
Learning samples	1	B102	$C3^2$	27.2	9.074	0.068	2.69	(3)	3
	2	B102	$C3^3$	10.9	21.66	36.6	0.51	(1)	1
	3	B102	$C4+5^1$	13.74	13.64	15.9	1.01	(2)	2
	4	B107	$C3^2$	18.4	0.18	0.025	2.82	(4)	4
	5	B108	$C3^3$	8.48	19.27	28.62	0.57	(1)	1
	6	B108	$C4+5^1$	13.98	12.44	15.72	0.83	(2)	2
	7	B110	$C4+5^1$	10.58	12.03	14.01	1.07	(2)	2
	8	B112	$C4+5^1$	11.91	9.16	0.77	1.93	(3)	3
	9	B115	$C4+5^1$	15.0	16.7	31.56	0.65	(2)	2
	10	B123	$C3^3$	12.5	15.13	22.12	1.27	(2)	2
	11	B130	$C3^2$	14.3	11.9	15.39	1.21	(2)	2
	12	B202	$C4+5^1$	12.2	1.5	1.28	1.4	(4)	4
	13	B205	$C3^1$	13.4	5.68	0.81	1.35	(3)	3
	14	B205	$C3^3$	20.26	9.45	1.89	1.83	(3)	3
	15	B205	$C4+5^1$	12.78	12.34	12.71	1.14	(2)	2
	16	B205	$C4+5^2$	12.8	14.91	33.42	0.44	(1)	1
	17	B206	$C3^3$	13.26	6.35	0.98	1.58	(3)	3
	18	B206	$C4+5^1$	14.2	13.84	21.34	0.92	(2)	2
	19	B206	$C4+5^2$	11.3	15.8	34.37	0.52	(1)	1
	20	B210	$C3^1$	21.89	8.35	0.58	1.37	(3)	3
	21	B210	$C4+5^1$	14.32	11.57	18.72	1.01	(2)	2
	22	B215	$C3^3$	13.6	12.96	15.61	0.98	(2)	2
	23	B215	$C4+5^1$	14.03	13.78	28.23	1.0	(2)	2
	24	B217	$C3^3$	17.98	11.53	20.73	0.84	(2)	2
	25	B217	$C4+5^1$	13.89	12.24	14.53	1.23	(2)	2
Prediction samples	26	B219	$C3^3$	14.74	12.32	23.79	0.84	(2)	2
	27	H186	$C4+5^2$	18.65	5.13	0.48	1.15	(3)	3
	28	W4	$C4+5^2$	19.9	1.09	0.003	3.84	(4)	4

[a]y^* = reservoir classification (1—excellent, 2—good, 3—average, 4—poor) determined by the well test. In y^*, numbers in parentheses are not input data but are used for calculating R(%).

Sample No.

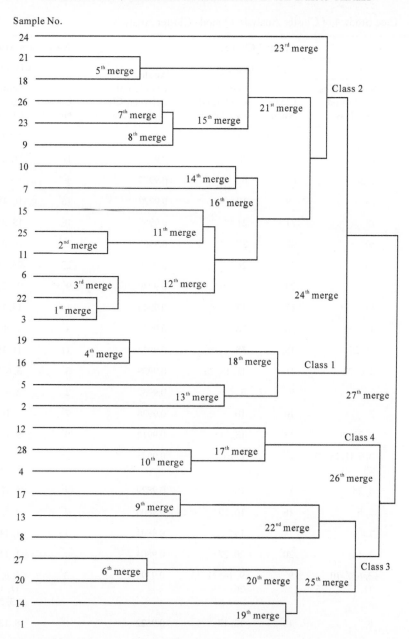

FIGURE 7.8   Case Study 4 of cluster analysis: Q-mode cluster pedigree chart.

one class. Users can see the order of dependence between samples in the cluster pedigree chart. Four classes can be divided in Figure 7.8 (classes with even numbers), based on which the class value of each sample is filled in the y column of Table 7.10. It is found from the comparison of y with y^* that the coincidence rate reaches to 28 / 28 = 100%.

TABLE 7.11   Case Study 4 of Cluster Analysis: Q-mode Cluster Analysis

| | Merged Classes | | | | | New Class After Merging | |
| | Class G_p | | Class G_q | | Analog | Class G_{p*} | |
Merge No.	Class No. p	Sample Nos. Contained by Class G_p	Class No. q	Sample Nos. Contained by Class G_q	Coefficient Between Class G_p and Class G_q	Class No. $p*$	Sample Nos. Contained by Class G_{p*}
1	3	3	22	22	0.9999	3^*	3, 22
2	11	11	25	25	0.9990	11^*	11, 25
3	3^*	3, 22	6	6	0.9975	3^*	3, 6, 22
4	16	16	19	19	0.9970	16^*	16, 19
5	18	18	21	21	0.9965	18^*	18, 21
6	20	20	27	27	0.9965	20^*	20, 27
7	23	23	26	26	0.9959	23^*	23, 26
8	9	9	23^*	23, 26	0.9949	9^*	9, 23, 26
9	13	13	17	17	0.9946	13^*	13, 17
10	4	4	28	28	0.9943	4^*	4, 28
11	11^*	11, 25	15	15	0.9942	11^*	11, 15, 25
12	3^*	3, 6, 22	11^*	11, 15, 25	0.9938	3^*	3, 6, 11, 15, 22, 25
13	2	2	5	5	0.9937	2^*	2, 5
14	7	7	10	10	0.9928	7^*	7, 10
15	9^*	9, 23, 26	18^*	18, 21	0.9912	9^*	9, 18, 21, 23, 26
16	3^*	3, 6, 11, 15, 22, 25	7^*	7, 10	0.9845	3^*	3, 6, 7, 10, 11, 15, 22, 25
17	4^*	4, 28	12	12	0.9829	4^*	4, 12, 28
18	2^*	2, 5	16^*	16, 19	0.9806	2^*	2, 5, 16, 19
19	1	1	14	14	0.9804	1^*	1, 14
20	1^*	1, 14	20^*	20, 27	0.9804	1^*	1, 14, 20, 27
21	3^*	3, 6, 7, 10, 11, 15, 22, 25	9^*	9, 18, 21, 23, 26	0.9794	3^*	3, 6, 7, 9, 10, 11, 15, 18, 21, 22, 23, 25, 26
22	8	8	13^*	13, 17	0.9622	8^*	8, 13, 17
23	3^*	3, 6, 7, 9, 10, 11, 15, 18, 21, 22, 23, 25, 26	24	24	0.9601	3^*	3, 6, 7, 9, 10, 11, 15, 18, 21, 22, 23, 24, 25, 26
24	2^*	2, 5, 16, 19	3^*		0.9391	2^*	

(Continued)

TABLE 7.11   Case Study 4 of Cluster Analysis: Q-mode Cluster Analysis—cont'd

| | | Merged Classes | | | | New Class After Merging | |
| | Class G_p | | Class G_q | | Analog | Class G_{p*} | |
Merge No.	Class No. p	Sample Nos. Contained by Class G_p	Class No. q	Sample Nos. Contained by Class G_q	Coefficient Between Class G_p and Class G_q	Class No. $p*$	Sample Nos. Contained by Class G_{p*}
25	1^*	1, 14, 20, 27	8^*	3, 6, 7, 9, 10, 11, 15, 18, 21, 22, 23, 24, 25, 26 8, 13, 17	0.8887	1^*	2, 3, 5, 6, 7, 9, 10, 11, 15, 16, 18, 19, 21, 22, 23, 24, 25, 26 1, 8, 13, 14, 17, 20, 27
26	1^*	1, 8, 13, 14, 17, 20, 27	4^*	4, 12, 28	0.8420	1^*	1, 4, 8, 12, 13, 14, 17, 20, 27, 28
27	1^*	1, 4, 8, 12, 13, 14, 17, 20, 27, 28	2^*	2, 3, 5, 6, 7, 9, 10, 11, 15, 16, 18, 19, 21, 22, 23, 24, 25, 26	0.6013	1^*	1, 2, 3, 4, 5, 6, 7, 8, 9, 10, 11, 12, 13, 14, 15, 16, 17, 18, 19, 20, 21, 22, 23, 24, 25, 26, 27, 28

2. *Sample reduction.* When the analog coefficient $r \geq 0.9998$ (this value can be adjusted by users) between two samples, it is possible to eliminate one of the samples to make the number of samples minus 1. From Table 7.11 it can be seen that $r = 0.9999 > 0.9998$ between sample No. 3 and sample No. 22 in the first merging, which indicates one of the two samples (e.g., sample No. 3) is probably eliminated from the original data when data analysis (e.g., MRA, BPNN, C-SVM, R-SVM, DTR, and BAC) is performed, leading the number of samples minus 1, from 28 to 27. Is this method of sample reduction reasonable or not? Now we will validate it using MRA calculation.

Using 25 learning samples, the calculation results by MRA are expressed by Equation (6.54) as follows:

$$y = 3.2701 + 0.00535x_1 - 0.10083x_2 - 0.015233x_3 + 0.2511x_4$$

The corresponding residual variance is 0.087777, multiple correlation coefficient is 0.9551, $\overline{R}(\%)$ is 2 (based on y which is taken as integer). y is shown to depend on the 4 parameters in decreasing order: x_2, x_4 x_3, and x_1.

Now using 24 learning samples: After sample No. 3 is reduced, the calculation results by MRA are as follows:

$$y = 3.2688 + 0.00537x_1 - 0.10109x_2 - 0.015103x_3 + 0.25211x_4 \tag{7.42}$$

The corresponding residual variance is 0.08807, the multiple correlation coefficient is 0.95495, and $\overline{R}(\%)$ is 2.08 (based on y which is taken as integer). y is shown to depend on the four parameters in decreasing order: x_2, x_4, x_3, and x_1.

TABLE 7.12　Case Study 4 of Cluster Analysis: R-mode Cluster Analysis

| | | Merged Classes | | | | New Class After Merging | |
| | | Class G_p | | Class G_q | | Analog | Class G_p | |
Merge No.	Class No. p	Parameter Nos. Contained by Class G_p	Class No. q	Parameter Nos. Contained by Class G_q	Coefficient Between Class G_p and Class G_q	Class No. p^*	Parameter Nos. Contained by Class G_{p^*}
1	2	2	3	3	0.8759	2^*	2, 3
2	1	1	2^*	2, 3	0.7233	1^*	1, 2, 3
3	1^*	1, 2, 3	4	4	0.1789	1^*	1, 2, 3, 4

From comparison of the two calculation results (mainly the residual variance and $\overline{R}(\%)$) of MRA before and after sample No.3 is reduced, it is found that there is very little difference between the two calculation results, showing that the sample reduction probably succeeds.

Table 7.11 shows that (a) as mentioned, $p < q$, and keep p as the number of new class; (b) for Class p or q, the number without superscript * contains only one sample, whereas the number with superscript * contains two or more samples; and (c) the same Class p^* possibly occurs two or more times due to its more than one merger, and this Class p^* contains more or more samples.

7.5.3.2. R-mode Cluster Analysis

The calculation results from R-mode cluster analysis are shown in Table 7.12 and Figure 7.9. Now the following two understandings have been gained through analysis:

1. *Parameter cluster.* Figure 7.9 illustrates a cluster pedigree chart of four parameters (shale content, porosity, permeability, permeability variation coefficient). It illustrates the whole process that four classes (i.e., four parameters) are merged step by step into one class. Users can see the order of dependence between parameters in the cluster pedigree chart.
2. *Parameter-dimension reduction.* It is possible to perform parameter-dimension reduction in this case study. The reason is that the variety of analog coefficients listed in Table 7.12 is consistent with the results of MRA used in this case study [see Equation (6.54)]. Table 7.12 shows that x_2 (porosity) or x_3 (permeability) is the first parameter for dimension reduction,

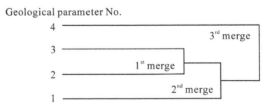

FIGURE 7.9　Case Study 4 of cluster analysis: R-mode cluster pedigree chart.

whereas Equation (6.54) shows that x_2 (porosity) is the first parameter for dimension reduction. If x_2 (porosity) is eliminated, leading the number of parameters to be minus 1, resulting in 4 reduced to 3, that is, the studied 5-D (plus y) problem is reduced to a 4-D (plus y) problem. Is this method of parameter reduction reasonable or not? Now it will be validated by MRA calculation.

Using four parameters (x_1, x_2, x_3, x_4), the calculation results by MRA are expressed by Equation (6.54) as follows:

$$y = 3.2701 + 0.00535x_1 - 0.10083x_2 - 0.015233x_3 + 0.2511x_4$$

The corresponding residual variance is 0.087777, the multiple correlation coefficient is 0.9551, and $\bar{R}(\%)$ is 2 (based on y, which is taken as integer). y is shown to depend on the four parameters in decreasing order: x_2, x_4, x_3, and x_1.

Eliminating x_2, i.e., using three parameters (x_1, x_3, x_4), the calculation results by MRA are expressed as follows:

$$y = 2.5448 - 0.0057101x_1 - 0.045692x_3 + 0.40067x_4 \tag{7.43}$$

The corresponding residual variance is 0.17221, multiple correlation coefficient is 0.90983, and $\bar{R}(\%)$ is 3 (based on y, which is taken as integer). y is shown to depend on the three parameters in decreasing order: x_3, x_4, and x_1.

It is found from comparison of the two calculation results (mainly the residual variance and $\bar{R}(\%)$) of MRA before and after x_1 is reduced that the difference between the two is larger, so it is impossible to perform parameter dimension reduction.

Table 7.12 demonstrates that (a) as mentioned, $p < q$, and keep p as the number of new class; (b) for Class p or q, the number without superscript * contains only one parameter, whereas the number with superscript * contains two or more parameters; and (c) the same Class p^* possiblely occurs two or more times due to its more than one merger, and this Class p^* contains more or more parameters.

It is noted in Figure 7.9 that the definition of the geological parameter number is 1, shale content; 2, porosity; 3, permeability; and 4, permeability variation coefficient.

7.5.4. Summary and Conclusions

In summary, using data for the reservoir classification in the C3 and C4+5 low-permeability reservoir in the Baibao Oilfield, i.e., the four parameters (shale content, porosity, permeability, permeability variation coefficient) of 28 layers, two cluster pedigree charts are gained by the data mining tool (Q-mode and R-mode cluster analyses). One chart is for 28 layers (Figure 7.8), indicating the order of dependence between layers; the classification coincidence rate reaches 100%, and the results coincide with practicality. The results are provided for the five algorithms of BAYSD, C-SVM, R-SVM, BPNN, and MRA (see Section 6.3). The other chart is for four parameters (Figure 7.9), indicating the order of dependence between the parameters. These two cluster pedigree charts are called *mined knowledge*. Therefore, this method can be spread to the cluster analysis for low-permeability reservoir classification in other oilfields. Moreover, it is found that (a) Q-mode cluster analysis can be used for layer sample reduction, indicating that Q-mode cluster analysis can serve as a pioneering sample-reduction tool; whereas (b) R-mode cluster analysis (using analog coefficient) cannot be used for parameter dimension reduction. The two discoveries are also called *mined knowledge*. As for parameter-dimension reduction, either MRA or BAYSD can serve as a pioneering dimension-reduction tool.

It can be seen from the preceding four Case Studies 1, 2, 3, and 4 (Table 7.13) that (a) Q-mode cluster analysis provides a cluster pedigree chart for samples, and the classification coincidence rate is passable; (b) if MRA is passed for validation, Q-mode cluster analysis can also serve as a pioneering sample-reduction tool; (c) R-mode cluster analysis provides a cluster pedigree chart for parameters; and (d) if MRA is passed for validation, R-mode cluster analysis (using a correlation coefficient) can also serve as a pioneering dimension-reduction tool. The so-called "pioneering tool" is whether it succeeds or needs a nonlinear tool (C-SVM, R-SVM, or BPNN) for the second validation so as to determine how many samples and how many parameters can be reduced. Why does it need a second validation? Because of the complexities of geosciences rules; the correlations between various classes of geosciences data are nonlinear in most cases. In general, therefore, C-SVM or R-SVM is adopted when nonlinearity is strong, whereas BPNN, DTR, or BAC is adopted when nonlinearity is not strong. The two linear algorithms of cluster analysis and MRA can serve as auxiliary tools, cooperating with major tools (BPNN, C-SVM, R-SVM, DTR, and BAC) for data mining.

TABLE 7.13 Application Values of Q-mode and R-mode Cluster Analyses in Four Case Studies

| | Control Information Input | | Calculation Results | | | | |
| | | | Q-Mode Cluster Analysis | | | R-Mode Cluster Analysis | |
Case Study	Data Normalization Method	Cluster Statistic	Cluster Pedigree Chart for Sample	Classification Coincidence Rate (%)	Sample Reduction	Cluster Pedigree Chart for Parameter	Parameter Reduction (Dimension Reduction)
Case Study1	Standard difference standardization	Analog coefficient	Figure 7.2	77	Can reduce sample No. 13	Figure 7.3	Cannot perform parameter-dimension reduction
Case Study2	Maximum difference normalization	Analog coefficient	Figure 7.4	70	Might can reduce sample No. 16	Figure 7.5	Cannot perform parameter-dimension reduction
Case Study3	Maximum difference standardization	Correlation coefficient	Figure 7.6	81	Might can reduce sample No. 12	Figure 7.7	Can perform parameter-dimension reduction (eliminate x_1)
Case Study4	Maximum difference standardization	Analog coefficient	Figure 7.8	100	Might can reduce sample No. 3	Figure 7.9	Cannot perform parameter-dimension reduction

EXERCISES

7-1. Are there any similarities or differences among the data used by a regression/classification algorithm and a cluster analysis algorithm?

7-2. What are the different purposes of Q-Mode cluster analysis and R-mode cluster analysis?

7-3. In this book, there are three kinds of data normalization (standard difference standardization, maximum difference standardization, maximum difference normalization), three cluster statistics (distance coefficient, analog coefficient, correlation coefficient), and three class-distance methods (the shortest-distance method, the longest-distance method, and the weighted mean method) for each cluster statistic. Hence, there are nine clustering methods from among which users can choose. How can we choose the best one in each application?

7-4. Q-mode cluster analysis and R-mode cluster analysis arc applied to Case Studies 1, 2, 3, and 4. What are the solution accuracies of the two algorithms?

References

Everitt, B.S., Landau, S., Leese, M., Stahl, D., 2011. Cluster Analysis, fifth ed. Wiley, Chichester, UK.

Li, H., Zhao, Y., 1998. Mathematical Petroleum Geology. Petroleum Industry Press, Beijing, China (in Chinese).

Zhao, X., 1992. Conspectus of Mathematical Petroleum Geology. Petroleum Industry Press, Beijing, China (in Chinese).

This chapter introduces the calculating techniques of *Kriging* as well as their applications in geosciences. For each technique, the applying ranges and conditions, basic principles, calculation methods, and application sample are provided. In each application sample, the calculation results and analyses are given. Though each application sample is small, it reflects the whole process of calculations to benefit readers in understanding and mastering the applied techniques.

Kriging is an approach to geostatistics. The differences between geostatistics and conventional statistics are (a) from methodology, conventional statistics uses conventional mathematics, whereas geostatistics uses a new combination of conventional statistics and geology; (b) from applying ranges, conventional statistics is applicable to multidisciplines, including geology, whereas geostatistics is applicable only to geology; and (c) from the space structure and randomicity of studied parameters, geostatistics is much superior to conventional statistics, because geostatistics can furthest utilize various information in geosciences, interpolate in both whole and local, and gain accurate results.

In one word, Kriging is a marginal *discipline*, taking variogram as a basic tool, selecting various appropriate methods, and performing optimal linear unbiased interpolation estimates on the space structure and randomicity of parameters.

Section 8.1 (preprocessing) introduces Case Study 1, the preprocessing of geostatistics for a parameter discretely distributed on a plane (x, y), with which we explain how to use coordinate transformation to perform the preprocessing of geostatistics. The results are then used in Sections 8.2, 8.4, and 8.5.

Section 8.2 (experimental variogram) introduces Case Study 2, an experimental variogram of a parameter discretely distributed on a plane (x, y), with which we explain how to use the distance-weighted method to perform the calculation of an experimental variogram. The results are used in Section 8.3.

Section 8.3 (optimal fitting of experimental variogram) introduces Case Study 3, the optimal fitting of an experimental variogram of a parameter discretely distributed on a plane (x, y), with which we explain how to use a spherical model to perform the calculation of the optimal fitting of an experimental variogram. The results are used in Sections 8.4 and 8.5.

Section 8.4 (cross-validation of Kriging) introduces Case Study 4, the cross-validation of a parameter discretely distributed on a plane (x, y), with which we explain how to use the method of the nearest point in a sector to perform the calculation of the cross-validation.

Section 8.5 (applications of Kriging) introduces Case Study 5, application of Kriging of a parameter discretely distributed on a plane (x, y), with which we explain how to use the method of the plane nearest the point to perform the application of Kriging.

Furthermore, this chapter introduces the five steps of the Kriging operation: preprocessing (see Section 8.1), experimental variogram (see Section 8.2), optimal fitting of experimental variogram (see Section 8.3), cross-validation of Kriging (see Section 8.4), and applications of Kriging (see Section 8.5). The contents of these five steps seem to be a connecting link between the preceding and the following, which form an integer. The ultimate purpose is to perform the cross-validation and applications of Kriging (see Figure 8.1). One case study is given for each of these steps. The five case studies are based on an original data, i.e., on the plane equidistant network with 80 nodes (i.e., 63 grids), on which there are values of a parameter at 41 nodes of 80 nodes, which form 41 samples. Different samples include different (x, y) and parameter value. In each step of the Kriging operation, data input and results output are given.

For more concrete discussion, Figure 8.1 illustrates the order of the five steps of the Kriging operation, i.e., preprocessing, experimental variogram, optimal fitting of experimental variogram, cross-validation of Kriging, and applications of Kriging. In fact, the first three steps are run, then the fourth step, achieving the Kriging estimated value on the known point. If the Kriging estimated values on these known points are qualified in the fourth step, the fifth step is run, achieving the Kriging estimated value at each grid center. Here the following should be stressed: (a) the results from the first step will be used in Steps 2, 4, and 5, (b) the results from the second step will be used in Step 3, and (c) the results from the third step will be used in Steps 4 and 5.

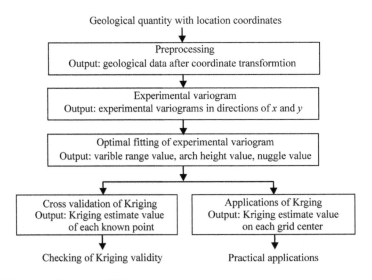

FIGURE 8.1   Calculation flowchart of Kriging.

Using the data of these 41 samples, the optimal fitting of the experimental variogram is obtained by the data mining tool of Kriging. This optimal fitting is believed to be qualified by cross-validation of Kriging. Therefore, this optimal fitting is called *mined knowledge*. In the prediction process, this knowledge is applied to calculate the estimate values of the parameter at 63 grid centers. Though the case study is small, it is a benefit to understand the method and practical applications. Therefore, Kriging can be spread to the applications in geosciences. The effect of its optimal linear unbiased interpolation estimate is superior to the general mathematics interpolation, so it can be widely applied.

8.1. PREPROCESSING

The *preprocessing* of geostatistics is used to perform coordinate system transformation, which prepares for the calculations of an experimental variogram (see Section 8.2) and Kriging (see Sections 8.4 and 8.5) and then makes the simplest statistical analysis to the parameter, such as mean value, variance, and histogram.

8.1.1. Calculation Method

The study object of Kriging is the parameter with location coordinates. For example, in 2-D space, let a parameter value be v_n, which is located at (x_n, y_n), $n = 1, 2, ..., N_{total}$. The study method of Kriging is of orientation, i.e., to conduct the study in a designated direction. Usually, one study is carried out in the direction parallel to the x axis and one in the direction parallel to the y axis, thus achieving geostatistical results in these two directions. If the study is conducted in other directions, the old coordinate system (x, y) should be rotated in an angle to a new coordinate system (x^*, y^*), on which the study is conducted in the directions of the x^* axis and the y^* axis. For example, let α be an included angle (units of angle) of this direction with the x positive axis; a new coordinate system (x^*, y^*) can be formed by anticlockwise rotation of the old coordinate system (x, y) in an angle of α, with the following Equations (8.1) and (8.2), and carrying out geostatistics in the directions parallel to x^* and y^*, respectively, on this new coordinate system, which is definitely consistent with the geostatistics carried out in the direction parallel to α direction and $\alpha + 90$ direction on the old coordinate system (Figure 8.2).

$$\theta = \alpha \cdot \pi / 180 \tag{8.1}$$

where θ is a direction angle, the included angle with the x positive axis, radian; α is a direction angle, the included angle with the x positive axis, angle; and π is pi, $\pi \approx 3.14159$.

$$\begin{cases} x^* = x \cdot \cos\theta + y \cdot \sin\theta \\ y^* = y \cdot \cos\theta - x \cdot \sin\theta \end{cases} \tag{8.2}$$

where (x^*, y^*) is the new coordinate system; (x^*, y^*) is the new coordinate system; and the other symbols have been defined in Equation (8.1).

For a parameter value v_n ($n = 1, 2, ..., N_{total}$), its mean value, variance, and the number of intervals of histogram can be calculated by Equations (8.3), (8.4), and (8.5), respectively.

$$\bar{v} = \frac{1}{N_{total}} \sum_{n=1}^{N_{total}} v_n \tag{8.3}$$

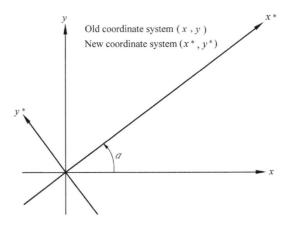

FIGURE 8.2   Coordinate system transformation.

where $\bar{v}$ is the mean value of a parameter; v_n is the nth value of the parameter, $n = 1, 2, ...,$ N_{total}; and N_{total} is the number of values of the parameter.

$$S^2 = \frac{1}{N_{\text{total}}} \sum_{n=1}^{N_{\text{total}}} (v_n - \bar{v})^2 \tag{8.4}$$

where S^2 is the variance of a parameter and the other symbols have been defined in Equation (8.3).

$$I_{\text{hist}} = \text{INT}\left(\frac{v_{\text{max}} - v_{\text{min}}}{I_{\text{intr}}}\right) + 1 \tag{8.5}$$

where I_{hist} is the number of intervals of a parameter-distribution histogram; INT is an integral function; I_{intr} is the interval length of the parameter distribution histogram; v_{max} is the maximum in v_n ($n = 1, 2, ..., N_{\text{total}}$); and v_{min} is the minimum of the histogram on the v axis. This v_{min} might be the minimum in v_n ($n = 1, 2, ..., N_{\text{total}}$) if this minimum is the tidy number that satisfies users, or it might not be the minimum in v_n ($n = 1, 2, ..., N_{\text{total}}$) if this minimum is the number that does not satisfy users or the untidy number.

8.1.2. Case Study 1: The Preprocessing of Geostatistics for a Parameter Discretely Distributed On a Plane (x, y)

Though the example is simple, it is very typical and is widely representative. So it is used as a case study in each section of this chapter. The explanation of this example makes readers forthrightly master all technologies of Kriging and flexibly apply to their own geosciences. The specific data of this example are illustrated in Figure 8.3 (Wang and Hu, 1988; Shi, 1999) and Table 8.1.

In Figure 8.3, the coordinate system uses relative numbers, i.e., mathematical coordinates; in practical applications the geodetic coordinate can be used too. The network is

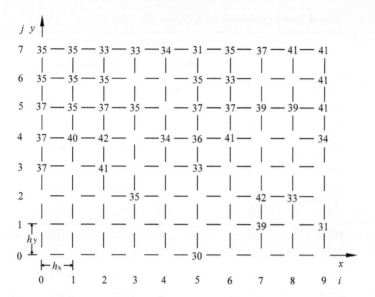

FIGURE 8.3 A discretely plane-distributed parameter.

TABLE 8.1 Old and New Coordinates for a Parameter

		Location				
		Old Coordinate		New Coordinate[a]		
Sample No.	Location Name	x	y	x^*	y^*	Parameter v
1	W5-0	5	0	4.096	−2.868	30
2	W7-1	7	1	6.308	−3.196	39
3	W9-1	9	1	7.946	−4.343	31
4	W3-2	3	2	3.605	−0.082	35
5	W7-2	7	2	6.881	−2.377	42
6	W8-2	8	2	7.700	−2.950	33
7	W0-3	0	3	1.721	2.457	37
8	W2-3	2	3	3.359	1.310	41
9	W5-3	5	3	5.816	−0.410	33
10	W0-4	0	4	2.294	3.277	37
11	W1-4	1	4	3.113	2.703	40
12	W2-4	2	4	3.933	2.129	42
13	W4-4	4	4	5.571	0.982	34

(Continued)

TABLE 8.1 Old and New Coordinates for a Parameter—cont'd

		Location				
		Old Coordinate		New Coordinate[a]		
Sample No.	Location Name	x	y	x^*	y^*	Parameter v
14	W5-4	5	4	6.390	0.409	36
15	W6-4	6	4	7.209	−0.165	41
16	W9-4	9	4	9.667	−1.886	34
17	W0-5	0	5	2.868	4.096	37
18	W1-5	1	5	3.687	3.522	35
19	W2-5	2	5	4.506	2.949	37
20	W3-5	3	5	5.325	2.375	35
21	W5-5	5	5	6.964	1.228	37
22	W6-5	6	5	7.783	0.654	37
23	W7-5	7	5	8.602	0.081	39
24	W8-5	8	5	9.421	−0.493	39
25	W9-5	9	5	10.24	−1.066	41
26	W0-6	0	6	3.441	4.915	35
27	W1-6	1	6	4.261	4.341	35
28	W2-6	2	6	5.080	3.768	35
29	W5-6	5	6	7.537	2.047	35
30	W6-6	6	6	8.356	1.473	33
31	W9-6	9	6	10.81	−0.247	41
32	W0-7	0	7	4.015	5.734	35
33	W1-7	1	7	4.834	5.160	35
34	W2-7	2	7	5.653	4.587	33
35	W3-7	3	7	6.472	4.013	33
36	W4-7	4	7	7.292	3.440	34
37	W5-7	5	7	8.111	2.866	31
38	W6-7	6	7	8.930	2.293	35
39	W7-7	7	7	9.749	1.719	37
40	W8-7	8	7	10.57	1.145	41
41	W9-7	9	7	11.39	0.572	41

[a]new coordinate (x^*, y^*) is formed by anticlockwise rotating in $35°$ based on old coordinate (x, y), i.e., calculated out with Equations (8.1) and (8.2) when $\alpha = 35°$.

partitioned into 10×8, i.e., 80 nodes (63 grids). At 41 of 80 nodes there are measured values (values of parameter); there are no values at the other nodes. Network partition depends on the number and form distribution of parameter values. In practical applications, most of the measured values do not fall at the nodes, so Figure 8.3 is designed only for the convenience of drawing.

In Table 8.1, the location of a parameter is expressed by the old coordinate system and the new coordinate system. It is possible to carry out geostatistics in the direction parallel to x and y using the location of (x, y) while carrying out geostatistics in the direction parallel to x^* and y^* using the location of (x^*, y^*). The two coordinate systems share the same origin, but the included angle between x and x^* axes or y and y^* axes is 35° (Figure 8.2).

In Table 8.1, when $\alpha = 0$, the coordinate system (called Example 1) is not rotated; when $\alpha = 35$, the coordinate system (called Example 2) is anticlockwise rotated in 35°.

There are 41 values for the parameter, with 30 of the minimum, 42 of the maximum, 36.37 of mean value, and 10.23 of variance. The basic data, which consist of a histogram, are 30 of the minimum and 44 of the maximum for the parameter, and seven intervals with equal length of 2. The occurrence count and the frequency for the parameter values in each interval are given. These results are not related to coordinate system transformation. The only difference between Example 1 and Example 2 is that the same value of the parameter has different coordinate values due to the coordinate transformation (Table 8.1).

8.2. EXPERIMENTAL VARIOGRAM

8.2.1. Applying Ranges and Conditions

8.2.1.1. *Applying Ranges*

The calculation of an experimental variogram is performed to a coordinate-transformed parameter (see Section 8.1), achieving the important properties of the affecting range, space distribution continuity, and anisotropy for the parameter and preparing for the optimal fitting of the experimental variogram (see Section 8.3).

8.2.1.2. *Applying Conditions*

For the parameter, the number of observations is large enough, the observed values are as precise as possible, and the observed values have been preprocessed (see Section 8.1).

8.2.2. Basic Principles

Since the calculation of an experimental variogram is relatively special, the calculation formulas will not be listed for the moment, but the example is used for explanation first.

Two types of calculation formula of experimental variogram are available: non-distance-weighted and distance-weighted. The basic lag distance of each type is given by users or by automatic calculation.

Taking Figure 8.3 as an example, let the basic lag distance h_x in x direction and the basic lag distance h_y in y direction be 1, and the non-distance-weighted calculation is performed.

8.2.2.1. Experimental Variogram in x Direction

The experimental variogram to be calculated is $\gamma(ih_x)$, $i = 1, 2, ..., 9$. Observing upward and rightward all lines on the x axis and lines parallel to the x axis with jh_y ($j = 1, 2, ..., 7$) of distance, it is found that there are 24 "data pairs" with h_x of distance: (42, 33), (37, 40), (40, 42), (34, 36), (36, 41), (37, 35), (35, 37), (37, 35), (37, 37), (37, 39), (39, 39), (39, 41), (35, 35), (35, 35), (35, 33), (35, 35), (35, 33), (33, 33), (33, 34), (34, 31), (31, 35), (35, 37), (37, 41), and (41, 41).

By cumulating the difference square of the above 24 data pairs and then dividing by the double number of data pairs, we obtain the experimental variogram value:

$$\gamma(h_x) = \frac{1}{2 \times 24}\left[(42 - 33)^2 + (37 - 40)^2 + ... + (37 - 41)^2 + (41 - 41)^2\right] = 4.104$$

As mentioned, the basic lag distance h_x in x direction is 1, and the corresponding experimental variogram value $\gamma(h_x) = 4.104$. If the lag distance is $2h_x$, the question is, how much is the corresponding experimental variogram value $\gamma(2h_x)$? Similarly, observing upward and rightward all lines on the x axis and lines parallel to the x axis with jh_y ($j = 1, 2, ..., 7$) of distance, it is found that there are 20 data pairs with $2h_x$ of distance: (39, 31), (37, 41), (37, 42), (42, 34), (34, 41), (37, 37), (35, 35), (35, 37), (37, 39), (37, 39), (39, 41), (35, 35), (35, 33), (35, 33), (33, 34), (33, 31), (34, 35), (31, 37), (35, 41), and (37, 41).

By cumulating the difference square of the above 20 data pairs and then dividing by the double number of data pairs, the experimental variogram value is obtained:

$$\gamma(2h_x) = \frac{1}{2 \times 20}\left[(39 - 31)^2 + (37 - 41)^2 + ... + (35 - 41)^2 + (37 - 41)^2\right] = 8.400$$

Similarly, $\gamma(3h_x) = 12.083$, $\gamma(4h_x) = 9.472$, $\gamma(5h_x) = 6.094$, $\gamma(6h_x) = 9.411$, $\gamma(7h_x) = 12.568$, $\gamma(8h_x) = 12.562$, and $\gamma(9h_x) = 11.591$ can be calculated.

8.2.2.2. Experimental Variogram in y Direction

The experimental variogram to be calculated is $\gamma(jh_y)$, $j = 1, 2, ..., 7$. Observing rightward and upward all lines on the y axis and lines parallel to the y axis with ih_x (1, 2, ..., 9) of distance, it is found that there are 22 data pairs with h_y of distance: (37, 37), (37, 37), (37, 35), (35, 35), (40, 35), (35, 35), (35, 35), (41, 42), (42, 37), (37, 35), (35, 33), (33, 36), (36, 37), (37, 35), (35, 31), (41, 37), (37, 33), (33, 35), (39, 42), (34, 41), (41, 41), and (41, 41).

By cumulating the difference square of the above 22 data pairs and then dividing by the double number of data pairs, we obtain the experimental variogram value:

$$\gamma(h_y) = \frac{1}{2 \times 22}\left[(37 - 37)^2 + (37 - 37)^2 + ... + (41 - 41)^2 + (41 - 41)^2\right] = 4.250$$

As mentioned, the basic lag distance h_y in y direction is 1, and the corresponding experimental variogram value $\gamma(h_y) = 4.250$. If the lag distance is $2h_y$, the question is, how much is the experimental variogram $\gamma(2h_y)$? Similarly, observing rightward and upward all lines on the y axis and lines parallel to the y axis with ih_x ($j = 1, 2, ..., 9$) of distance, it is found that there are 18 data pairs with $2h_y$ of distance: (37, 37), (37, 35), (37, 35), (40, 35), (35, 35), (41, 37), (42, 35), (37, 33), (35, 33), (33, 37), (36, 35), (37, 31), (41, 33), (37, 35), (39, 37), (39, 41), (34, 41), and (41, 41).

By cumulating the difference square of the above 18 data pairs and then dividing by a double number of data pairs, the experimental variogram value is obtained:

$$\gamma(2h_y) = \frac{1}{2 \times 18}\left[(37-37)^2 + (37-35)^2 + \dots + (34-41)^2 + (41-41)^2\right] = 8.222$$

Similarly, $\gamma(3h_y) = 10.900$, $\gamma(4h_y) = 17.333$, $\gamma(5h_y) = 24.200$, $\gamma(6h_y) = 18.429$, and $\gamma(7h_y) = 7.000$ can be calculated.

8.2.3. Distribution Characteristics of Parameter Described by Experimental Variogram

The following introduces the calculation method. It is suggested that readers carefully read the basic principles in Section 8.3.2, since the calculation of an experimental variogram is exceptional.

Assume there are a certain number of discretely distributed values of a parameter on the plane (x, y) (Figure 8.3). To study its distribution characteristics, we calculate the experimental variogram of this parameter.

The calculation of an experimental variogram is directional, i.e., we calculate in two mutually perpendicular directions, e.g., in the directions parallel to the x axis and y axis, respectively. If calculations are in another two mutually perpendicular directions, by rotating the coordinate system at an angle to form a new coordinate system, calculations are in the directions parallel to the new x axis and the new y axis, respectively (see Section 8.1).

Two types of calculation formula of experimental variogram are available: non-distance-weighted and distance-weighted.

8.2.3.1. The Non-Distance-Weighted Calculation Formula

When a parameter's location (x, y) falls at the node composed by basic lag distance, it is possible to use a non-distance-weighted calculation formula (Figure 8.3).

1. Calculation in x direction.

$$\gamma_x\left(ih_x\right) = \frac{1}{2N(ih_x)} \sum_{k=1}^{N(ih_x)} \left[v\left(x_k, y_k\right) - v\left(x_k + ih_x, y_k\right)\right]^2 \quad (i = 1, 2, \dots, N_x) \tag{8.6}$$

where γ_x is the experimental variogram in x direction (non-distance-weighted); h_x is the basic lag distance in x direction; ih_x is lag distance in x direction, $i = 1, 2, \dots, N_x$, in which N_x is the number of effective lag distance on x axis, and ih_x is the basic distance when $i = 1$; v is the parameter with location coordinate (x, y); (x_k, y_k), $(x_k + ih_x, y_k)$ are two points with v value on any lines (including x axis) parallel to the x axis, and the distance between the two points is ih_x, then the v value on these two points is called "a pair of data v," which meets the condition for calculation, $k = 1, 2, \dots, N(ih_x)$; k is the ordinal of "a pair of data v," which meets the condition for calculation; and $N(ih_x)$ is the count of "a pair of data v," which meets the condition for calculation.

2. Calculation in y direction.

$$\gamma_y(jh_y) = \frac{1}{2N(jh_y)} \sum_{k=1}^{N(jh_y)} \left[v(x_k, y_k) - v(x_k, y_k + jh_y)\right]^2 \quad (i = 1, 2, ..., N_y) \tag{8.7}$$

where γ_y is the experimental variogram in y direction (non-distance-weighted); h_y is basic lag distance in y direction; jh_y is lag distance in y direction, $j = 1, 2, ..., N_y$, in which N_y is the number of effective lag distance on y axis, and jh_y is the basic distance when $j = 1$; v is parameter with location coordinate (x, y); (x_k, y_k), $(x_k, y_k + jh_y)$ are two points with v value on any lines (including y axis) parallel to y axis, and the distance between the two points is jh_y, then the v value on these two points is called "a pair of data v," which meets the condition for calculation, $k = 1, 2, ..., N(jh_y)$; k is the ordinal of "a pair of data v," which meets the condition for calculation; and $N(jh_y)$ is the count of "a pair of data v," which meets the condition for calculation.

8.2.3.2. The Distance-Weighted Calculation Formula

When a parameter's location (x, y) does not all fall at the node composed by basic lag distance, a distance-weighted calculation formula should be used (Figure 8.3).

1. Calculation in x direction.

$$\gamma_x^*(ih_x) = \frac{1}{2\sum_{k=1}^{N(ih_x)} d_k} \sum_{k=1}^{N(ih_x)} d_k \left[v(x_k, y_k) - v(x_k + \Delta x_k, y_k + \Delta y_k)\right]^2 \quad (i = 1, 2, ..., N_x)$$

$$\tag{8.8}$$

where γ_x^* is the experimental variogram in x direction (distance-weighted); (x_k, y_k), $(x_k + \Delta x_k, y_k + \Delta y_k)$ is two points with v value that meet the condition expressed by Equations (8.9) and (8.10), then the v value on the two points is called "a pair of data v," which meets the condition for calculation, $k = 1, 2, ..., N(ih_x)$; Δx_k is the difference of x coordinate values between the previous two points, $\Delta x_k \geq 0$; Δy_k is the difference of y coordinate values between the previous two points, $\Delta y_k \geq 0$; d_k is the distance between the previous two points, $d_k = \sqrt{(\Delta x)^2 + (\Delta y)^2}$; and the other symbols have been defined in Equation (8.6).

$$\beta_x = \text{tg}^{-1}\left(\frac{\Delta y_k}{\Delta x_k}\right) \leq \beta_t \tag{8.9}$$

where β_x is the included angle between the line linking the previous two points and x positive axis, radian; β_t is the tolerance of the preceding included angle, radian; and the other symbols have been defined in Equation (8.8).

$$ih_x/2 < d_k \leq ih_x \quad (I = 1, 2, ..., N_x) \tag{8.10}$$

where all symbols have been defined in Equations (8.6) and (8.8).

2. Calculation in y direction.

$$\gamma_y^*(jh_y) = \frac{1}{2\sum_{k=1}^{N(jh_y)} d_k} \sum_{k=1}^{N(jh_y)} d_k[v(x_k, y_k) - v(x_k + \Delta x_k, y_k + \Delta y_k)]^2 \quad (i = 1, 2, ..., N_y) \quad (8.11)$$

where γ_y^* is the experimental variogram in y direction (distance-weighted); (x_k, y_k), $(x_k + \Delta x_k, y_k + \Delta y_k)$ is two points with v value that meet the conditions expressed by Equations (8.12) and (8.13), then the v value on the two points is called "a pair of data v," which meets the condition for calculation, $k = 1, 2, ..., N(jh_y)$; and the other symbols have been defined in Equations (8.7) and (8.8).

$$\beta_y = tg^{-1}\left(\frac{\Delta x_k}{\Delta y_k}\right) \leq \beta_t \quad (8.12)$$

where β_y is the included angle between the line linking the previous two points and y positive axis, radian; and the other symbols have been defined in Equations (8.8) and (8.9).

$$ih_y/2 < d_k \leq ih_y \quad (j = 1, 2, ..., N_y) \quad (8.13)$$

By explaining the examples in Section 8.2 (basic principles) and the preceding formulas, we have so far distinctly introduced the calculations for experimental variograms. Here the following should be stressed: (a) if the locations of a parameter all fall at the nodes composed by basic lag distance, then the results obtained from non-distance-weighted calculation formulas (8.6) and (8.7) and the results obtained from distance-weighted calculation formulas (8.8) and (8.11) are the same, which can be seen from the expressions of the formulas, i.e., Equations (8.6) and (8.7) are the special case of Equations (8.8) and (8.11), respectively; and (b) in practical applications, it is impossible that the locations of a parameter all fall at the nodes composed by basic lag distance as the parameter discretely distributes, so it is unadvisable to use a non-distance-weighted calculation formula, but it is advisable to use a distance-weighted calculation formula. Hence, the calculation formula used in the corresponding software is distance-weighted, i.e., do not use Equations (8.6) and (8.7), but use Equations (8.8) and (8.11).

The symbol $\sum$ in the preceding formula has the following meanings except for the concept of cumulative summation, that is, to select "a pair of data v" under the conditions of (a) the included angle between the line linking two points and x or y positive axis is less than a tolerance (for example, $10°$, i.e., $\pi / 18$) so as to ensure this linking line is approximately parallel to x or y axis; see Equation (8.9) or (8.12), and (b) the distance between two points $\in$ (lag distance / 2, lag distance); see Equation (8.10) or (8.13).

3. To calculate basic lag distance. The basic lag distances h_x and h_y can be assigned by users or figured out by the programs.

The calculation way for h_x is to calculate the distance of "a pair of data v," which meets the condition of Equation (8.9), and select the minimum value in the nonzero distance. It can be regarded as h_x after an integral operation is taken.

The calculation way for h_y is to calculate the distance of "a pair of data v," which meets the condition of Equation (8.12), and select the minimum value in the nonzero distance. It can be regarded as h_y after an integral operation is taken.

TABLE 8.2   Experimental Variogram Values in x and y Directions in a Coordinate System without Rotation

		x Direction				y Direction	
i	Lag Distance ih_x	"A Pair of Data v" Count N_i	Experimental Variogram Value[a] $\gamma(ih_x)$	j	Lag Distance jh_y	"A Pair of Data v" Count N_j	Experimental Variogram Value[a] $\gamma(jh_y)$
1	1	24	4.104	1	1	22	4.250
2	2	20	8.400	2	2	18	8.222
3	3	18	12.083	3	3	15	10.900
4	4	18	9.472	4	4	6	17.333
5	5	16	6.094	5	5	5	24.200
6	6	28	9.411	6	6	7	18.429
7	7	22	12.568	7	7	3	7.000
8	8	16	12.562				
9	9	11	11.591				

[a]in calculation, basic lag distance $h_x = h_y = 1$, tolerance angle $\beta_t = 10°$.

8.2.4. Case Study 2: Experimental Variogram for a Parameter Discretely Distributed On a Plane (x, y)

The data used in the case study are the results in Section 8.1 (preprocessing). For the sake of convenience for readers to understand, the results in the old coordinate system are taken, that is x, y, and v in Table 8.1; see Figure 8.3 and the preceding Example 1($\alpha = 0$).

When the basic lag distances h_x and h_y are input by users (Example 1: $\alpha = 0$), $h_x = 1$, $h_y = 1$; when the basic lag distances h_x and h_y are calculated by the program (Example 2: $\alpha = 35$), $h_x = 1$, $h_y = 1$. Therefore, the calculation results when h_x and h_y are assigned to 1 by users are the same as that when h_x and h_y are calculated by the program. The difference is that in the calculation result of Example 2, there appear related intermediate results in the process of calculation of basic lag distances by the program.

Table 8.2 lists the experimental variogram values.

The map of experimental variogram in x and y directions can be drawn using the lag distance and experimental variogram values in Table 8.2. Three major distribution characteristics (the affecting range, space distribution continuity, and anisotropy) of a parameter can be gained by the experimental variogram map.

8.3. OPTIMAL FITTING OF EXPERIMENTAL VARIOGRAM

8.3.1. Applying Ranges and Conditions

8.3.1.1. Applying Ranges

When the discrete values of an experimental variogram for a parameter have been calculated (see Section 8.2), curve fitting will be made to these discrete values (Table 8.2 or Figure 8.4) so as to intensify the important understanding to the affecting range, space

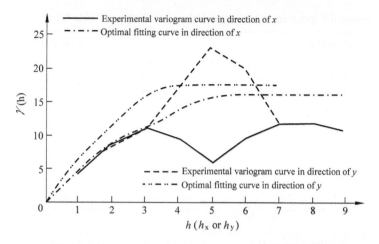

FIGURE 8.4   Experimental variogram curve and optimal fitting curve in x and y directions.

distribution continuity, and anisotropy of this parameter and to prepare for the Kriging calculation (see Sections 8.4 and 8.5).

8.3.1.2. Applying Conditions

There are at least three discrete values of experimental variogram obtained in Section 8.2, and most of the discrete values increase as the lag distance increases.

8.3.2. Basic Principles

To correctly reflect the important properties of the parameter, an optimal fitting function is made based on the discrete values of an experimental variogram so as to make up the following defects of experimental variograms (Figure 8.4): (a) some discrete values decrease as the lag distance increases, which does not accord with the whole rule; and (b) a straight line links the discrete points, which leads to loss of accuracy.

To avoid the first defect, it is required to delete those discrete points that decrease as the lag distance increases before fitting, which will be realized by the program (Tables 8.2 and 8.3). To avoid the second defect, the optimal fitting curve of the spherical model is selected, which is widely applied here and will be introduced in detail in Section 8.3.3.

Is it possible to adopt the least-squares method mentioned in Section 2.2.3 to realize the optimal fitting of an experimental variogram? The answer is no. The reason is that the study object is the parameter with a location coordinate; it is impossible to fully consider the geological characteristics that the first three or four points in the experimental variogram curve are much more important than the last few points by simply adopting the least-squares method (Table 8.3).

Generally, the optimal fitting of an experimental variogram must meet the following three requirements: (1) to select reasonable data and weighted value so as to reach the optimal fitting at least square, (2) to make sure that the fitting results are uniquely determined so as to avoid subjectively made errors, and (3) to not need to repeatedly adjust the calculation since the calculation method is not difficult. The spherical model and its calculation method introduced in Section 8.3.3 absolutely follow these three requirements.

TABLE 8.3 Incremental Experimental Variogram Values Selected from Table 8.2

	x Direction				y Direction		
i	Lag Distance ih_x	"A Pair of Data v" Count N_i	Experimental Variogram Value $\gamma(ih_x)$	j	Lag Distance jh_y	"A Pair of Data v" Count N_j	Experimental Variogram Value $\gamma(jh_y)$
1	1	24	4.104	1	1	22	4.250
2	2	20	8.400	2	2	18	8.222
3	3	18	12.083	3	3	15	10.900
7	7	22	12.568	4	4	6	17.333
				5	5	5	24.200

TABLE 8.4 Experimental Variogram Values in Table 8.3 Used by Optimal Fitting

	x Direction				y Direction		
i	Lag Distance H_i	"A pair of Data v" Count N_i	Experimental Variogram Value γ_i	i	Lag Distance H_i	"A Pair of Data v" Count N_i	Experimental Variogram Value γ_i
1	1	24	4.104	1	1	22	4.250
2	2	20	8.400	2	2	18	8.222
3	3	18	12.083	3	3	15	10.900
4	7	22	12.568	4	4	6	17.333
				5	5	5	24.200

8.3.3. Optimal Fitting of Experimental Variogram Implemented by Spherical Model

The following introduces the calculation method. It is required to filter the discrete values of experimental variogram (Table 8.2) at first, i.e., to delete those discrete values that do not increase as the lag distance increases. Those discrete values that are left are the ones that increase, and they can be regarded as the basic data for the optimal fitting (Table 8.3). For the sake of the expression of the calculation formula, Table 8.3 is rewritten to Table 8.4.

Using the spherical model, the obtained optimal fitting curves of the experimental variogram are expressed as follows (Wang and Hu, 1988; Sun et al., 1997; Shi, 1999):

$$\gamma(H) = \begin{cases} 0 & \text{when } H = 0 \\ C_0 + C\left(\frac{3H}{2a} - \frac{H^3}{2a^3}\right) & \text{when } 0 < H \le a \\ C_0 + C & \text{when } H \ge a \end{cases} \tag{8.14}$$

where γ is a fitting value of experimental variogram, expressing γ_x or γ_y; H is a lag distance in x or y direction; C_0 is a nuggle value, calculated by Equation (8.15) or (8.28); a is a variable range value, calculated by Equation (8.15) or (8.29); and C is an arch height value, calculated by Equation (8.15) or (8.30).

8.3.3.1. To Calculate C_0, a, C, and C_b by Formula

The following can be obtained by using weighted polynomial regression (Wang and Hu, 1988; Shi, 1999):

$$\begin{cases} C_0 = b_0 \\ a = \sqrt{\frac{-b_1}{3b_2}} \\ C = \frac{2}{3}b_1 a \\ C_b = C + C_0 \end{cases} \tag{8.15}$$

where b_0, b_1, and b_2 can be calculated by Equation (8.18) and the other symbols have been defined in Equation (8.14).

C_b in Equation (8.15) is the base table value, which is the sum of arch height value C and nuggle value C_0.

The sum of "a pair of data v" in x or y direction is

$$N = \sum_{i=1}^{I} N_i \tag{8.16}$$

where N_i is the count of "a pair of data v" corresponding to the ith experimental variogram value in x or y direction, which plays a role of weight for fitting; and i is the number of experimental variogram values in x or y direction.

Three cases for b_0, b_1, and b_2 in Equation (8.15) are discussed next.

1. In the case of $b_0 \geq 0$ and $b_1 b_2 < 0$, let

$$B_1 = L_{11}L_{22} - (L_{12})^2 \tag{8.17}$$

where L_{11}, L_{22}, and L_{12} can be calculated by Equation (8.19).

$$\begin{cases} b_1 = (L_{1\gamma}L_{22} - L_{2\gamma}L_{12})/B_1 \\ b_2 = (L_{2\gamma}L_{11} - L_{1\gamma}L_{12})/B_1 \\ b_0 = \left(\overline{\gamma} - b_1\overline{H} - b_2\overline{H}^3\right) \end{cases} \tag{8.18}$$

where $\overline{\gamma}$ is the weighted mean of experimental variogram in x or y direction [see Equation (8.21)]; $\overline{H}$ is the weighted mean of lag distance in x or y direction [see Equation (8.22)]; $\overline{H}^3$ is the weighted mean of lag distance to the third power in x or y direction [see Equation (8.23)]; and the other symbols will be defined in Equations (8.19) and (8.20).

$$\begin{cases} L_{11} = \sum_{i=1}^{I} N_i H_i^2 - \left(\sum_{i=1}^{I} N_i H_i \right)^2 \Big/ N \\[2mm] L_{12} = \sum_{i=1}^{I} N_i H_i^4 - \left(\sum_{i=1}^{I} N_i H_i \right) \left(\sum_{i=1}^{I} N_i H_i^3 \right) \Big/ N \\[2mm] L_{22} = \sum_{i=1}^{I} N_i H_i^6 - \left(\sum_{i=1}^{I} N_i H_i^3 \right)^2 \Big/ N \end{cases} \tag{8.19}$$

where H_i is the lag distance corresponding to the ith experimental variogram value in x or y direction, and the other symbols have been defined in Equation (8.16).

$$\begin{cases} L_{1\gamma} = \sum_{i=1}^{I} \gamma_i N_i H_i - \left(\sum_{i=1}^{I} N_i H_i \right) \left(\sum_{i=1}^{I} \gamma_i N_i \right) \Big/ N \\[2mm] L_{2\gamma} = \sum_{i=1}^{I} \gamma_i N_i H_i^3 - \left(\sum_{i=1}^{I} N_i H_i^3 \right) \left(\sum_{i=1}^{I} \gamma_i N_i \right) \Big/ N \end{cases} \tag{8.20}$$

where γ_i is the ith experimental variogram value in x or y direction and all the other symbols have been defined in Equations (8.16) and (8.19).

The weighted mean of the experimental variogram in x or y direction is expressed as

$$\overline{\gamma} = \left(\sum_{i=1}^{I} \gamma_i N_i \right) \Big/ N \tag{8.21}$$

where all symbols have been defined in Equations (8.18), (8.20), and (8.16).

The weighted mean of lag distance in x or y direction is expressed as

$$\overline{H} = \left(\sum_{i=1}^{I} H_i N_i \right) \Big/ N \tag{8.22}$$

where all symbols have been defined in Equations (8.18), (8.19), and (8.16).

The weighted mean of lag distance to the third power in x or y direction is expressed as

$$\overline{H}^3 = \left(\sum_{i=1}^{I} H_i^3 N_i \right) \Big/ N \tag{8.23}$$

where all symbols have been defined in Equations (8.18), (8.19) and (8.16).

2. In the case of $b_0 < 0$, if $b_0 < 0$ calculated by Equation (8.18), then $b_0 = 0$ is artificially given, and b_1 and b_2 should be recalculated by Equation (8.25). Let

$$B_2 = L'_{11} L'_{22} - \left(L'_{12} \right)^2 \tag{8.24}$$

where L'_{11}, L'_{22}, and L'_{12} can be calculated by Equation (8.26).

$$
\begin{cases}
b_1 = \left(L'_{1\gamma}L'_{22} - L'_{2\gamma}L'_{12}\right)\big/B_2 \\[2mm]
b_2 = \left(L'_{2\gamma}L'_{11} - L'_{1\gamma}L'_{12}\right)\big/B_2 \\[2mm]
b_0 = 0
\end{cases}
\tag{8.25}
$$

where all symbols of the right-hand side are defined in Equations (8.26), (8.27), and (8.24).

$$
\begin{cases}
L'_{11} = \sum_{i=1}^{I} N_i H_i^2 \\[3mm]
L'_{12} = \sum_{i=1}^{I} N_i H_i^4 \\[3mm]
L'_{22} = \sum_{i=1}^{I} N_i H_i^6
\end{cases}
\tag{8.26}
$$

where all symbols of the right-hand side have been defined in Equations (8.16) and (8.19).

$$
\begin{cases}
L'_{1\gamma} = \sum_{i=1}^{I} \gamma_i N_i H_i \\[3mm]
L'_{2\gamma} = \sum_{i=1}^{I} \gamma_i N_i H_i^3
\end{cases}
\tag{8.27}
$$

where all symbols of the right-hand side have been defined in Equations (8.20), (8.16), and (8.19).

3. In the case of $b_0 \geq 0$ and $b_1 b_2 \geq 0$. Obviously, in this case, the formula fails. The reason is that if $b_1 b_2 > 0$, it is impossible to calculate a and C in Equation (8.15); and if $b_1 b_2 = 0$, that is, $b_1 = 0$ or $b_2 = 0$. Furthermore, when $b_1 = 0$ or $b_2 = 0$, if $b_1 = 0$, then $a = 0$ from Equation (8.15), which makes it impossible to calculate Equation (8.14), and if $b_2 = 0$, then $a = \infty$ from Equation (8.15), which makes Equation (8.14) a linear model. At this time, the following rough calculation can be employed.

8.3.3.2. Rough Calculation of C_0, a, C and C_b

When the calculation of C_0, a, and C by formula fails, the rough calculation can be employed to roughly calculate C_0, a, and C (Sun et al., 1997; Shi, 1999).

C_0 is the y-coordinate value of the intersection point of the line linking the first two discrete points of experimental variogram with the y axis, i.e.,

$$
C_0 = y_1 - H_1 (y_2 - y_1)/(H_2 - H_1)
\tag{8.28}
$$

where C_0 is the nuggle value, taking $C_0 = 0$ if $C_0 < 0$; γ_1, γ_2 are y-coordinate values on the first two discrete points of experimental variogram; H_1, H_2 are x-coordinate values on the first two discrete points of the experimental variogram.

a is the *x*-coordinate value on the last but one discrete point of the experimental variogram, i.e.,

$$a = H_{I-1} \tag{8.29}$$

where *a* is the variable range value.

C is *y*-coordinate value on the last but one discrete point of the experimental variogram minus the nuggle value C_0, i.e.,

$$C = \gamma_{I-1} - C_0 \tag{8.30}$$

where *C* is the arch height value.

Base table value is derived from Equation (8.30):

$$C_b = \gamma_{I-1} = C + C_0 \tag{8.31}$$

where C_b is the base table value.

Thus, nuggle value C_0, variable range value *a*, and arch height value *C* have been achieved. Substituting them into Equation (8.14), the optimal fitting curves of the experimental variogram are obtained. This optimal fitting incarnates the use of N_i that plays a role of weight; see Equations (8.19)–(8.23), (8.26), and (8.27). Generally, the first three or four points of N_i are relatively larger, whereas the last few of N_i are relatively smaller, which fully considers the geological characteristics that the first three or four points are much more important than the last few points. If the first three or four points of N_i are found to be too small, which might result from the too-small basic lag distance (h_x or h_y), it is necessary to augment this basic lag distance in the calculation of the experimental variogram.

8.3.4. Case Study 3: Optimal Fitting of Experimental Variogram for a Parameter Discretely Distributed on a Plane (*x*, *y*)

The data used in the case study come from the results in Section 8.2 (experimental variogram); see Table 8.2.

For the *x* direction, calculation of C_0, *a*, and *C* using the formula succeeds; whereas for the *y* direction, calculation of C_0, *a*, and *C* using the formula fails, so rough calculation is used to calculate C_0, *a*, and *C*.

For the *x* direction, according to the principle of increasing γ, four γ values (Table 8.3) of nine γ values (Table 8.2) are selected to the optimal fitting calculation. $b_1 = 4.746$, $b_2 = -0.058$, $b_0 = -0.595$ are calculated by Equation (8.18). Because $b_0 < 0$, $b_0 = 0$ is artificially given, b_1 and b_2 are recalculated, and $b_1 = 4.446$ and $b_2 = -0.054$ are calculated by Equation (8.25). Because $b_1 b_2 < 0$, the calculation by using formula succeeds, attaining $C_0 = 0$ of nuggle value, $a = 5.236$ of variable range value, and $C = 15.52$ of arch height value. $C_b = 15.52$ of base table value can be calculated by Equation (8.31). Substituting C_0, *a*, and *C* into Equation (8.14), the following spherical model formula of optimal fitting in the *x* direction is obtained:

$$\gamma(H) = \begin{cases} 0 & \text{when} \quad H = 0 \\ 15.52\left[\frac{3H}{2\times 5.236} - \frac{H^3}{2\times(5.236)^3}\right] & \text{when} \quad 0 < H \leq 5.236 \\ 15.52 & \text{when} \quad H > 5.236 \end{cases} \tag{8.32}$$

TABLE 8.5   Experimental Variogram Values and Their Corresponding Optimal fitting Values in x and y Directions

		x Direction				y Direction	
i	Lag Distance H_i	Experimental Variogram Value $\gamma(H_i)$	Optimal Fitting Value $\gamma(H_i)$	i	Lag Distance H_i	Experimental Variogram Value $\gamma(H_i)$	Optimal Fitting Value $\gamma(H_i)$
1	1	4.104[a]	4.392	1	1	4.250[a]	6.540
2	2	8.400[a]	8.460	2	2	8.222[a]	12.003
3	3	12.083[a]	11.879	3	3	10.900[a]	15.867
4	4	9.472	14.325	4	4	17.333[a]	17.333
5	5	6.094	15.473	5	5	24.200[a]	17.333
6	6	9.411	15.520	6	6	18.429	17.333
7	7	12.568[a]	15.520	7	7	7.000	17.333
8	8	12.562	15.520				
9	9	11.591	15.520				

[a]*the experimental variogram values used in optimal fitting calculation are from Table 8.3 or Table 8.4.*

Nine γ values of 4.392, 8.460, ..., 15.520 (Table 8.5) are obtained by substituting $H = 1, 2, ..., 9$ (Table 8.2) in Equation (8.32). The optimal fitting curve in the x direction in Figure 8.4 is composed of these nine points and the starting point (0, 0).

For the y direction, according to the principle of increasing γ, five γ values (Table 8.3) of seven γ values (Table 8.2) are selected to optimal fitting calculation. $b_1 = 2.561$, $b_2 = 0.076$, $b_0 = 1.823$ are calculated by Equation (8.18). Because $b_0 > 0$, but $b_1 b_2 > 0$, the formula cannot be used, but rough calculation can be. C_0, a, and C are calculated from Equations (8.28), (8.29), and (8.30), respectively, attaining $C_0 = 0.278$ of nuggle value, $a = 4$ of variable range value, and $C = 17.06$ of arch height value. $C_b = 17.338$ of base table value can be calculated by Equation (8.31). Substituting C_0, a, and C into Equation (8.14), the following spherical model formula of optimal fitting in y direction is obtained:

$$\gamma(H) = \begin{cases} 0 & \text{when } H = 0 \\ 0.278 + 17.06\left[\frac{3H}{2\times 4} - \frac{H^3}{2\times 4^3}\right] & \text{when } 0 < H \leq 4 \\ 0.278 + 17.06 & \text{when } H > 4 \end{cases} \qquad (8.33)$$

Seven γ values of 6.540, 12.003, ..., 17.333 (Table 8.5) are obtained by substituting $H = 1, 2, ..., 7$ (Table 8.2) in Equation (8.33). The optimal fitting curve in the y direction in Figure 8.4 is composed of these seven points and the starting point (0, 0).

Figure 8.5 illustrates the geometric meaning of variable range value a, nuggle value C_0, arch height value C, and base table value C_b (i.e., $C + C_0$).

Table 8.6 collects a, C, C_0, and C_b in x and y directions.

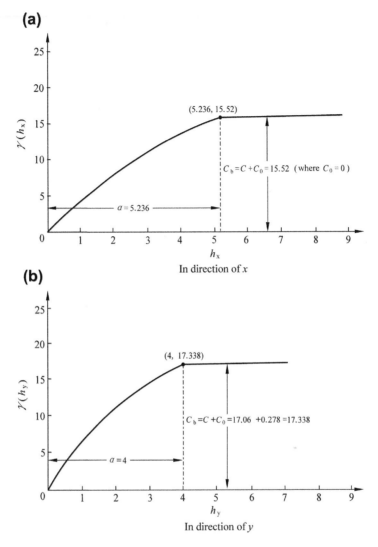

FIGURE 8.5 Geometric meaning of variable range value a, nuggle value C_0, arch height value C, and base table value C_b (i.e., $C + C_0$).

TABLE 8.6 The Values of Variable Range, Arch Height, Nuggle, and Base Table in x and y Directions

Direction	Variable Range a	Arch Height C	Nuggle C_0	Base Table C_b
x	5.236	15.52	0	15.52
y	4	17.06	0.278	17.338

8.4. CROSS-VALIDATION OF KRIGING

8.4.1. Applying Ranges and Conditions

8.4.1.1. Applying Ranges

When the known value of a coordinate-transformed parameter (see Section 8.1) and the optimal fitting of the experimental variogram (see Section 8.3) have been obtained, *Kriging* is used to calculate the estimation value of this parameter. Analysis is made to the variance and deviation between the estimation value and known value so as to check the validity that Kriging is used to this parameter and to confirm whether Kriging will be applied (see Section 8.5).

8.4.1.2. Applying Conditions

For the parameter, the number of observations is large enough, the observed values are as precise as possible, and the optimal fitting of its experimental variogram must have been attained (see Section 8.3).

8.4.2. Basic Principles

Two types of general Kriging are available: *point* Kriging and *block* Kriging. The former estimates the parameter at a point; the latter estimates the parameter at the center of a block. Generally, a smaller amount of calculation is needed for point Kriging; a larger amount of calculation is required for block Kriging, but it is more practical (Wang and Hu, 1988; Sun et al., 1997; Shi, 1999). This book introduces only point Kriging, which is strongly practicable for finer network partition. This section introduces the validation for point Kriging.

Assume there is a parameter on the plane, the value of which is v_n and location is (x_n, y_n), $n = 1, 2, \ldots, N_{\text{total}}$. Now point Kriging is used to estimate the value of this parameter, designated as v_n^*, and then the variance and deviation between v_n^* and v_n are calculated to check the validity of Kriging. This is called the *cross-validation* of point Kriging.

In the cross-validation of point Kriging, the estimation values v_n^* of each parameter are linearly expressed by several v_n, i.e.,

$$\begin{cases} v_n^* = \sum_{k=1}^{N_{\text{sp}}} \lambda_k v_k \\ \sum_{k=1}^{N_{\text{sp}}} \lambda_k = 1 \end{cases} \tag{8.34}$$

where λ_k is the weight of the Kriging estimate value, which meets the Kriging equations about λ_k ($k = 1, 2, \ldots, N_{\text{sp}}$) using a Lagrange multiplier to solve the conditional extremum; see Equation (8.40). But N_{sp} is defined in this way such that taking the location of v_n (x_n, y_n) as an origin, a rectangular coordinate system with four quadrants is formed, and each quadrant is divided into N_s equi-parts. Thus, every quadrant is composed of N_s sectors with equi-included-angle, and the whole rectangular coordinate system is composed of $4N_s$ sectors with equi-included-angle (Figure 8.6). In each sector we search for v_k that is the nearest

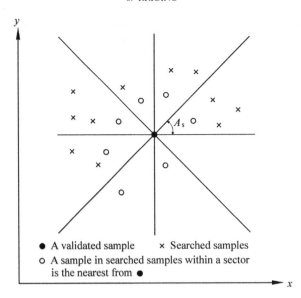

FIGURE 8.6   Sketch map of quadrant equi-partition in equal-angle and searching in a sector.

from the origin. Because one v_k can only appear in one sector, and there might not be v_k in some sectors, the total number of v_k ($k \neq n$) is nearest to v_n, designated as N_{sp}, $N_{sp} \leq 4N$. The included angle in each sector is expressed as:

$$A_s = (90/N_s)(\pi/180) \tag{8.35}$$

In Figure 8.6, $N_s = 2$, $4N_s = 8$, $A_s = \pi/4$, $N_{sp} = 7$, so $N_{sp} < 4N_s$.

Is the Kriging weight the best in Equation (8.34)? For example, people often use the concept of the reciprocal of distance square to form weight and make the sum of each weight equal to 1. It cannot be denied that this is one of the better interpolation approaches, but it is a pure mathematical concept. As mentioned at the beginning of this chapter, it is not as good as Equation (8.34), which combines mathematical concept with geological concept.

In practical applications, N_s can be taken to be 1, 2, 3, and 4. The size of N_s depends on the quantity and distribution status of the samples. Generally, a larger N_s is taken when the samples are more and distribute evenly; on the contrary, a smaller N_s is taken.

8.4.3. Cross-Validation of Kriging Implemented by Nearest Point in Sector

Kriging uses two sets of data: the coordinate-transformed geological data (Table 8.1) and variable range values, arch height values, and nuggle values in x and y directions (Table 8.6). The first set of data covers location name LN_n, location in the new coordinate system (x_n, y_n), and the value of a parameter v_n, $n = 1, 2, ..., N_{total}$; whereas the second set of data covers variable range values, arch height values, and nuggle values in x and y directions in the new coordinate system, designated as a_x, C_x, C_{0_x}, a_y, C_y, C_{0_y}, respectively.

The following introduces the calculation method. Calculation will be run in the following nine steps:

1. To calculate a, C, and C_0. In Kriging, variable range value a, arch height value C, nuggle value C_0, and base table value C_b are calculated by

$$\begin{cases} a = a_x \\ C = \left(C_x + C_y\right)/2 \\ C_0 = \left(C_{0_x} + C_{0_y}\right)/2 \\ C_b = C + C_0 \end{cases} \tag{8.36}$$

2. To divide x, y, and a by a proportional factor. To prevent later matrix operation from overlarge rounding error and even floating-point overflow, particularly when a geodetic coordinate is used for the sample location, x, y, and a, which have distance concepts, are divided by a proportional factor D_y, i.e.,

$$\begin{cases} x_n/D_y \Rightarrow x_n \\ y_n/D_y \Rightarrow y_n \\ a/D_y \Rightarrow a \end{cases} \tag{8.37}$$

where D_y is defined in this way, first to calculate the maximum and minimum of x_n and y_n, i.e., x_{max} and x_{min}, y_{max} and y_{min}, and designate $D_x = x_{max} - x_{min}$, $D_y = y_{max} - y_{min}$. If $D_x < D_y$, then $D_y \Rightarrow D_x$, so $D_x \geq D_y$. Finally, D_y is defined by

$$D_y = \begin{cases} 1 & \text{when } D_x \leq 10^2 \\ 10^2 & \text{when } 10^2 < D_x \leq 10^3 \\ 10^3 & \text{when } 10^3 < D_x \leq 10^4 \\ 10^4 & \text{when } 10^4 < D_x \leq 10^5 \\ 10^5 & \text{when } D_x > 10^5 \end{cases} \tag{8.38}$$

3. $1 \Rightarrow i$ (sample number).
4. To search the nearest sample in each sector.

   As for Sample i, to search Sample j ($j \neq i$) that is the nearest to Sample i and the distance $<a$ in each sector (see Section 8.2.2) and to store LN_j, x_j, y_j, v_j, and D_j of the selected Sample j, of which D_j is the distance between Sample j and Sample i. $j \in [1, N_{total}]$, but $j \neq i$, and j in different sectors are different. LN_j, x_j, y_j, v_j, and D_j take "its own seat" in the sectors it is in. Sector number $\in [1, 4N_s]$ (see Section 8.2.2). Because probably no samples are available in some sectors, the count of selected samples j corresponding to Sample i is $N_{sp} \leq 4N_s$. Thus, a new set of data is formed by sequentially arranging the sectors according to sector number and skipping the possible sectors where samples are not available:

$$LN'_k, x'_k, y'_k, v'_k, D_k \quad (k = 1, 2, ..., N_{sp}) \tag{8.39}$$

This set of data is used to generate an augmented matrix for Kriging equations.

5. To generate a coefficient matrix and right-hand-side vector in Kriging equations. For Sample i, $(N_{sp} + 1)$ equations, called *Kriging equations*, can be formed; see Equation (8.40). The matrix of Kriging equations is expressed as follows (Wang and Hu, 1988; Shi, 1999):

$$K \cdot \lambda = M_2 \tag{8.40}$$

where K, λ, and M_2 are expressed by Equations (8.41), (8.43), and (8.44), respectively.

$$K = \begin{bmatrix} C_{11} & C_{12} & \cdots\cdots & C_{1N_{sp}} & 1 \\ C_{21} & C_{22} & \cdots\cdots & C_{2N_{sp}} & 1 \\ \cdots & \cdots & \cdots\cdots & \cdots & 1 \\ C_{N_{sp}1} & C_{N_{sp}2} & \cdots\cdots & C_{N_{sp}N_{sp}} & 1 \\ 1 & 1 & \cdots\cdots & 1 & 0 \end{bmatrix} \tag{8.41}$$

where C_{kl} $(k = 1, 2, \ldots, N_{sp}; l = 1, 2, \ldots, N_{sp})$ is the following function of the distance H_{kl} between the samples expressed by Equation (8.39).

$$C_{kl} = C_b - \gamma(H_{kl}) \tag{8.42}$$

where C_b is a base table value [see Equation (8.36)] and γ is the optimal fitting of the experimental variogram [see Equation (8.14)].

It must be pointed out that H_{kl} should be calculated according to the principle of the nesting structure, i.e.,

$$H_{kl} = \sqrt{(x'_k - x'_l)^2 + K^2(y'_k - y'_l)^2}$$

where (x'_k, y'_k) is the location of the kth sample [see Equation (8.39)]; (x'_l, y'_l) is the location of the lth sample [see Equation (8.39)]; and K is the nesting structure coefficient, $K = a_x/a_y$.

Because $H_{kl} = H_{lk}$, $C_{kl} = C_{lk}$, and obviously matrix K expressed in Equation (8.41) is a symmetrical matrix,

$$\lambda = \begin{bmatrix} \lambda_1 \\ \lambda_2 \\ \cdots \\ \lambda_{N_{sp}} \\ -\mu \end{bmatrix} \tag{8.43}$$

where λ_k is the weight of Kriging estimate value, $k = 1, 2, \ldots, N_{sp}$, and μ is a Lagrange multiplier to solve the conditional extremum.

λ in Equation (8.43) is the solution of the Kriging equations (8.40).

$$M_2 = \begin{bmatrix} C_{01} \\ C_{02} \\ \cdots \\ C_{0N_{sp}} \\ 1 \end{bmatrix} \tag{8.44}$$

where C_{0k} is the function of the distance D_k shown in Equation (8.39) between Sample i and the nearest sample in sector:

$$C_{0k} = C_b - \gamma(D_k) \quad (k = 1, 2, \ldots, N_{sp})$$ (8.45)

where all symbols of the right-hand side have been defined in Equation (8.42).

6. To solve Kriging equations. Gauss elimination is used to solve Kriging Equation (8.40). In the solution process, the $(N_{sp} + 1) \times (N_{sp} + 1)$ matrix K and the $(N_{sp} + 1) \times 1$ vector M_2 are composed into a $(N_{sp} + 1) \times (N_{sp} + 2)$ augmented matrix. The solved unknown is expressed by Equation (8.43).

7. To calculate the Kriging estimate value and its variance and deviation. The Kriging estimate value of Sample i is

$$v_i^* = \sum_{k=1}^{N_{sp}} \lambda_k v_k'$$ (8.46)

where λ_k and v_k' are expressed by Equations (8.43) and (8.39), respectively.

The Kriging variance of Sample i is

$$\sigma_i^2 = C_b - \sum_{k=1}^{N_{sp}} \lambda_k C_{0k} + \mu$$ (8.47)

The Kriging deviation of Sample i is

$$e_i = v_i - v_i^*$$ (8.48)

8. $i + 1 \Rightarrow i$, and turn to Step 4 if $i \leq N$.

9. Print. Print n, LN_n, v_n, v_n^*, σ_n^2, and e_n, and print LN_k of Sample k related to Sample n, where Sample k is the nearest to Sample n in sector. Here $n = 1, 2, \ldots, N_{total}$; $k = 1, 2, \ldots, N_{sp}$. It should be noted that different n might have different N_{sp}.

For the sake of analysis, the mean of v, v^*, σ^2, and e are also printed:

$$\begin{cases} \bar{v} = \dfrac{1}{N_{total}} \displaystyle\sum_{n=1}^{N_{total}} v_n \\[2em] \bar{v}^* = \dfrac{1}{N_{total}} \displaystyle\sum_{n=1}^{N_{total}} v_n^* \\[2em] \bar{\sigma}^2 = \dfrac{1}{N_{total}} \displaystyle\sum_{n=1}^{N_{total}} \sigma_n^2 \\[2em] \bar{e} = \dfrac{1}{N_{total}} \displaystyle\sum_{n=1}^{N_{total}} e_n \end{cases}$$ (8.49)

8.4.4. Case Study 4: Cross-Validation of Experimental Variogram for a Parameter Discretely Distributed On a Plane (x, y)

The data used in the case study are the results in Section 8.1 (preprocessing). For the sake of convenience in readers' understanding, the results in the old coordinate system are taken, that is x, y, and v in Table 8.1; see Example 1 ($\alpha = 0$). In addition, the results in Section 8.3 (optimal fitting of experiment variogram) are used, that is, a, C, and C_0 in Table 8.6; see Figures 8.3 and 8.5.

Assume that each quadrant is composed of one sector, i.e., a quadrant is equal to a sector. Table 8.7 lists estimate values, variance values, and deviation values of each sample, and Table 8.8 lists the location names of other samples in four sectors (quadrants), which are the nearest from the origin of each sample.

TABLE 8.7   Cross-Validation Results of Kriging

		Parameter			
Sample No. i	Location Name	Known Value v_i	Kriging Estimate Value v_i^*	Kriging Variance σ_i^2	Kriging Deviation e_i
1	w5-0	30	37.27	14.08	−7.27
2	w7-1	39	36.67	6.52	2.33
3	w9-1	31	35.03	11.42	−4.03
4	w3-2	35	36.55	8.44	−1.55
5	w7-2	42	35.31	4.84	6.69
6	w8-2	33	36.82	5.00	−3.82
7	w0-3	37	37.18	7.71	−0.18
8	w2-3	41	38.74	4.87	2.26
9	w5-3	33	37.49	5.84	−4.49
10	w0-4	37	37.60	3.45	−0.60
11	w1-4	40	38.21	3.69	1.79
12	w2-4	42	38.92	3.35	3.08
13	w4-4	34	35.53	5.05	−1.53
14	w5-4	36	35.96	3.10	0.04
15	w6-4	41	37.89	4.37	3.11
16	w9-4	34	38.20	6.14	−4.20
17	w0-5	37	35.80	3.45	1.20
18	w1-5	35	37.31	3.10	−2.31
19	w2-5	37	37.16	3.10	−0.16

(Continued)

TABLE 8.7 Cross-Validation Results of Kriging—cont'd

		Parameter			
Sample No. i	Location Name	Known Value v_i	Kriging Estimate Value v_i^*	Kriging Variance σ_i^2	Kriging Deviation e_i
20	w3-5	35	35.60	5.05	−0.60
21	w5-5	37	35.62	3.35	1.38
22	w6-5	37	37.38	3.10	−0.38
23	w7-5	39	36.98	4.60	2.02
24	w8-5	39	39.04	4.14	−0.04
25	w9-5	41	37.80	3.45	3.20
26	w0-6	35	35.80	3.45	−0.80
27	w1-6	35	35.00	3.10	0.00
28	w2-6	35	34.92	3.35	0.08
29	w5-6	35	33.72	3.42	1.28
30	w6-6	33	35.97	3.35	−2.97
31	w9-6	41	40.80	3.74	0.20
32	w0-7	35	35.00	5.71	0.00
33	w1-7	35	34.29	4.11	0.71
34	w2-7	33	34.29	4.11	−1.29
35	w3-7	33	33.75	4.47	−0.75
36	w4-7	34	32.62	4.63	1.38
37	w5-7	31	34.64	4.11	−3.64
38	w6-7	35	33.71	4.11	1.29
39	w7-7	37	38.16	4.47	−1.16
40	w8-7	41	39.18	4.63	1.82
41	w9-7	41	41.00	5.71	0.00
	Mean	36.37	36.56	4.82	−0.19

Cross-validation is mainly used to make analysis of the validity that point Kriging is applied to the specific geological data, which should be conducted before point Kriging is applied to this set of geological data. According to the location, variance, and deviation of each sample, a plane variance contour and a deviation contour are drawn so as to observe error distribution status. When there are different numbers of sectors in each quadrant, the two kinds of contours are different, based on which to confirm the appropriate number of sectors by comparison. In addition, it is possible to obtain different kinds of contours by

TABLE 8.8   Other Samples in Four Sectors (Quadrants) that are the Nearest from the Origin of Each Sample

		The Nearest Other Samples			
Sample No. i	Location Name	The 1st Sector	The 2nd Sector	The 3rd Sector	The 4th Sector
1	w5-0	w7-1	w3-2		
2	w7-1	w8-2	w7-2	w5-0	
3	w9-1	w8-2	w7-1		
4	w3-2	w5-3	w2-3	w5-0	
5	w7-2	w8-2	w5-3	w5-0	w7-1
6	w8-2	w9-4	w6-4	w7-2	w9-1
7	w0-3	W1-4	w0-4	w3-2	
8	w2-3	w4-4	w2-4	w0-3	w3-2
9	w5-3	w6-4	w5-4	w3-2	w7-2
10	w0-4	w1-4	w0-5	w0-3	
11	w1-4	w2-4	w1-5	w0-4	w2-3
12	w2-4	w3-5	w2-5	w1-4	w2-3
13	w4-4	w5-4	w3-5	w2-4	w5-3
14	w5-4	w6-4	w5-5	w4-4	w5-3
15	w6-4	w7-5	w6-5	w5-4	w7-2
16	w9-4	w9-5	w8-2	w9-1	
17	w0-5	w1-5	w0-6	w0-4	
18	w1-5	w2-5	w1-6	w0-5	w1-4
19	w2-5	w3-5	w2-6	w1-5	w2-4
20	w3-5	w5-5	w2-6	w2-5	w4-4
21	w5-5	w6-5	w5-6	w4-4	w5-4
22	w6-5	w7-5	w6-6	w5-5	w6-4
23	w7-5	w8-5	w6-6	w6-5	w9-4
24	w8-5	w9-5	w8-7	w7-5	w9-4
25	w9-5	w9-6	w8-5	w9-4	
26	w0-6	w1-6	w0-7	w0-5	
27	w1-6	w2-6	w1-7	w0-6	w1-5
28	w2-6	w3-7	w2-7	w1-6	w2-5
29	w5-6	w6-6	w5-7	w4-4	w5-5
30	w6-6	w7-7	w6-7	w5-6	w6-5

(Continued)

TABLE 8.8   Other Samples in Four Sectors (Quadrants) that are the Nearest from the Origin of Each Sample—cont'd

		The Nearest Other Samples			
Sample No. i	Location Name	The 1st Sector	The 2nd Sector	The 3rd Sector	The 4th Sector
31	w9-6	w9-7	w8-5	w9-5	
32	w0-7	w1-7	w0-6		
33	w1-7	w2-7	w0-7	w1-6	
34	w2-7	w3-7	w1-7	w2-6	
35	w3-7	w4-7	w2-7	w3-5	
36	w4-7	w5-7	w3-7	w5-6	
37	w5-7	w6-7	w4-7	w5-6	
38	w6-7	w7-7	w5-7	w6-6	
39	w7-7	w8-7	w6-7	w7-5	
40	w8-7	w9-7	w7-7	w9-6	
41	w9-7	w8-7	w9-6		

different schemes through accepting or rejecting the geological data; thus suitable data for an appropriate scheme are selected.

By the way, in calculation results, for sample No. 27 (w1-6), sample No. 32 (w0-7), and sample No. 41 (w9-7), their deviations are all zero, which results from the fact that the value of each sample is equal to the value of the nearest sample in the respective sector.

8.5. APPLICATIONS OF KRIGING

8.5.1. Applying Ranges and Conditions

8.5.1.1. Applying Ranges

When the known value of a coordinate-transformed parameter (see Section 8.1) and the optimal fitting of an experimental variogram (see Section 8.3) have been obtained, Kriging is used to calculate the estimate value of this parameter at each grid center on the plane.

8.5.1.2. Applying Conditions

For the parameter, the number of observations is large enough, the observed values are as precise as possible, and the optimal fitting of its experimental variogram must have been attained (see Section 8.3).

8.5.2. Basic Principles

This section introduces the application of point Kriging. Assume there is a parameter on the plane, the value of which is v_n and location is (x_n, y_n), $n = 1, 2, ..., N_{total}$. To estimate the distribution of this parameter on the plane, we divide the plane into a certain number of rectangular grids according to the studied area. First $x_{min} = \min\{x_n\}$, $x_{max} = \max\{x_n\}$, $y_{min} = \min\{y_n\}$, and $y_{max} = \max\{y_n\}$ are calculated, then the number of grids N_x and N_y in x and y directions are calculated.

$$
\begin{cases}
N_x = \text{INT}\left(\dfrac{x_{max} - x_{min}}{\Delta x} + 0.5\right) \\[2mm]
N_y = \text{INT}\left(\dfrac{y_{max} - y_{min}}{\Delta y} + 0.5\right)
\end{cases}
\tag{8.50}
$$

where INT is integral function; Δx is grid spacing in x direction; and Δy is grid spacing in y direction.

Thus, the studied area is a rectangle with the length of $N_x\Delta x$ in x direction and the length of $N_y\Delta y$ in y direction, i.e., the area is composed of N_xN_y rectangular grids (Δx, Δy).

Now the question is how to solve v^* at each grid center based on the known x_n, y_n, and v_n, i.e.,

$$
\begin{cases}
v^* = \displaystyle\sum_{k=1}^{N_d} \lambda_k v_k \\[2mm]
\displaystyle\sum_{k=1}^{N_d} \lambda_k = 1
\end{cases}
\tag{8.51}
$$

where λ_k is the weight of the Kriging estimate value, which meet the Kriging equations about λ_k ($k = 1, 2, ..., N_{sp}$) using the Lagrange multiplier to solve the conditional extremum; see Equation (8.56). N_d is the count of samples that are used in calculating v^* value at a grid center to be evaluated. N_d samples are the nearest from the grid center in all samples. In practical applications, N_d can be taken to be 6. The size of N_d depends on the quantity and distribution status of the samples. Generally, a larger N_d is taken when the samples are more and distribute evenly; on the contrary, a smaller N_d is taken.

8.5.3. Applications of Kriging Implemented by Nearest Point in Plane

Kriging uses two sets of data: the coordinate-transformed geological data (Table 8.1) and variable range values, arch height values, and nuggle values in x and y directions (Table 8.6). The first set of data covers location name LN_n, location in the new coordinate system (x_n, y_n), and the value of a parameter v_n, $n = 1, 2, ..., N_{total}$; whereas the second set of data covers variable range values, arch height values, and nuggle values in x and y directions in the new coordinate system, designated as a_x, C_x, C_{0x}, a_y, C_y, C_{0y}, respectively.

The following introduces the calculation method. Calculation will be run in 12 steps:

1. To calculate a, C, and C_0. In Kriging, variable range value a, arch height value C, nuggle value C_0, and base table value C_b are calculated by Equation (8.36).
2. To divide x, y, and a by a proportional factor. To prevent later matrix operation from overlarge rounding error and even floating-point overflow, particularly when a geodetic coordinate is used for sample location, x, y, and a, which have distance concept, are divided by a proportional factor D_y; see Equations (8.37) and (8.38).

3. To generate plane network. The grid count N_x and N_y in x and y directions are calculated by Equation (8.50) based on the given grid spacing Δx and Δy, to be $N_x N_y$ of grid count in total.

4. $1 \Rightarrow n_y$ in y direction.

5. $1 \Rightarrow n_x$ in x direction.

6. To search N_d nearest samples on plane. The coordinate of a grid center located on (n_x, n_y) is expressed as

$$
\begin{cases}
x_0 = x_{\min} + n_x \Delta x - 0.5 \Delta x \\
y_0 = y_{\min} + n_y \Delta y - 0.5 \Delta y
\end{cases}
\tag{8.52}
$$

For the sake of unification, x_0 and y_0 are divided by a proportional factor D_y, respectively:

$$
\begin{cases}
x_0 / D_y \Rightarrow x_0 \\
y_0 / D_y \Rightarrow y_0
\end{cases}
\tag{8.53}
$$

The distance between the location of grid center (x_0, y_0) shown in Equation (8.52) and the location of each sample (x_n, y_n) is expressed as

$$
D_n = \sqrt{(x_n - x_0)^2 + (y_n - y_0)^2} \quad (n = 1, 2, ..., N_{\text{total}})
\tag{8.54}
$$

A new set of data is formed by selecting N_d samples that their D_n are smaller than others:

$$
x'_k, y'_k, v'_k, D_k \quad (k = 1, 2, ..., N_d)
\tag{8.55}
$$

This set of data is used to generate augmented matrix for Kriging equations.

7. To generate a coefficient matrix and a right-hand-side vector of Kriging equations. For the grid center (n_x, n_y), $(N_d + 1)$ equations, called *Kriging equations*, can be formed; see Equation (8.56).

The matrix of Kriging equations is expressed as follows (Wang and Hu, 1988; Shi, 1999):

$$
K \cdot \lambda = M_2
\tag{8.56}
$$

where K, λ, and M_2 are expressed by Equations (8.57), (8.59), and (8.60), respectively.

$$
K = \begin{bmatrix}
C_{11} & C_{12} & \cdots & C_{1N_d} & 1 \\
C_{21} & C_{22} & \cdots & C_{2N_d} & 1 \\
\cdots & \cdots & \cdots & \cdots & 1 \\
C_{N_d 1} & C_{N_d 2} & \cdots & C_{N_d N_d} & 1 \\
1 & 1 & \cdots & 1 & 0
\end{bmatrix}
\tag{8.57}
$$

where C_{kl} ($k = 1, 2, ..., N_d$; $l = 1, 2, ..., N_d$) is the following function of the distance H_{kl} between N_d samples expressed by Equation (8.55):

$$
C_{kl} C_b - \gamma(H_{kl})
\tag{8.58}
$$

where all symbols have been defined in Equation (8.42).

It must be pointed out that H_{kl} should be calculated according to the principle of nesting structure, i.e.,

$$H_{kl} = \sqrt{(x'_k - x'_l)^2 + K^2(y'_k - y'_l)^2}$$

where (x'_k, y'_k) is the location of the kth sample [see Equation (8.55)]; (x'_l, y'_l) is the location of the lth sample [see Equation (8.55)]; and K is the nesting structure coefficient, $K = a_x/a_y$.

Because $H_{kl} = H_{lk}$, $C_{kl} = C_{lk}$, and obviously matrix K expressed in Equation (8.57) is a symmetrical matrix,

$$\lambda = \begin{bmatrix} \lambda_1 \\ \lambda_2 \\ \cdots \\ \lambda_{N_d} \\ -\mu \end{bmatrix} \tag{8.59}$$

where all symbols have been defined in Equation (8.43).

λ in Equation (8.59) is the solution of the Kriging equations (8.59).

$$M_2 = \begin{bmatrix} C_{01} \\ C_{02} \\ \cdots \\ C_{0N_d} \\ 1 \end{bmatrix} \tag{8.60}$$

where C_{0k} is the function of the distance D_k between the grid center (x_0, y_0) expressed by Equation (8.53) and locations of N_d samples expressed by Equation (8.55):

$$C_{0k} = C_b - \gamma(D_k) \quad (k = 1, 2, ..., N_d) \tag{8.61}$$

where all symbols of the right-hand side have been defined in Equation (8.42).

8. To solve Kriging equations. Gauss elimination is used to solve Kriging equations (8.56). In the solution process, the $(N_d + 1) \times (N_d + 1)$ matrix K and the $(N_d + 1) \times 1$ vector M_2 are composed into a $(N_d + 1) \times (N_d + 2)$ augmented matrix. The solved unknown is expressed by Equation (8.59).

9. To calculate the Kriging estimate value and its variance. The Kriging estimate value and its variance at the grid center (n_x, n_y) are expressed by Equations (8.62) and (8.63).

$$v^* = \sum_{k=1}^{N_d} \lambda_k v'_k \tag{8.62}$$

where λ_k and v'_k are expressed by Equations (8.59) and (8.55), respectively.

$$\sigma^2 = C_b - \sum_{k=1}^{N_d} \lambda_k C_{0k} + \mu \tag{8.63}$$

10. $n_x + 1 \Rightarrow n_x$, turn to Step 6 if $n_x \le N_x$.
11. $n_y + 1 \Rightarrow n_y$, turn to Step 5 if $n_y \le N_y$.

12. To print the calculation results. Print n_x, n_y, $x_0 D_y$, $y_0 D_y$, v^*, and σ^2 at each grid center. The reason that x_0 and y_0 are multiplied by D_y is that they were once before divided by D_y. For the sake of analysis, the mean of v^* and σ^2 are also printed:

$$\begin{cases} \bar{v}^* = \frac{1}{N_x N_y} \sum_{n=1}^{N_x N_y} v_n^* \\ \bar{\sigma}^2 = \frac{1}{N_x N_y} \sum_{n=1}^{N_x N_y} \sigma_n^2 \end{cases} \tag{8.64}$$

8.5.4. Case Study 5: Kriging Application for a Parameter Discretely Distributed On a Plane (x, y)

The data used in the case study are the results in Section 8.1 (preprocessing). For the sake of reader understanding, the results in the old coordinate system are taken, that is, x, y, and v in Table 8.1; see Figure 8.3 and Example 1 ($\alpha = 0$). In addition, the results in Section 8.3 (optimal fitting of experiment variogram) are used, that is a, C, and C_0 in Table 8.6; see Figures 8.3 and 8.5.

Take $\Delta x = 1$ and $\Delta y = 1$. Letting the sample count N_d for determining the estimate value at grid center be 6, Table 8.9 lists the estimate values, variance values, and their mean at each grid center.

TABLE 8.9   Application Results of Kriging

Grid No. n	Grid (n_x, n_y)		Coordinates of a Grid Center		Kriging Estimate Value v_n^*	Kriging Variance σ_n^2
	n_x	n_y	x_0	y_0		
1	1	1	0.5	0.5	36.67	16.48
2	2	1	1.5	0.5	35.23	13.69
3	3	1	2.5	0.5	34.13	10.69
4	4	1	3.5	0.5	32.10	7.74
5	5	1	4.5	0.5	30.85	4.40
6	6	1	5.5	0.5	33.60	4.07
7	7	1	6.5	0.5	36.75	4.54
8	8	1	7.5	0.5	36.08	5.12
9	9	1	8.5	0.5	32.76	5.16
10	1	2	0.5	1.5	37.04	10.90
11	2	2	1.5	1.5	37.21	9.21

(Continued)

8. KRIGING

TABLE 8.9 Application Results of Kriging—cont'd

Grid No. n	Grid (n_x, n_y)		Coordinates of a Grid Center		Kriging Estimate Value v_n^*	Kriging Variance σ_n^2
	n_x	n_y	x_0	y_0		
12	3	2	2.5	1.5	35.72	5.40
13	4	2	3.5	1.5	33.33	4.55
14	5	2	4.5	1.5	32.63	5.81
15	6	2	5.5	1.5	35.73	5.38
16	7	2	6.5	1.5	39.62	2.99
17	8	2	7.5	1.5	37.55	2.60
18	9	2	8.5	1.5	31.87	2.80
19	1	3	0.5	2.5	37.62	5.09
20	2	3	1.5	2.5	39.01	4.71
21	3	3	2.5	2.5	38.13	2.83
22	4	3	3.5	2.5	34.89	3.81
23	5	3	4.5	2.5	32.63	4.01
24	6	3	5.5	2.5	36.32	4.24
25	7	3	6.5	2.5	40.89	3.74
26	8	3	7.5	2.5	37.63	3.36
27	9	3	8.5	2.5	32.70	3.72
28	1	4	0.5	3.5	38.05	2.55
29	2	4	1.5	3.5	40.36	2.53
30	3	4	2.5	3.5	39.51	2.76
31	4	4	3.5	3.5	35.33	3.70
32	5	4	4.5	3.5	33.70	2.42
33	6	4	5.5	3.5	36.52	2.51
34	7	4	6.5	3.5	40.63	3.66
35	8	4	7.5	3.5	37.92	5.09
36	9	4	8.5	3.5	34.57	3.77
37	1	5	0.5	4.5	36.90	2.13
38	2	5	1.5	4.5	38.58	2.13
39	3	5	2.5	4.5	38.26	2.38
40	4	5	3.5	4.5	35.05	2.96
41	5	5	4.5	4.5	34.98	2.48

(Continued)

TABLE 8.9 Application Results of Kriging—cont'd

Grid No. n	Grid (n_x, n_y)		Coordinates of a Grid Center		Kriging Estimate Value v_n^*	Kriging Variance σ_n^2
	n_x	n_y	x_0	y_0		
42	6	5	5.5	4.5	37.49	2.10
43	7	5	6.5	4.5	39.54	2.60
44	8	5	7.5	4.5	38.47	3.57
45	9	5	8.5	4.5	37.23	2.46
46	1	6	0.5	5.5	35.71	2.16
47	2	6	1.5	5.5	35.92	2.16
48	3	6	2.5	5.5	36.17	2.52
49	4	6	3.5	5.5	34.74	4.15
50	5	6	4.5	5.5	35.46	2.97
51	6	6	5.5	5.5	35.62	2.16
52	7	6	6.5	5.5	36.27	2.53
53	8	6	7.5	5.5	38.30	3.57
54	9	6	8.5	5.5	40.03	2.36
55	1	7	0.5	6.5	35.09	2.16
56	2	7	1.5	6.5	34.58	2.16
57	3	7	2.5	6.5	33.96	2.30
58	4	7	3.5	6.5	33.99	3.00
59	5	7	4.5	6.5	34.12	2.38
60	6	7	5.5	6.5	33.75	2.16
61	7	7	6.5	6.5	35.51	2.30
62	8	7	7.5	6.5	38.96	2.89
63	9	7	8.5	6.5	40.91	2.38
			Mean		36.20	4.05

Figure 8.3 shows that the parameter does not distribute evenly on the plane (x, y); in particular, no data are available in the lower-left corner, i.e., southwest in geographic concept, which covers 1/10 of the total area. So in this location the variance of estimate value at the grid center calculated by Kriging is over 7 (Table 8.9). The variance of estimate value at the grid center calculated by Kriging is smaller in the other locations in this figure, especially in the area with more data. If a contour is used to illustrate Kriging variance, it is prone to observe the distribution status of Kriging variance, thereby to know the accuracy distribution of Kriging estimate values.

8.6. SUMMARY AND CONCLUSIONS

In summary, this chapter introduced five steps of the Kriging operation: preprocessing (see Section 8.1), experimental variogram (see Section 8.2), optimal fitting of experimental variogram (see Section 8.3), cross-validation of Kriging (see Section 8.4), and applications of Kriging (see Section 8.5). The contents of these five steps seem to be connecting links between the preceding and the following, which form an integer. The ultimate purpose is to perform the cross-validation and applications of Kriging (see Figure 8.1). One case study is given for each step. The five case studies are based on original data, i.e., on the plane equidistant network with 80 nodes (i.e., 63 grids), there are values of a parameter at 41 nodes of 80 nodes, which form 41 samples. Different samples include different (x, y) and parameter value. In each step of the Kriging operation, data input and results output are given.

Using the data of these 41 samples, the optimal fitting of experimental variogram is obtained using the data mining tool of Kriging. This optimal fitting is believed to be qualified by cross-validation of Kriging. Therefore, this optimal fitting is called *mined knowledge*. In the prediction process, this knowledge is applied to calculate the estimate values of the parameter at 63 grid centers (Table 8.9). Though the case is small, it is beneficial to understand the method and practical applications via which Kriging can be spread to geoscience applications. The effect of Kriging's optimal linear unbiased interpolation estimate is superior to the general mathematics interpolation, so it can be widely applied.

EXERCISES

8-1. What are the advantage of geostatistics compared with conventional statistics?

8-2. How is Kriging organically composed of five operation steps?

8-3. Why is the coordinate system transformation required in the preprocessing of geostatistics (see Section 8.1)?

8-4. What are the actions of the experimental variogram (see Section 8.2)?

8-5. What are the actions of the optimal fitting of the experimental variogram (see Section 8.3)?

8-6. Why is the optimal fitting curve rather than the experimental variogram curve (Figure 8.4) provided for Kriging (see Sections 8.4 and 8.5)?

8-7. Is it possible to adopt the least-squares method mentioned in Section 2.2.3 to realize the optimal fitting of the experimental variogram?

8-8. What are the actions of the cross-validation of Kriging (see Section 8.4)?

8-9. What are the actions of the applications of Kriging (see Section 8.5)?

References

Shi, G., 1999. New Computer Application Technologies in Earth Sciences. Petroleum Industry Press, Beijing, China (in Chinese).

Sun, H., Kang, Y., Du, H., 1997. A Package of Practical Geostatistics. Geology Press, Beijing, China (in Chinese).

Wang, R., Hu, G., 1988. Linear Geostatistics. Geology Press, Beijing, China (in Chinese).

Data Mining and Knowledge Discovery for Geoscientists
http://dx.doi.org/10.1016/B978-0-12-410437-2.00009-6

This chapter introduces the techniques of other soft computing algorithms as well as their applications in geosciences. For each technique, the applying ranges and conditions, basic principles, calculation method, calculation flowchart, and case studies are provided. In each case study, the calculation flowchart, calculation results, and analyses are given. Though the case study is small, it reflects the whole process of calculations to benefit readers in understanding and mastering the techniques. Though one flowchart is drawn for a specific case study, it is actually all-purpose only if the specific case study is replaced by readers' own application samples.

Generally, *hard computing* means conventional calculation, that is, calculation results are affirmatory and accurate. However, in natural and social sciences, there exist some calculation problems that are not always completely affirmatory and accurate; thus *soft computing* is needed to solve these problems. Fuzzy mathematics, gray systems, and fractal geometry introduced in this chapter fall in the category of soft computing. As for linear programming, it essentially falls into the category of hard computing; however, since it is required to repeatedly adjust the mathematical model and constraint condition for adapting to the practicality in geosciences, linear programming is constrainedly arranged in soft computing.

Section 9.1 (fuzzy mathematics) introduces a simple case study of fuzzy integrated evaluation on an exploration drilling object, with which to explain how to adopt fuzzy integrated decisions to select exploration drilling objects. This is the application of fuzzy mathematics in geosciences. Concretely, if it is of significance to drill a well, we use two kinds of data for this well: the evaluation results that 10 experts made on 6 geological factors in terms of 5 comment ratings, and the weight of the 6 geological factors. Then we use data mining tool (fuzzy integrated decision), the fuzzy integrated decision value for drilling this well is obtained. The results coincide with practicality. This decision is called *mined knowledge*, which provides the basis for the decision to drill this well. Therefore, this method can be spread to the decision of drilling other exploration wells.

Section 9.2 (gray system) introduces four simple case studies, in which the first three pertain to space-time calculations conducted by gray prediction and the last pertains to geological object quality determined by gray integrated decision. In gray prediction there are two calculation schemes: Calculation Scheme 1, in which the original data are not processed; and Calculation Scheme 2, in which the original data are processed so that the minimum value is subtracted from each piece of data.

1. Simple Case Study 1 is an oil production prediction of an oil production zone in the Shengli Oilfield using gray prediction, with which we explain how to adopt gray prediction for time-domain prediction. Concretely, using the data of this oil production zone, i.e., annual oil production in five years from 1985 to 1989, a *whiting equation* (9.25) with its solution (9.26) for the annual oil production of the oil production zone is constructed by a data mining tool (gray prediction). The results coincide with practicality. This whiting equation with its solution is called *mined knowledge*. This knowledge is applied to predict the annual oil production in the coming 10 years from 1990 to 1999. Using Calculation Scheme 1, the prediction results basically agree with the actual production. Therefore, this method can be spread to the prediction of annual oil and gas production of other production zones.

2. Simple Case Study 2 is reservoir prediction of a well in the Liaohe Oilfield using gray prediction, with which we explain how to adopt gray prediction for the space-domain prediction. Concretely, during drilling a well, we use data of this well: the depths of the top of five reservoirs where oil and gas shows appear downward. Then we use data mining tool (gray prediction), a whiting equation with its solution for the reservoir top depth in that well is constructed. The results coincide with practicality. This whiting equation with its solution is called *mined knowledge*. This knowledge is applied to predict the depth of the top of the sixth reservoir. Using Calculation Scheme 2, the prediction results basically agree with the actual production. Therefore, this method can be spread to the prediction of reservoir top depth in drilling other wells.

3. Simple Case Study 3 is an oil production prediction of the Romashkino Oilfield using gray prediction, with which we explain how to adopt gray prediction for time-domain prediction. Concretely, using the data of the Romashkino Oilfield, i.e., annual oil production in the 39 years from 1952 to 1990, a whiting equation with its solution for the annual oil production of this oilfield is constructed by a data mining tool (gray prediction). There exists considerable difference between the calculation results and the real production. The whiting equation with its solution may be called *mined wrong "knowledge."* This wrong "knowledge" is applied to predict the annual oil production in the coming 15 years from 1991 to 2005. The calculation results are found to be considerably different from the actual production, too. Moreover, calculation was run using both Calculation Scheme 1 and Calculation Scheme 2. The difference of the calculation results of the two schemes is small, and Calculation Scheme 2 is a little bit better than Calculation Scheme 1. Here it is indicated that annual oil production of the Romashkino Oilfield can be successfully predicted (see Section 3.1.5) using an error

back-propagation neural network (BPNN; see Chapter 3). Why is gray prediction applicable to predict the annual oil production of the aforementioned oil production zone but not to predict the annual oil production of the Romashkino Oilfield? This is attributed to the annual oil production gradient variation of the learning sample: The annual oil production of that oil production zone is 428.8250–451.7581 (10^4t) for five years, whereas the annual oil production for 39 years of the Romashkino Oilfield is 200–8150 (10^4t). Obviously, it is impossible to adopt gray prediction, but we can adopt BPNN to predict annual oil and gas production of an oilfield when the annual production gradient is large, since the solution accuracy of BPNN is not restricted by a gradient.

4. The last case study is about the gray integrated evaluation of exploration drilling objects, with which we explain how to adopt gray integrated decisions to select exploration drilling objects. Concretely, using the data of four exploration drilling objects in an exploration area, i.e., for each exploration drilling object the evaluation results on six geological factors according to five comment ratings made by five experts as well as the weight of these six geological factors, the integrated evaluation for each exploration drilling object is obtained by a data mining tool (a gray integrated decision). The results coincide with practicality. This integrated evaluation is called *mined knowledge*, which provides evidence to decide the selection of the first exploration drilling object. Therefore, this method can be spread to the selection of exploration drilling objects in other exploration areas.

Section 9.3 (fractal geometry) introduces a case study, analysis of a fractal feature of a fault, with which to explain how to adopt fractal geometry for quantitative description of the relationship between fractal features of fault oil gas migration. Concretely, using the data of two structural layers in Dongying Sag in Shengli Oilfield, i.e., the structure map containing fault locations for each structural layer, the fractal dimension in the four areas of north, south, west, and east for each structural layer is obtained by a data mining tool (fractal geometry). The results coincide with practicality. The fractal dimension is called *mined knowledge*, which expresses the development extent of the fault in the sag, high in the north while low in the south, thus predicting the development extent of the reservoir in the sag, high in the north while low in the south. This understanding agrees with what it is in practice. Therefore, this method can be spread to the prediction of the reservoir development extent in other sags.

Section 9.4 (linear programming) introduces a simple case study, the best economic efficiency for an exploration drilling plan, with which to explain how to adopt the simplex method of linear programming to realize optimization of the *exploration and production* (E&P) scheme. Concretely, using the data of five oil-bearing traps in this area, including the success probability of exploration drilling for each oil-bearing trap, the oil revenue in two years for each well, as well as five geological exploration requirements, the object function, and its constraint condition in this area, are obtained by data mining tool (linear programming). The results coincide with practicality. This object function and its constraint condition are called *mined knowledge*, which is applied to calculate the optimal number of wells to be drilled in each oil-bearing trap. Therefore, this method can be spread to the optimization of the number of wells to be drilled in other areas.

9.1. FUZZY MATHEMATICS

9.1.1. Applying Ranges and Conditions

9.1.1.1. Applying Ranges

The common algorithm is one of fuzzy integrated decisions, which can be adopted under a certain E&P condition to determine the quality rating (e.g., excellent, good, average, poor, and very poor) of the studied geological object. Assuming that some geological factors related to the aforementioned geological object quality are used to describe their relationship with this geological object in terms of rating, it is possible to adopt fuzzy integrated decisions to determine the quality rating of this geological object. The geological objects studied can be the evaluation of reservoirs; identification of oil-, gas-, and water-bearing formations, and so on. Certainly, if these geological factors can be expressed in numerical forms, the fuzzy integrated decision can also be adopted to determine the quality rating of a geological object.

9.1.1.2. Applying Conditions

There must be sufficient geological factors affecting the geological object quality as well as the weight assignment and impact rating for the geological objects.

9.1.2. Basic Principles

Since the means of petroleum E&P as well as the data acquired are limited, the decision on a geological object is a fuzzy concept. There are three cases:

1. For an exploration drilling object, before *spudding* (Spudding is a special term in well drilling. Spudding means start drilling), no one expert affirms that there must be commercial oil flow or there must not be commercial oil flow, but experts often believe that both success and failure are possible. Obviously, it is not the two values of 0 to express failure or 1 to express success for the forecast of subsurface objects. That is to say, it is not the concept of success or failure, but the concept of "both are possible," in which all membership functions within the range of [0, 1].
2. The basis of a decision comes from the analysis of several geological factors.
3. The results of a decision by every expert are different. In light of the previous explanation, the fuzzy integrated decision can be adopted to calculate the success ratio of this exploration drilling object.

9.1.3. Geological Object Quality Determined by Fuzzy Integrated Decision

9.1.3.1. Calculation Method

Assuming that there are n geological factors affecting a geological object and m comment ratings, and letting E be the value of a fuzzy integrated decision, E is expressed as follows (Guo and Sun, 1992; Shi, 1999):

$$E = W \cdot R \cdot V^{\mathrm{T}} \tag{9.1}$$

where E is integrated evaluation; the value ranges between the first element and the last element of vector V. W is a $(1 \times n)$ matrix, called a *weight vector*, showing the weight of the related geological factors; see Equation (9.2). R is a $(n \times m)$ matrix, called a *fuzzy matrix*, showing the membership degree of every geological factor versus each comment rating; see Equation (9.3). V is a $(1 \times m)$ matrix, called a *comment vector*, showing a comment rating; see Equation (9.4). V^T is the transposed vector of V.

The weight vector is

$$W = (w_1, w_2, ..., w_n) \tag{9.2}$$

where w_i $(i = 1, 2, ..., n)$ is the weight of the ith geological factor for evaluating the geological object, $\sum_{i=1}^{n} w_i = 1$.

The fuzzy matrix is

$$R = \begin{bmatrix} r_{11} & r_{12} & ...... & r_{1m} \\ r_{21} & r_{22} & ...... & r_{2m} \\ ... & ... & ... & ... \\ r_{n1} & r_{n2} & ...... & r_{nm} \end{bmatrix} \tag{9.3}$$

where r_{ij} $(i = 1, 2, ..., n; j = 1, 2, ..., m)$ is the membership degree of the ith geological factor versus the jth comment rating, $\sum_{j=1}^{m} r_{ij} = 1$.

The comment vector

$$V = (v_1, v_2, ..., v_m) \tag{9.4}$$

where m is usually taken to be odd number. If $m = 3$, $V = (1, 0, -1)$, indicating three comment ratings of good, average, and poor, then $E \in [-1, 1]$ in Equation (9.1). If $m = 5$, $V = (2, 1, 0, -1, -2)$, indicating five comment ratings of excellent, good, average, poor, and very poor, then $E \in [-2, 2]$ in Equation (9.1).

It is easy for readers to understand Equations (9.2) and (9.4). As for Equation (9.3), it will be explained in the simple case study.

9.1.3.2. Simple Case Study: Fuzzy Integrated Evaluation of Exploration Drilling Object

Ten experts are invited to adopt fuzzy integrated decisions to evaluate drilling an exploratory well in an exploration area in China. They select six geological factors affecting the quality of well location based on petroleum system theory: hydrocarbon generation, migration, trapping, accumulation, cap rock, and preservation (Shi, 1999). Finally, the weight of these six geological factors is determined through collective discussion, i.e., the weight vector corresponding to Equation (9.2) is

$$W = (0.15, 0.10, 0.15, 0.20, 0.25, 0.15) \tag{9.5}$$

And the five comment ratings have also been determined: excellent, good, average, poor, and very poor. Thus the comment vector corresponding to Equation (9.4) is

$$V = (2, 1, 0, -1, -2) \tag{9.6}$$

TABLE 9.1   Summation of Comments from 10 Experts

Geological Factor	p_{ij} [Number of Experts for a Geological Factor (i) and a Comment Rating (j)]				
	Excellent	Good	Average	Poor	Very Poor
Hydrocarbon generation	3	4	3	0	0
Migration	2	4	2	1	1
Trapping	2	3	2	2	1
Accumulation	5	3	2	0	0
Cap rocks	1	2	3	3	1
Preservation	2	2	3	1	2

Afterward, the 10 experts make evaluations on the aforementioned six geological factors in terms of five comment ratings. Table 9.1 lists the summation of the comments.

The numbers in Table 9.1 indicate the number of experts for a geological factor versus a comment rating, designated as statistic value p_{ij} ($i = 1, 2, ..., 6; j = 1, 2, ..., 5$), forming statistics matrix P. By normalization, p_{ij} is transformed to fuzzy matrix element r_{ij} corresponding to Equation (9.3) as follows:

$$r_{ij} = p_{ij} / \sum_{j=1}^{5} p_{ij} \quad (i = 1, 2, ..., 6; j = 1, 2, ..., 5) \tag{9.7}$$

So the fuzzy matrix corresponding to Equation (9.3) is

$$R = \begin{bmatrix} 0.3 & 0.4 & 0.3 & 0 & 0 \\ 0.2 & 0.4 & 0.2 & 0.1 & 0.1 \\ 0.2 & 0.3 & 0.2 & 0.2 & 0.1 \\ 0.5 & 0.3 & 0.2 & 0 & 0 \\ 0.1 & 0.2 & 0.3 & 0.3 & 0.1 \\ 0.2 & 0.2 & 0.3 & 0.1 & 0.2 \end{bmatrix} \tag{9.8}$$

Substituting Equations (9.5), (9.8), and (9.6) into Equation (9.1), the fuzzy integrated decision value $E = 0.495$ is obtained. Since 2, 1, 0, −1, and −2 in Equation (9.6) indicate excellent, good, average, poor, and very poor, respectively, but value E ranges between 1 and 0, slightly leaning to 0, the integrated evaluation result of this exploration drilling object approaches to average, or falls in the range between good and average.

If there are several exploration drilling objects, the fuzzy integrated decision can be made under the same evaluation standard (weight vector W and comment vector V), respectively, achieving their fuzzy integrated decision values, which will be ranked according to quality so as to reach the aim of optimal selection.

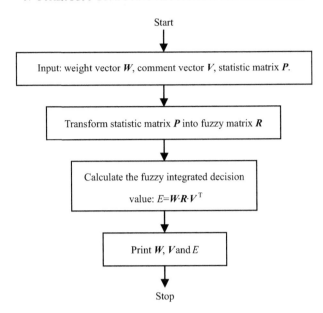

FIGURE 9.1 Calculation flowchart of fuzzy integrated decision.

1. Calculation flowchart (Figure 9.1).
2. Results and analyses. Calculation results give 0.495 of the fuzzy integrated decision value of this exploration drilling object, which falls in the range between good and average in the five comment ratings, where 2 is excellent, 1 is good, 0 is average, −1 is poor, and −2 is very poor, or is approximately defined to be average, because 0.495 is closer to 0 but a little bit far from 1. In fact, this conclusion can be reflected in Table 9.1 or Equation (9.8); no more than quantification and accuracy as a fuzzy integrated decision.

In summary, if it is of significance to drill a well, we use two kinds of data for this well: the evaluation results that 10 experts made on 6 geological factors in terms of 5 comment ratings, and the weight of the 6 geological factors. Then we use data mining tool (fuzzy integrated decision), the fuzzy integrated decision value for drilling this well is obtained. The results coincide with practicality. This decision is called *mined knowledge*, which provides the basis for the decision to drill this well. Therefore, this method can be spread to the decision of drilling other exploration wells.

9.2. GRAY SYSTEMS

9.2.1. Applying Ranges and Conditions

9.2.1.1. *Applying Ranges*

The following are the common projects:

1. *Space-time calculations using gray prediction.* Assume that the values of an E&P index in the past are known; then gray prediction can be adopted to calculate the value of this

index in the future. This is the gray prediction of a time domain. Assume that the value of a known parameter distributed in local space is known; then the gray prediction can be adopted to calculate the value of this parameter distributed in the adjacent space. This is the gray prediction of a space domain. Both adopt the absolutely same gray prediction algorithm, and the difference is just whether the prediction series is of time or space.

2. *Geological object quality determined by gray integrated decisions.* Gray integrated decisions can be adopted to determine the quality rating (e.g., excellent, good, average, poor, or very poor) of the several studied geological objects under a certain E&P condition. Assuming that some geological factors related to the aforementioned geological object quality are used to describe their relationship with this geological object in terms of rating, it is possible to adopt gray integrated decisions to determine the quality rating of this geological object. The geological objects can be the evaluation of reservoirs, identification of oil-, gas-, and water-bearing formations, and so on. Certainly, if these geological factors can be expressed in numerical forms, the gray integrated decision can also be adopted to determine the quality rating of a geological object.

9.2.1.2. Applying Conditions

For gray prediction, it is required that the original data distribute steadily and smoothly in time or space domain and the amount of data is not less than 5, whereas for gray integrated decisions, it is required that there are sufficient geological factors affecting the geological object quality as well as the weight assignment and impact rating for the geological objects.

9.2.2. Basic Principles

Based on gray theory, a number that is completely confirmed is called a *white number,* whereas a number that is not completely confirmed is called a *black number,* and the number of which probable range is known but the exact value is unknown is called a *gray number.* Hence, it is inferred that the system in which all information is composed of white numbers is a *white system,* the system in which all information is composed of black numbers is a *black system,* and the system between the two is a *gray system.* Therefore, there exist three types of numbers in the gray system: white numbers, black numbers, and gray numbers. In light of the need for scientific research and production, it is necessary to conduct the study on some gray numbers. The specific way is by applying the gray processing method to turn these gray numbers to white numbers that have enough reliability.

9.2.2.1. Gray Prediction

The specific way of gray prediction is to turn a white number into a new white number in a proper form, to construct a whiting equation for the new white number, to solve the whiting equation by obtaining the prediction of new white numbers, and to revert this prediction value to the original white number, which was the gray number but now is the white number.

9.2.2.2. Gray Integrated Decision

Since the means of petroleum E&P as well as the data acquired are limited, the decision on several geological objects is a fuzzy concept. Here are three cases:

1. For several exploration drilling objects, not an expert before spudding will affirm that there must be commercial oil flow or there must not be commercial oil flow, but experts often believe that both success and failure are possible. Obviously, it is not the concept of either success or failure but the concept that "both are possible" that makes it a gray system.
2. The basis of a decision comes from analysis of several geological factors.
3. The decision results by every expert are different. In the light of the preceding explanation, it is possible to adopt fuzzy integrated decisions to determine the quality rating of exploration drilling objects.

9.2.3. Space-time Calculations Conducted by Gray Prediction

9.2.3.1. Calculation Method

Here we introduce the model GM(1, 1) (Deng, 1986; Shi, 1999), which expresses a whiting one-order ordinary differential equation (a whiting equation):

$$\frac{dX_1}{dt} + aX_1 = U \tag{9.9}$$

where X_1 is the data column $X_1(t)$ $(t = 1, 2, ..., n)$ generated by cumulating the original data column $X_0(t)$ $(t = 1, 2, ..., n)$; see Equation (9.10). t is the ordinal of time or space, $t = 1, 2, ..., n$. a, U are two equation coefficients that can be calculated out from Equation (9.11).

Gray prediction runs in the following six steps:

1. Calculate X_1.

$$X_1(t) = \sum_{k=1}^{t} X_0(k) \quad (t = 1, 2, ..., n) \tag{9.10}$$

2. Calculate a and U by least-squares method:

$$\begin{bmatrix} a \\ U \end{bmatrix} = (B^\mathrm{T}B)^{-1}B^\mathrm{T}Y \tag{9.11}$$

where B is $(n-1) \times 2$ matrix expressed by Equation (9.12); Y is $(n-1) \times 1$ matrix expressed by Equation (9.13).

$$B = \begin{bmatrix} -[X_1(1) + X_1(2)]/2 & 1 \\ -[X_1(2) + X_1(3)]/2 & 1 \\ \cdots & \cdots \\ -[X_1(n-1) + X_1(n)]/2 & 1 \end{bmatrix} \tag{9.12}$$

$$Y = [X_0(2), X_0(3), \cdots X_0(n)]^{\mathrm{T}} \tag{9.13}$$

3. Data column by gray prediction $\hat{X}_1$. Substituting a and U calculated by Equation (9.11) into Equation (9.9), the solution of Equation (9.9) is

$$\left. \begin{array}{l} \hat{X}_1(t) = [X_1(1) - U/a]\exp[-a(t-1)] + U/a \\[2mm] (t = 1, 2, ..., n, n+1, n+2, ..., n+m) \end{array} \right\} \tag{9.14}$$

where $t = n+1, n+2, ..., n+m$ are gray prediction. The number of prediction m depends on what is needed.

4. Revert data column $\hat{X}_1$ into data column $\hat{X}_0$ by reverse cumulation. Since X_1 is resulted from cumulation based on X_0, whereas $\hat{X}_1$ is resulted from the solution of ordinary differential equation based on X_1, $\hat{X}_1$ should be reverted to $\hat{X}_0$ by reverse cumulation, i.e.,

$$\left. \begin{array}{l} \hat{X}_0(1) = \hat{X}(1) \\[2mm] \hat{X}_0(t) = \hat{X}_1(t) - \hat{X}_1(t-1) \\[2mm] (t = 2, ..., n, n+1, n+2, ..., n+m) \end{array} \right\} \tag{9.15}$$

$\hat{X}_0(n+1), \hat{X}_0(n+2), ..., \hat{X}_0(n+m)$ are m gray prediction values related to time or space when $t = n+1, n+2, ..., n+m$.

5. Calculate prediction validation error $R(\%)$ when $t = n$. The relative residual between $\hat{X}_0(t)$ and the original data $X_0(t)$ ($t = 1, 2, ..., n$) is used to express the reliability of m prediction values $\hat{X}_0(t)$ ($t = n+1, n+2, ..., n+m$). Nevertheless, the key relative residual is the relative residual when $t = n$, i.e.,

$$R(\%) = \left| \frac{\hat{X}_0(n) - X_0(n)}{X_0(n)} \right| \times 100 \tag{9.16}$$

Note: $X_0(n)$ in the preceding equation must be reverted to original data when Calculation Scheme 2 is implemented.

Generally, the smaller the relative residual $R(\%)$, the more accurate the prediction result $\hat{X}_0(t)$ ($t = n+1, n+2, ..., n+m$) will be.

6. Calculate association degree D. Letting the residual between $\hat{X}_0$ and X_0 be $\varepsilon(t)$,

$$\varepsilon(t) = \left| \hat{X}_0(t) - X_0(t) \right| \quad (t = 1, 2, ..., n) \tag{9.17}$$

Note: $X_0(t)$ in the preceding equation must be reverted to original data column when Calculation Scheme 2 is implemented.

In data column $\varepsilon(t)$, the minimum value $\varepsilon_{\min}$ and the maximum value $\varepsilon_{\max}$ are

$$\varepsilon_{\min} = \min_t \{\varepsilon(t)\} \tag{9.18}$$

$$\varepsilon_{\max} = \max_t \{\varepsilon(t)\} \tag{9.19}$$

Letting the association coefficient between $\hat{X}_0$ and X_0 be $L(t)$,

$$L(t) = \frac{\varepsilon_{\min} + K\varepsilon_{\max}}{\varepsilon(t) + K\varepsilon_{\max}} \quad (t = 1, 2, \ldots, n) \tag{9.20}$$

where K is resolving coefficient, can be taken as 0.5.

Thus, the association degree D between $\hat{X}_0$ and X_0 has been obtained as follows:

$$D = \frac{1}{n} \sum_{t=1}^{n} L(t) \tag{9.21}$$

Generally, the larger the association degree D is, the more accurate the prediction results $\hat{X}_0(t)$ $(t = n + 1, n + 2, \ldots, n + m)$ will be.

In application, $\hat{X}_0$, $R(\%)$ and D can be calculated out by calculation of the preceding six steps, which is called Calculation Scheme 1. Another way is to transform $X_0(t)$, i.e., the minimum value is subtracted from each $X_0(t)$, followed by calculation of the six steps. Certainly, it is necessary to revert the aforementioned transformation, i.e., the preceding minimum value must be added to $X_0(t)$ and $\hat{X}_0(t)$ at the end of Step 4. At this time, another $\hat{X}_0$, $R(\%)$, and D are obtained, which is called Calculation Scheme 2. Which is better, $\hat{X}_0$ from Calculation Scheme 1 or $\hat{X}_0$ from Calculation Scheme 2? This question can be answered by comparing their $R(\%)$ and D. Generally, the smaller the $R(\%)$ and the larger the D, the more accurate $\hat{X}_0$ will be. But sometimes it depends on specific application. Practice shows that it is better to adopt Calculation Scheme 1 for some application fields, whereas it is better to adopt Calculation Scheme 2 for other application fields. The following Simple Case Studies 1 and 2 will prove this conclusion.

9.2.3.2. Simple Case Study 1: Oil Production Prediction of an Oil Production Zone in Shengli Oilfield Using Gray Prediction

The annual oil production in five years from 985 to 1989 (10^4t) of an oil production zone in Shengli Oilfield in China is $X_0(t)$, $t = 1, 2, \ldots, 5$ [see Equation (9.22)]. Now gray prediction is adopted to calculate the annual oil production in the coming 10 years from 1990 to 1999 by Calculation Scheme 1 (Sun and Fan, 1996; Shi, 1999).

$$X_0 = (444.2043, 451.7581, 442.7612, 437.3083, 428.8250) \tag{9.22}$$

X_0 is transformed into X_1 by Equation (9.10) as follows:

$$X_1 = (444.2043, 895.9624, 1338.7236, 1776.0320, 2204.8569) \tag{9.23}$$

a and U are calculated out by Equation (9.11) as follows:

$$a = 0.0168722; \; U = 462.667511 \tag{9.24}$$

Substituting a and U into Equation (9.9), the whiting equation is constructed:

$$\frac{dX_1}{dt} + 0.0168722X_1 = 462.667511 \tag{9.25}$$

From Equation (9.14), the solution of white Equation (9.25) is

$$\hat{X}_1(t) = \left(444.2043 - \frac{462.667511}{0.0168722}\right)\exp[-0.0168722(t-1)]$$
$$+ \frac{462.667511}{0.0168722} \quad (t = 1, 2, ..., 5, 6, 7, ...15) \tag{9.26}$$

where $t = 6, 7, ..., 15$ is gray prediction, and the number of predictions is 10.

By Equation (9.26), $\hat{X}_1$ is calculated:

$$\hat{X}_1 = (444.2043, 895.5586, 1339.3633, 1775.7402, 2204.8164,$$
$$2626.7148, 3041.5547, 3449.4531, 3850.5293, 4244.8945, \tag{9.27}$$
$$4632.6602, 5013.9395, 5388.8398, 5757.4668, 6119.9277)$$

By Equation (9.15), $\hat{X}_1$ expressed by Equation (9.27) is transformed into $\hat{X}_0$:

$$\hat{X}_0 = (444.2043, 451.3543, 443.8047, 436.3770, 429.0762,$$
$$421.8984, 414.8398, 407.8984, 401.0762, 394.3652, \tag{9.28}$$
$$387.7656, 381.2793, 374.9004, 368.6270, 362.4609)$$

By Equation (9.16), the relative residual $R(\%)$ is calculated:

$$R(\%) = \left|\frac{\hat{X}_0(5) - X_0(5)}{X_0(5)}\right| \times 100 = \left|\frac{429.0762 - 428.8250}{428.8250}\right| \times 100 = 0.05871 \tag{9.29}$$

It can be seen from Equation (9.29) that since $R(\%)$ is quite small, the prediction values expressed in Equation (9.28) are of high reliability. It is known from Equation (9.28) that the annual oil production in the coming 10 years from 1990 to 1999 is predicted to be 421.8984, 414.8398, 407.8984, 401.0762, 394.3652, 387.7656, 381.2793, 374.9004, 368.6270, and 362.4609 $(10^4 t)$, respectively.

By calculating Equations (9.17)–(9.21), the association degree D is

$$D = 0.5860 \tag{9.30}$$

1. Calculation flowchart (Figure 9.2).
2. Results and analyses. It must be pointed out that the original data used in this gray prediction and the prediction values are all verified production, i.e., cumulative production of new wells and the production attained from stimulation the same year (Sun and Fan, 1996; Shi, 1999). The gray prediction is feasible under this definition of the annual production; otherwise, the gray prediction is hard to describe, this complicated condition in the case of artificial factors interfere, such as increase of new production wells and measures of stimulation for the existing production wells.

In the preceding, no calculation results are given for Calculation Scheme 2, but the calculation results for Calculation Scheme 1 are given.

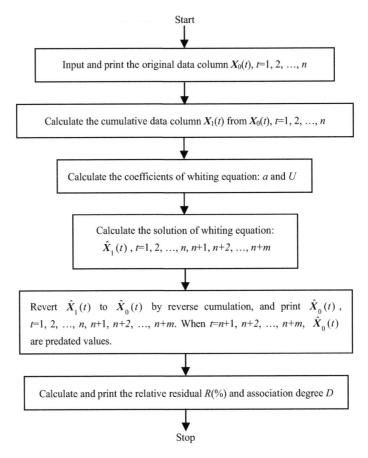

FIGURE 9.2   Calculation flowchart of gray prediction.

$R(\%)$ and D of the two calculation schemes are as follows:

For Calculation Scheme 1, $R(\%) = 0.05853$, $D = 0.5862$;
For Calculation Scheme 2, $R(\%) = 0.8510$, $D = 0.6663$.

Though the association degree D calculated by Calculation Scheme 2 is larger than that by Calculation Scheme 1, $0.6663 > 0.5862$, which shows that its prediction result is more accurate, but the relative residual $R(\%)$ is much larger than that by Calculation Scheme 1, $0.8510\% \gg 0.05853\%$, which shows its prediction result is not as good as that by Calculation Scheme 1. Consequently, Calculation Scheme 1 should be adopted in application, i.e., not to process the original data. Here the conclusion is that if the original data are processed, the accuracy for the gray prediction results will be lower.

In summary, using the data of an oil production zone, i.e., annual oil production in five years from 1985 to 1989, a whiting equation (9.25) with its solution (9.26) for the annual oil production of this oil production zone is constructed by a data mining tool (gray prediction). The results coincide with practicality. This whiting equation with its solution is called *mined knowledge*. This knowledge is applied to predict the annual oil production in the coming 10 years from 1990 to 1999. By using Calculation Scheme 1, the prediction results basically agree with the actual production. Therefore, this method can be spread to the prediction of annual oil and gas production of other production zones.

9.2.3.3. Simple Case Study 2: Reservoir Prediction of a Well in Liaohe Oilfield Using Gray Prediction

Five reservoirs with oil and gas shows are discovered during exploration drilling in a well in Liaohe Oilfield in China. The tops of the reservoirs are at the depth of 1490, 1508, 1516, 1525, and 1535 (m), respectively. Now try to adopt gray prediction, and applying Calculation Scheme 2 in which the five depths are processed that the minimum value 1490 is subtracted from each depth, $X_0(t)$ ($t = 1, 2, ..., 5$) is obtained [see Equation (9.31)], to calculate the top depth of the sixth reservoir (Qi, 1996; Shi, 1999).

$$X_0 = (0, 18, 26, 35, 45) \tag{9.31}$$

X_0 is transformed into X_1 by Equation (9.10) as follows:

$$X_1 = (0, 18, 44, 79, 124) \tag{9.32}$$

a and U are calculated out by Equation (9.11) as follows:

$$a = -0.2891642, \ U = 16.324923 \tag{9.33}$$

Substituting a and U into Equation (9.9), the whiting equation is constructed:

$$\frac{dX_1}{dt} - 0.2891642 X_1 = 16.324923 \tag{9.34}$$

From Equation (9.14), the solution of white Equation (9.34) is

$$\hat{X}_1(t) = \left(0 + \frac{16.324923}{0.2891642}\right) \exp[0.2891642(t-1)] - \frac{16.324923}{0.2891642} \right\} \quad (t = 1, 2, ...5, 6) \tag{9.35}$$

where $t = 6$ is gray prediction, and the number of prediction is 1.

By Equation (9.35), $\hat{X}_1$ is calculated:

$$\hat{X}_1 = (0, 18.9302, 44.2078, 77.9613, 123.0328, 183.2172) \tag{9.36}$$

By Equation (9.15), $\hat{X}_1$ expressed by Equation (9.36) is transformed into $\hat{X}_0$:

$$\hat{X}_0 = (1490, 1508.9302, 1515.2776, 1523.7535, 1535.0714, 1550.1843) \tag{9.37}$$

By Equation (9.16), the relative residual $R(\%)$ is calculated:

$$R(\%) = \left| \frac{\hat{X}_0(5) - X_0(5)}{X_0(5)} \right| \times 100 = \left| \frac{1535.0714 - 1535}{1535} \right| \times 100 = 0.004652 \tag{9.38}$$

It can be seen from Equation (9.38) that since $R(\%)$ is quite small, so the prediction values expressed in Equation (9.37) are of high reliability. It is known from Equation (9.37) that the sixth reservoir top is predicted to be at the depth of 1550.1843 (m).

By calculating Equations (9.17)–(9.21), the association degree D is

$$D = 0.6190 \qquad\qquad (9.39)$$

1. Calculation flowchart (Figure 9.2). It is same as the previous Simple Case Study 1.
2. Results and analyses. In the preceding, no calculation results are given for Calculation Scheme 2, but the calculation results for Calculation Scheme 1 are given.

$R(\%)$ and D of the two calculation schemes and the real error $(RERR)$ are as follows:

For Calculation Scheme 1, $R(\%) = 0.03157$, $D = 0.4696$, $RERR = 4.3438$ (m), i.e., the depth of the sixth reservoir top is predicted to be 1543.6562 (m), but real depth is 1548 (m).

For Calculation Scheme 2, $R(\%) = 0.004652$, $D = 0.6190$, $RERR = 2.1843$ (m), i.e., the depth of the sixth reservoir top is predicted to be 1550.1843 (m), but real depth is 1548 (m).

It is found from the comparison that Calculation Scheme 2 has small relative residual $R(\%)$, large association degree D, and small real error $RERR$, which indicates that the prediction result of Calculation Scheme 2 is obviously better than that of Calculation Scheme 1. Therefore, Calculation Scheme 2 should be adopted, i.e., to process the original data. Here comes the conclusion that the processing of original data will improve the accuracy of gray prediction results.

In summary, during drilling a well, we use data of this well: the depths of the top of five reservoirs where oil and gas shows appear downward. Then we use data mining tool (gray prediction), a whiting equation with its solution for the reservoir top depth in that well is constructed. The results coincide with practicality. This whiting equation with its solution is called *mined knowledge*. This knowledge is applied to predict the depth of the top of sixth reservoir. Using Calculation Scheme 2, the prediction results basically agree with the actual production. Therefore, this method can be spread to the prediction of reservoir top depth in drilling other wells.

This simple case study shows that it is necessary to process the original data, whereas the aforementioned Simple Case Study 1 shows that it is not necessary to process the original data. This indicates that when users adopt gray prediction, the original data need to be processed for the test.

9.2.3.4. Simple Case Study 3: Oil Production Prediction of Romashkino Oilfield Using Gray Prediction

Simple Case Study 2 in Section 3.1.5 used the annual oil production in 39 years from 1952 to 1990 of the Romashkino Oilfield in Russia (Table 3.4). BPNN was adopted to calculate the annual oil production in the coming 15 years from 1991 to 2005, and the calculation results were filled into Table 3.4. Now Table 3.4 is used as input data for Simple Case Study 3 of this chapter. For the sake of algorithm comparisons, Table 3.4 is wholly copied into Table 9.2, and the results of gray prediction are also filled into Table 9.2.

TABLE 9.2   Predicted Oil Production of the Romashkino Oilfield Using Gray Prediction

Annual Oil Production (10^4 t)[a]

Sample type	Sample No.	x (Year)[a]	Real Production[b] y^*	BPNN	Gray Prediction Calculation Scheme 1	Calculation Scheme 2
Learning samples	1	1952	200	204	200	200
	2	1953	300	244	4507	4513
	3	1954	500	426	4524	4529
	4	1955	800	803	4541	4546
	5	1956	1400	1295	4558	4563
	6	1957	1900	1848	4575	4579
	7	1958	2400	2465	4592	4596
	8	1959	3050	3154	4609	4613
	9	1960	3800	3892	4626	4630
	10	1961	4400	4616	4643	4647
	11	1962	5000	5264	4661	4664
	12	1963	5600	5806	4678	4681
	13	1964	6000	6255	4696	4698
	14	1965	6500	6654	4713	4715
	15	1966	6900	7041	4731	4733
	16	1967	7350	7416	4748	4750
	17	1968	7600	7729	4766	4767
	18	1969	7900	7931	4784	4785
	19	1970	8150	8030	4802	4802
	20	1971	8100	8068	4820	4820

Annual Oil Production (10^4 t)[a]

Sample Type	Sample No.	x (Year)[a]	Real Production[b] y^*	BPNN	Gray Prediction Calculation Scheme 1	Calculation Scheme 2
Learning samples	28	1979	6090	6490	4966	4964
	29	1980	5450	5970	4984	4982
	30	1981	4900	5441	5003	5000
	31	1982	4350	4927	5021	5018
	32	1983	3900	4445	5040	5037
	33	1984	3450	4005	5059	5055
	34	1985	3100	3609	5078	5074
	35	1986	2700	3256	5097	5093
	36	1987	2450	2943	5116	5111
	37	1988	2200	2667	5135	5130
	38	1989	2000	2424	5154	5149
	39	1990	1900	2211	5173	5168
Prediction samples	40	1991	(1770)	2023	5193	5187
	41	1992	(1680)	1858	5212	5206
	42	1993	(1590)	1712	5232	5225
	43	1994	(1500)	1584	5251	5245
	44	1995	(1410)	1472	5271	5264
	45	1996	(1320)	1373	5290	5283
	46	1997	(1230)	1285	5310	5303
	47	1998	(1180)	1207	5330	5322

(Continued)

TABLE 9.2　Predicted Oil Production of the Romashkino Oilfield Using Gray Prediction—cont'd

Sample type		Annual Oil Production (10^4t)[a]				
		Real Production[b]			y	
					Gray Prediction	
Sample No.	x (Year)[a]	y^*	BPNN	Calculation Scheme 1	Calculation Scheme 2	Sample Type
21	1972	8050	8072	4838	4838	
22	1973	8000	8049	4856	4856	
23	1974	7900	7990	4874	4873	
24	1975	7800	7871	4892	4891	
25	1976	7600	7670	4910	4909	
26	1977	7250	7367	4929	4927	
27	1978	6750	6967	4947	4945	

		Annual Oil Production (10^4t)[a]				
		Real Production[b]			y	
					Gray Prediction	
Sample No.	x (Year)[a]	y^*	BPNN	Calculation Scheme 1	Calculation Scheme 2	
48	1999	(1150)	1139	5350	5342	
49	2000	(1110)	1078	5370	5362	
50	2001	(1090)	1024	5390	5381	
51	2002	(1070)	975	5410	5401	
52	2003	(1050)	932	5430	5421	
53	2004	(1040)	894	5450	5441	
54	2005	(1020)	860	5471	5461	

[a]x (year) is used by BPNN but not by gray prediction.
[b]in y^*, numbers in parentheses are not input data but are used for calculating R(%).

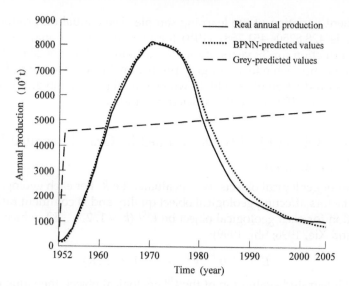

FIGURE 9.3   Comparison between real, BPNN-predicted and gray-predicted values of oil production in the Romashkino Oilfield.

For Calculation Schemes 1 and 2 of gray prediction, the relative residual $R(\%)$ calculated by Equation (9.16) is 172.31 and 172.03, respectively, and the association degree D calculated by Equation (9.21) is 0.53651 and 0.53686, respectively. It seems that the results of Calculation Scheme 2 are better than those of Calculation Scheme 1, that is, $R(\%)$ a little bit smaller and D a little bit larger. However, $R(\%)$ of both schemes are stupendously large, which indicates that the two schemes have very low reliability for prediction values. Therefore, the gray prediction is not applicable to predict oil production in the Romashkin Oilfield, but only BPNN is applicable (Table 9.2, Figure 9.3). Since the calculation results of the two schemes are very close, only the curve of Calculation Scheme 2 is plotted on Figure 9.3, which is the only difference between Figure 9.3 and Figure 3.4.

In summary, using the data of the Romashkino Oilfield, i.e., annual oil production in 39 years from 1952 to 1990, a whiting equation with its solution for the annual oil production of this oilfield is constructed by a data mining tool (gray prediction). There exists considerable difference between the calculation results and the real production. The whiting equation with its solution may be called *mined wrong "knowledge."* This wrong "knowledge" is applied to predict the annual oil production in the coming 15 years from 1991 to 2005. The calculation results are found to be considerably different from the actual production, too. Moreover, calculation was run using both Calculation Scheme 1 and Calculation Scheme 2. The difference of the calculation results of the two schemes is small, and Calculation Scheme 2 is a little bit better than Calculation Scheme 1.

Here it is indicated that annual oil production of the Romashkino Oilfield can be successfully predicted (see Section 3.1.5) by using BPNN. Why is gray prediction applicable to predict the annual oil production of the aforementioned oil production zone but not to predict the annual oil production of the Romashkino Oilfield? This is attributed to the annual oil production gradient variation of the learning sample: The annual oil production of that oil production zone is 428.8250–451.7581 (10^4t) for five years [see Equation (9.22)], whereas the annual oil production for 39 years of the Romashkino Oilfield is 200–8150 (10^4t) (Table 9.2). Obviously, it is impossible to adopt gray prediction, but we can adopt BPNN to predict annual oil and gas production of an oilfield when annual production gradient is large, since the solution accuracy of BPNN is not restricted by gradient.

9.2.4. Geological Object Quality Determined by Gray Integrated Decision

9.2.4.1. *Calculation Method*

Let the number of geological objects to be evaluated be K. For each geological object, there are n geological factors affecting geological object quality and m comment ratings. Let the integrated evaluation for each geological object be $E^{(k)}$ ($k = 1, 2, ..., K$), which is expressed as follows (Wang and Xie, 1996; Shi, 1999):

$$E^{(k)} = B^{(k)} \cdot V^{T} \quad (k = 1, 2, ..., K) \tag{9.40}$$

where $E^{(k)}$ is the integrated evaluation of the kth geological object, the value ranges between the first element and the last element of comment vector V. $B^{(k)}$ is a $(1 \times m)$ matrix, called an *integrated evaluation vector* of the kth geological object; see Equation (9.41). V is a $(1 \times m)$ matrix, called a *comment vector* to express a comment rating; see Equation (9.51). V^{T} is the transposed vector of V.

The integrated evaluation vector of the kth geological object is

$$B^{(k)} = \left(b_1^{(k)}, b_2^{(k)}, \cdots, b_m^{(k)}\right) \quad (k = 1, 2, ..., K) \tag{9.41}$$

and

$$B^{(k)} = W \cdot R^{(k)} \tag{9.42}$$

where W is a $(1 \times n)$ matrix, called a *weight vector*, commonly used for each geological object and denoting the weight related to geological factors; see Equation (9.43). $R^{(k)}$ is a $(n \times m)$ matrix, called an *evaluation weight matrix* of the kth geological object; see Equation (9.44).

A weight vector commonly used for each geological object is

$$W = (w_1, w_2, \cdots w_n) \tag{9.43}$$

where w_i ($i = 1, 2, ..., n$) is the weight of the ith geological factor for evaluating each geological object, $\sum_{i=1}^{n} w_i = 1$.

The evaluation weight matrix of the kth geological object is

$$R^{(k)} = \begin{bmatrix} r_{11}^{(k)} & r_{12}^{(k)} & \cdots & r_{1m}^{(k)} \\ r_{21}^{(k)} & r_{22}^{(k)} & \cdots & r_{2m}^{(k)} \\ \cdots & \cdots & \cdots & \cdots \\ r_{n1}^{(k)} & r_{n2}^{(k)} & \cdots & r_{nm}^{(k)} \end{bmatrix} \quad (k = 1, 2, ..., K) \tag{9.44}$$

where $r_{ij}^{(k)}$ is expressed as follows:

$$r_{ij}^{(k)} = \frac{\eta_{ij}^{(k)}}{\sum\limits_{j=1}^{m} \eta_{ij}^{(k)}} \quad (i = 1, 2, \cdots n; \ j = 1, 2, \cdots, m; \ k = 1, 2, \cdots, K) \quad (9.45)$$

where $\eta_{ij}^{(k)}$ is the evaluation coefficient of the kth geological object for the ith geological factor versus the jth comment rating; see Equation (9.64).

The evaluation coefficient of the kth geological object for the ith geological factor versus the jth comment rating is

$$\eta_{ij}^{(k)} = \sum\limits_{l=1}^{L} f_j\left(d_{il}^{(k)}\right) \quad (i = 1, 2, \cdots n; \ j = 1, 2, \cdots, m; \ k = 1, 2, \cdots, K) \quad (9.46)$$

where f_j is the jth *whiting function*; see Equations (9.48)–(9.50). $d_{il}^{(k)}$ is the comment of the lth expert on the ith geological factor to the kth geological object; and L is the number of experts.

Assume that the comment rating is divided into good, average, and poor, expressed by 5, 3, and 1, respectively, i.e., $m = 3$ and $j = 1, 2, 3$. Experts can also use 5, 4, 3, 2, and 1 to stand for excellent, good, average, poor, and very poor. In a word, experts can use five numbers to assign $d_{il}^{(k)}$, but there are still three whiting functions available.

For the kth geological object, $d_{il}^{(k)}$ can form a comment matrix as follows:

$$\mathbf{D}^{(k)} = \begin{bmatrix} d_{11}^{(k)} & d_{12}^{(k)} & \cdots & d_{1L}^{(k)} \\ d_{21}^{(k)} & d_{22}^{(k)} & \cdots & d_{2L}^{(k)} \\ \cdots & \cdots & \cdots & \cdots \\ d_{n1}^{(k)} & d_{n2}^{(k)} & \cdots & d_{nL}^{(k)} \end{bmatrix} \quad (k = 1, 2, ..., K) \quad (9.47)$$

where the lth column ($l = 1, 2, ..., L$) expresses the comment of the lth expert on n geological factors.

As mentioned in Section 9.2.2, the gray number is a number of which probable range is known but the exact value is unknown. It is designated as $\otimes$. To describe the gray number, the following three classes of whiting functions are often used, i.e., $m = 3$ (Wang and Xie, 1996; Shi, 1999):

1. The first whiting function. $j = 1$ (good), gray number $\otimes_1 \in [0, \infty]$, whiting function is f_1; see Equation (9.48) and Figure 9.4.

$$f_1\left(d_{il}^{(k)}\right) = \begin{cases} \frac{d_{il}^{(k)}}{d_1} & d_{il}^{(k)} \in [0, d_1] \\ 1 & d_{il}^{(k)} \in [d_1, \infty] \\ 0 & \text{other cases} \end{cases} \quad (9.48)$$

2. The second whiting function. $j = 2$ (average), gray number $\otimes_2 \in [0, 2d_2]$, the dwhiting function is f_2; see Equation (9.49) and Figure 9.5.

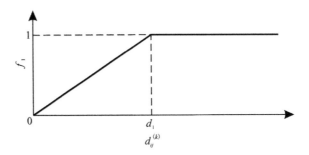

FIGURE 9.4   Class 1 whiting function.

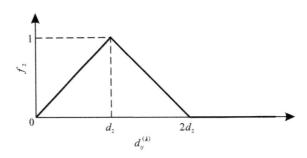

FIGURE 9.5   Class 2 whiting function.

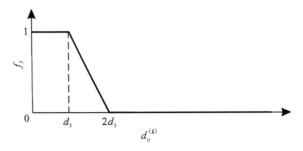

FIGURE 9.6   Class 3 whiting function.

$$
f_2\left(d_{il}^{(k)}\right) = \begin{cases} \dfrac{d_{il}^{(k)}}{d_2} & d_{il}^{(k)} \in [0, d_2] \\[2mm] \dfrac{2d_2 - d_{il}^{(k)}}{d_2} & d_{il}^{(k)} \in [d_2, 2d_2] \\[2mm] 0 & \text{other cases} \end{cases}
\tag{9.49}
$$

3. The third whiting function. $j = 3$ (poor), gray number $\otimes_3 \in [0, 2d_3]$, whiting function is f_3; see Equation (9.50) and Figure 9.6.

$$f_3\left(d_{il}^{(k)}\right) = \begin{cases} 1 & d_{il}^{(k)} \in [0, d_3] \\ \frac{2d_3 - d_{il}^{(k)}}{d_3} & d_{il}^{(k)} \in [d_3, 2d_3] \\ 0 & \text{other cases} \end{cases} \tag{9.50}$$

The three inflexions of d_1, d_2, and d_3 for the previous three classes of whiting functions are called a *threshold*, which can be determined in terms of a criterion or experience; or they can take the mean of the maximum value and the minimum value of $D^{(k)}$ ($k = 1, 2, ..., K$) as d_1 and d_3, respectively, and can take $(d_1 + d_3)/2$ as d_2. These three thresholds are commonly used for each geological object.

If three more, that is, $m > 3$, whiting functions are needed, usually to subdivide Class 2 whiting function into several ones, e.g., $d_1 > d_{2_1} > d_{2_2} > \cdots > d_{2_Q} > d_3$, at this time there are $2 + Q$ whiting functions in total. If Q is an odd number, then $m = 2 + Q$. And $d_{2_Q} = d_3 + \frac{d_1 - d_3}{Q+1}(Q - q + 1)$, where $q = 1, 2, \cdots, Q$

Thus, the calculation of $B^{(k)}$ in Equation (9.40) has been explained, and only V is left. The comment vector is

$$V = (v_1, v_2, \cdots, v_m) \tag{9.51}$$

where m is usually taken to be an odd number. If $m = 3$, $V = (5, 3, 1)$ expresses three comment ratings of good, average, and poor, then $E^{(k)} \in [5, 1]$ in Equation (9.40). If $m = 5$, $V = (9, 7, 5, 3, 1)$ expresses five comment ratings of excellent, good, average, poor, and very poor, then $E^{(k)} \in [9, 1]$ in Equation (9.40).

It must be noted that V, W, f_j ($j = 1, 2, ..., m$) in the preceding symbols are commonly used for each geological object so as to ensure the integrated evaluation $E^{(k)}$ can be obtained under the same condition. The other symbols are all related to k, that is, related to a specific geological object.

Now the calculation process is summarized:

1. Input V, W, and $D^{(k)}$ ($k = 1, 2, ..., K$). See Equations (9.51), (9.43), and (9.47).
2. Calculate the threshold d_j of f_j ($j = 1, 2, ..., m$). See Equations (9.48), (9.49), and (9.50).
3. $1 \Rightarrow k$. Calculate the first geological object.
4. Calculate $\eta_{ij}^{(k)}$ and $r_{ij}^{(k)}$ ($i = 1, 2, ..., n$; $j = 1, 2, ..., m$), and print $R^{(k)} = (r_{ij}^{(k)})$. See Equations (9.46), (9.45), and (9.44).
5. Calculate and print $B^{(k)}$. See Equation (9.42).
6. Calculate and print $E^{(k)}$. See Equation (9.40).
7. $k + 1 \Rightarrow k$. If $k > K$, the calculation is finished; otherwise, turn to (4) to calculate the next geological object.

9.2.4.2. Simple Case Study: Gray Integrated Evaluation of Exploration Drilling Objects

There are four exploration drilling objects in an exploration area of China, one of which is going to be selected for drilling the first exploratory well. For this purpose, five experts are invited to make an evaluation on this by using gray integrated decisions. Based on petroleum system theory, they select six geological factors that affect the quality of the four exploration drilling objects. These six geological factors are hydrocarbon generation, migration, trapping, accumulation, cap rock, and preservation (Shi, 1999), the weight of which was determined after the discussion by the five experts, i.e., the weight vector corresponding to Equation (9.43) is

$$W = (0.15, 0.10, 0.15, 0.20, 0.25, 0.15) \qquad (9.52)$$

Three comment ratings of good, average, and poor are defined too. So the comment vector corresponding to Equation (9.51) is

$$V = (5, 3, 1) \qquad (9.53)$$

As mentioned previously, the experts can use 5, 4, 3, 2, and 1 to express the comment rating of excellent, good, average, poor, and very poor. The comments the five experts made for the four exploration drilling objects are listed in Tables 9.3A, 9.3B, 9.3C, and 9.3D, respectively. It is seized from the preceding that $K = 4$, $n = 6$, $m = 3$, $L = 5$.

TABLE 9.3A Expert Comments for the First Exploration Drilling Object

| Geological Factor | $d_{il}^{(1)}$ (Comment on the ith Geological Factor from the lth Expert) | | | | |
	Expert 1	Expert 2	Expert 3	Expert 4	Expert 5
Hydrocarbon generation	5	4	5	4	4
Migration	4	5	4	5	5
Trapping	5	4	5	4	5
Accumulation	4	5	4	5	4
Cap rocks	5	4	5	4	5
Preservation	4	5	4	5	4

TABLE 9.3B Expert Comments for the Second Exploration Drilling Object

| Geological Factor | $d_{il}^{(2)}$ (Comment on the ith Geological Factor from the lth Expert) | | | | |
	Expert 1	Expert 2	Expert 3	Expert 4	Expert 5
Hydrocarbon generation	4	3	4	3	4
Migration	4	4	4	5	4
Trapping	3	3	3	4	3
Accumulation	4	4	4	3	4
Cap rocks	5	5	5	4	5
Preservation	3	3	3	4	3

TABLE 9.3C Expert Comments for the Third Exploration Drilling Object

Geological Factor	$d_{il}^{(3)}$ (Comment on the ith Geological Factor from the lth Expert)				
	Expert 1	Expert 2	Expert 3	Expert 4	Expert 5
Hydrocarbon generation	3	2	2	3	2
Migration	2	3	2	3	2
Trapping	2	2	1	2	2
Accumulation	2	2	2	2	3
Cap rocks	3	2	2	3	2
Preservation	2	2	2	2	1

TABLE 9.3D Expert Comments for the Fourth Exploration Drilling Object

Geological Factor	$d_{il}^{(4)}$ (Comment on the ith Geological Factor from the lth Expert)				
	Expert 1	Expert 2	Expert 3	Expert 4	Expert 5
Hydrocarbon generation	1	2	1	2	1
Migration	2	1	2	1	1
Trapping	1	2	1	1	2
Accumulation	1	2	2	1	2
Cap rocks	1	1	2	1	1
Preservation	1	2	1	1	2

The following is the specific calculation process of the first exploration drilling object ($k = 1$).

After the comment matrix corresponding to Equation (9.47) on the four exploration drilling objects are input, the maximum value and the minimum value of the four matrixes (Tables 9.3A, 9.3B, 9.3C, 9.3D) are calculated to be 5.0 and 4.0 when $k = 1$, 5.0 and 3.0 when $k = 2$, 3.0 and 1.0 when $k = 3$, and 2.0 and 1.0 when $k = 4$, respectively. So $d_1 = \frac{1}{4} \sum_{k=1}^{4}$ (maximun)$_k = 3.75$, $d_3 = \frac{1}{4} \sum_{k=1}^{4}$ (minimun)$_k = 2.25$, and $d_2 = d_3 + \frac{d_1 - d_3}{2} = 3.0$.

It is known from Equations (9.46), (9.45), and (9.44) that

$$R^{(1)} = \begin{bmatrix} 0.6000 & 0.3200 & 0.0800 \\ 0.6429 & 0.3000 & 0.0571 \\ 0.6429 & 0.3000 & 0.0571 \\ 0.6000 & 0.3200 & 0.0800 \\ 0.6429 & 0.3000 & 0.0571 \\ 0.6000 & 0.3200 & 0.0800 \end{bmatrix} \qquad (9.54)$$

From Equation (9.42),

$$\boldsymbol{B}^{(1)} = (0.6214, 0.3100, 0.0686) \qquad (9.55)$$

From Equation (9.40), $\boldsymbol{E}^{(1)} = 4.106$.

Similarly, $\boldsymbol{E}^{(2)} = 3.693$, $\boldsymbol{E}^{(3)} = 2.779$, and $\boldsymbol{E}^{(4)} = 2.660$ can be obtained by calculation. Consequently, the first exploration drilling object is optimal, in the range between excellent and good and leaning to good.

1. Calculation flowchart (Figure 9.7).
2. Results and analyses. Though it seems that the quality of the first exploration drilling object is the best from Tables 9.3A to 9.3D, which list the rank of the four exploration drilling objects in terms of quality, gray integrated decisions can give the rank order in

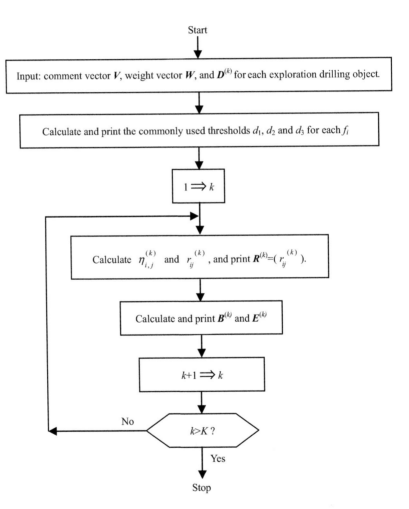

FIGURE 9.7   Calculation flowchart of gray integrated decision.

quantitative form, and the best one has quantitative integrated evaluation. If the quality of the exploration drilling object to be evaluated cannot be determined directly, it is more necessary to adopt gray integrated decisions for selection.

In summary, using the data of four exploration drilling objects in an exploration area, i.e., for each exploration drilling object the evaluation results on six geological factors according to five comment ratings made by five experts, as well as the weight of these six geological factors, the integrated evaluation to each exploration drilling object is obtained by data mining tool (gray integrated decision). The results coincide with practicality. This integrated evaluation is called *mined knowledge*, which provides evidence to decide the selection of the first exploration drilling object. Therefore, this method can be spread to the selection of exploration drilling objects in other exploration areas.

9.3. FRACTAL GEOMETRY

9.3.1. Applying Ranges and Conditions

9.3.1.1. Applying Ranges

Network overlay is often adopted for quantitative description of geological structure features such as fault and fracture to provide the evidence for the analysis of hydrocarbon migration accumulation.

9.3.1.2. Applying Conditions

In a studied geological volume, the distribution of geological structure features that reflect the basic status of this geological volume should be known.

9.3.2. Basic Principles

The meaning of fractal geometry is that each component of a geological volume is similar to the volume in a given form. If the volume or each component is partitioned into several grids, then fractal dimension D can be used to quantitatively describe the complexity of fractal geometry (Zhang et al., 1996; Chen and Chen, 1998; Shi, 1999):

$$N = \frac{C}{L^D} \tag{9.56}$$

where N is the number of grids with some geological feature; C is a proportional constant; L is a line degree of grid (plane or volume); and D is fractal dimension.

Natural logarithm is taken at both sides of Equation (9.56), attaining

$$\ln(N) = \ln(C) - D\ln(L) \tag{9.57}$$

Obviously, $\ln(N)$ is linearly related to $\ln(L)$, and D is the slope of a straight line.

Assume there is a group (L, N). Different D values can be calculated from different values of (L, N). The larger the value of D, the stronger the geological structure feature, whereas the smaller the value of D, the weaker the geological structure feature.

The question is, how to calculate the D value?

9.3.3. Geological Structure Features Described by Fractal Geometry

9.3.3.1. Calculation Method

Calculation is conducted in the following four steps:

1. *Calculate the samples (L, N) by network overlay.* In the case of two dimensions, the square grid with the side length L is used to overlay the whole studied geological volume or the other components; in the case of three dimensions, the cube grid with the side length L is used to partition the whole studied geological volume or the other components. The number of grids that contain the geological structure features such as fault or fracture is obtained by statistics, designated as N. The side length is altered step by step, similarly attaining the corresponding N. Let each side be $L_j(j = 1, 2 \cdots, n)$ and the corresponding statistical number of grids be $N_j(j = 1, 2\cdots, n)$. This is a group of samples (L, N) that is obtained by network overlay. This method can be performed by manpower or by computers if conditions allow.

2. Calculate fractal dimension D by least squares. Based on Equation (9.57), referring to Equation (2.16) in Section 2.2.3, taking $\ln(L)$ as x, $\ln(N)$ as y, i.e., $\ln(L_j)$ stands for x_i, $\ln(N_j)$ stands for y_i, which are substituted into Equation (2.17), a and b are attained. Obviously, $\ln(C) = b, C = \exp(b); -D = a, D = -a$. Therefore, in the light of the preceding calculation based on a group of samples (L, N), not only can fractal dimension D be calculated, but also the expression of Equation (9.56) is more specified.

3. Calculate the correlation coefficient of (L, N). The correlation coefficient r_{LN} of (L, N) can be obtained by referring to Equation (2.31) in Section 2.2.4.

$$r_{LN} = \left| \sum_{j=1}^{n} \frac{\left[\ln(L_j) - \overline{L} \right] \cdot \left[\ln(N_j) - \overline{N} \right]}{\sigma_L \sigma_N} \right| \tag{9.58}$$

where

$$\left. \begin{aligned} \overline{L} &= \frac{1}{n} \sum_{j=1}^{n} \ln(L_j) \\[2mm] \overline{N} &= \frac{1}{n} \sum_{j=1}^{n} \ln(N_j) \end{aligned} \right\} \tag{9.59}$$

$$\left. \begin{aligned} \sigma_L &= \sqrt{\sum_{j=1}^{n} \left[\ln(L_j) - \overline{L} \right]^2} \\[2mm] \sigma_N &= \sqrt{\sum_{j=1}^{n} \left[\ln(N_j) - \overline{N} \right]^2} \end{aligned} \right\} \tag{9.60}$$

Generally, since there exists similarity for geological structure features (Shi, 1999), the correlation coefficient calculated by Equation (9.58) approaches 1. Certainly, to ensure the accuracy of the correlation coefficient, it is necessary to conduct correlation validation.

4. Conduct correlation validation. Based on the number of samples n, and the given significance level α (e.g., 0.10, 0.05, 0.02, 0.01, 0.001), critical value r_α of a correlation coefficient can be found in Table 9.4 (A compiling group in Chongqing University, 1991; Shi, 1999).

TABLE 9.4　Critical Value r_α of Correlation Coefficient $[P\,(|r|>r_a)=\alpha]$

$n-2$	r_α				
	$\alpha=0.10$	$\alpha=0.05$	$\alpha=0.02$	$\alpha=0.01$	$\alpha=0.001$
1	0.98769	0.99692	0.999507	0.999877	0.9999988
2	0.90000	0.95000	0.98000	0.99000	0.99900
3	0.8054	0.8783	0.93433	0.95873	0.99116
4	0.7293	0.8114	0.8822	0.91720	0.97406
5	0.6694	0.7545	0.8329	0.8745	0.95074
6	0.6215	0.7067	0.7887	0.8343	0.92493
7	0.5822	0.6664	0.7498	0.7977	0.8982
8	0.5494	0.6319	0.7155	0.7646	0.8721
9	0.5214	0.6021	0.6851	0.7348	0.8471
10	0.4973	0.5760	0.6581	0.7079	0.8233
11	0.4762	0.5529	0.6339	0.6835	0.8010
12	0.4575	0.5324	0.6120	0.6614	0.7800
13	0.4409	0.5139	0.5923	0.6411	0.7603
14	0.4259	0.4974	0.5742	0.6226	0.7420
15	0.4124	0.4821	0.5577	0.6055	0.7246
16	0.4000	0.4683	0.5425	0.5897	0.7084
17	0.3887	0.4555	0.5285	0.5751	0.6932
18	0.3783	0.4438	0.5155	0.5614	0.6787
19	0.3687	0.4329	0.5034	0.5487	0.6652
20	0.3598	0.4227	0.4921	0.5368	0.6524
25	0.3233	0.3809	0.4451	0.4869	0.5974
30	0.2960	0.3494	0.4093	0.4487	0.5541
35	0.2746	0.3246	0.3810	0.4182	0.5189

(Continued)

TABLE 9.4   Critical Value r_α of Correlation Coefficient $[P\,(|r|>r_a) = \alpha]$—cont'd

			r_α		
$n-2$	$\alpha=0.10$	$\alpha=0.05$	$\alpha=0.02$	$\alpha=0.01$	$\alpha=0.001$
40	0.2573	0.3044	0.3578	0.4032	0.4896
45	0.2428	0.2875	0.3384	0.3721	0.4648
50	0.2306	0.2732	0.3218	0.3541	0.4433
60	0.2108	0.2500	0.2948	0.3248	0.4078
70	0.1954	0.2319	0.2737	0.3017	0.3799
80	0.1829	0.2172	0.2565	0.2830	0.3568
90	0.1726	0.2050	0.2422	0.2673	0.3375
100	0.1638	0.1946	0.2331	0.2540	0.3211

If $r_{LN} > r_a$, that means correlation of (L, N) is significant, and it is feasible to use fractal dimension D value to describe the geological structure features.

Assume a studied geological volume is divided into m components. For each component, i.e., fractal object, n kinds of grid-line degree are used to divide them.

The known parameters:

$$L_j\left(j = 1,2\cdots,n\right); \;\; N_{ij}\left(i = 1,2,\cdots,m; \;\; j = 1,2,\cdots,n\right)$$

Calculation results:

$$D_i, r_{LN_i} \;\;\left(i = 1,2,\cdots,m\right)$$

r_{a_i} is obtained from Table 9.4.

9.3.3.2. Case Study: Relationship between the Fractal Features of Faults and Hydrocarbon Migration Accumulation

There are two structural layers (T_6 and T_2) in Dongying Sag in the Shengli Oilfield in China. Each layer is divided into two areas, north and south, followed by another two areas, west and east, with eight areas in total. For each area, that is, each fractal object, five kinds of line degree (L) with 0.5, 1, 1.5, 2, and 4 (km) are used for network overlay, achieving the corresponding number of grids (N) that contain faults; see Table 9.5. Afterward, the fractal dimension D, correlation coefficient r_{LN} of each area are calculated and filled in Table 9.5. Finally, critical value r_α of a correlation coefficient is obtained by looking up Table 9.4, and validation analysis is made, which is filled in Table 9.5 (Shen et al., 1996; Shi, 1999). The specific way to look up Table 9.4 is that because $n = 5$, $n - 2 = 3$, so look up r_a in the row of $n - 2 = 3$ in the table; and because eight r_{LN} are all larger than all r_a in this row, take the maximum one $r_a = 0.99116$ that corresponds to significance level $\alpha = 0.001$ to fill it in Table 9.5, and put "significant" on the column of validation analysis.

TABLE 9.5 Fractal Data and Results for Faults of Structural Layers T_2 and T_6

Structural Layers	Fractal Object	N					Calculation Results			Validation Analysis
		L_1 0.5 (km)	L_2 1 (km)	L_3 1.5 (km)	L_4 2 (km)	L_5 4 (km)	D	r_{LN}	r_α	
T_6	South area	1685	782	534	380	167	1.1034	0.99923	0.99116	Significant
	North area	1179	560	353	255	95	1.2023	0.99725	0.99116	Significant
	East area	1569	719	468	352	144	1.1363	0.99893	0.99116	Significant
	West area	1305	613	419	283	118	1.1496	0.99804	0.99116	Significant
T_2	South area	1409	715	426	300	149	1.0980	0.99885	0.99116	Significant
	North area	930	442	273	196	92	1.1187	0.99986	0.99116	Significant
	East area	1453	703	441	319	141	1.1232	0.99961	0.99116	Significant
	West area	886	454	258	177	100	1.0812	0.99549	0.99116	Significant

1. Calculation flowchart (Figure 9.8).
2. Results and analyses. The results (D, r_{LN}, r_a) of each area have been filled in Table 9.5. Obviously, $r_{LN} > r_a$ in each area, indicating the linear correlation between $\ln(L)$ and $\ln(N)$ in each area is highly significant and it is feasible to describe the fault feature in this sag using fractal dimension D value.

Table 9.5 shows that for layer T_6, D value (1.2023) in the north area is larger than D value (1.1034) in the south area, and D value (1.1363) in the east area is smaller than D value (1.1496) in the west area. For layer T_2, D value (1.1187) in the north area is also larger than D value (1.0980) in the south area, and D value (1.1232) in the east area is larger than D value (1.0812) in the west area, which indicates that the tectonic movement in this sag leads the faults to be high in the north and low in the south. Since faults tend to play an important role in hydrocarbon migration accumulation, the reservoir is predicted to be high in the north and low in the south in this sag, which is proved by reality. The size and productivity of reservoirs in this sag are higher in the north area than in the south.

In summary, using the data of two structural layers in Dongying Sag in the Shengli Oilfield, i.e., the structure map containing fault locations for each structural layer, the fractal dimension in the four areas of north, south, west, and east for each structural layer is obtained by data mining tool (fractal geometry). The results coincide with practicality. The fractal dimension is called *mined knowledge*, which expresses the development extent of the fault in the sag, high in the north while low in the south, thus predicting the development extent of the reservoir in the sag. This understanding agrees with what it is in practice. Therefore, this method can be spread to the prediction of reservoir development extent in other sags.

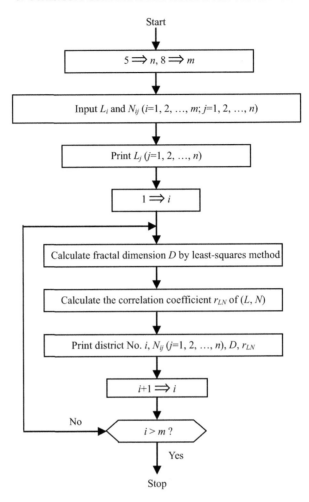

FIGURE 9.8   Calculation flowchart of fractal geometry.

9.4. LINEAR PROGRAMMING

9.4.1. Applying Ranges and Conditions

9.4.1.1. Applying Ranges

Linear programming is often adopted for the optimal programming under the constraint conditions for a given E&P scheme as well as the funds.

If an exploration or production plan is made in an area, including the constraint of the arrangement and the funds, what we have to work out is to determine the arrangement that should not only meet the constraint conditions but also minimize the expenditure.

9.4.1.2. *Applying Conditions*

There must be appropriate object function from which extremum is taken and sufficient constraint conditions without contradiction with each other.

9.4.2. Basic Principles

We adopt linear programming to deal with optimization, of which the mathematical model is the following two equations:

Object function:

$$\max \ z = c_1 x_1 + c_2 x_2 + \cdots + c_n x_n \tag{9.61}$$

The constraint conditions:

$$\begin{cases} a_{11} x_1 + a_{12} x_2 + \cdots + a_{1n} x_n = b_1 \\ a_{21} x_1 + a_{22} x_2 + \cdots + a_{2n} x_n = b_2 \\ \qquad \cdots \cdots \\ a_{m1} x_1 + a_{m2} x_2 + \cdots + a_{mn} x_n = b_m \\ \qquad x_1, x_2, \cdots, x_n \geq 0 \end{cases} \tag{9.62}$$

where $c_j(j = 1, 2, \cdots, n)$ is the value coefficient; $a_{ij}(i = 1, 2, \cdots, m, \ j = 1, 2, \cdots, n)$ is the constraint coefficient: $b_i(i = 1, 2, \cdots, m)$ is the resource coefficient, $b_i \geq 0$; otherwise, at both sides of constraint condition multiplied by -1 simultaneously. c_j, a_{ij} and b_i are all known numbers. $x_j(j = 1, 2, \cdots, n)$ is an unknown number, called a *decision variable*. Optimization is to adopt linear programming to calculate x_j, enabling Equation (9.61) to reach the maximum and meet Equation (9.62).

Equations (9.61) and (9.62) are the standard pattern for linear programming. In practical application, the mathematical models for various kinds of linear programming should be transformed to standard patterns before the solution is worked out. There are four cases for this kind of transformation:

1. To change the minimum object function to maximum object function. Only when $z' = -z$ will minz be changed to max z'.
2. To change the negative decision variable (e.g., x_k) to a nonnegative decision variable (e.g., x'_k) in the constraint conditions. It is available only when $x'_k = -x_k$.
3. To change the decision variable without constraint, e.g., x_k, to the variable with constraint. It is available only when $x_k = x'_k - x''_k$, where x'_k and x''_k are dual nonnegative slack variables.
4. To change the unequal sign to an equal sign in constraint conditions containing a_{ij}. For the sign of $<$ or $\leq$, add a nonnegative slack variable at the left-hand side; for the sign of $>$ or $\geq$, to subtract a nonnegative surplus variable at the left-hand side.

Obviously, any form of mathematical model can be transformed to the standard pattern. The following example is taken for explanation.

Assume a linear programming problem in a nonstandard pattern is:

$$\min z = -x_1 + 2x_2 - 3x_3$$

$$\begin{cases} x_1 + x_2 + x_3 \leq 7 \\ x_1 - x_2 + x_3 \geq 2 \\ -3x_1 + x_2 + 2x_3 = 5 \\ x_1 \geq 0, x_2 \leq 0, x_3 \quad \text{is in unconstraint} \end{cases}$$

The process to transform the problem from nonstandard pattern to standard pattern is: let $z' = -z$; let $x'_2 = -x_2$; let $x_3 = x_4 - x_5$, of which the slack variables $x_4 \geq 0$ and $x_5 \geq 0$; and add slack variable x_6 and subtract surplus variable x_7 at the left-hand side of the first and the second constraint inequality, respectively, of which $x_6 \geq 0$ and $x_7 \geq 0$.

Thus, this linear programming has been in standard pattern:

$$\max z' = x_1 + 2x'_2 + 3(x_4 - x_5) + 0x_6 + 0x_7$$

$$\begin{cases} x_1 - x'_2 + x_4 - x_5 + x_6 = 7 \\ x_1 + x'_2 + x_4 - x_5 - x_7 = 2 \\ -3x_1 - x'_2 + 2x_4 - 2x_5 = 5 \\ x_1, x'_2, x_4, x_5, x_6, x_7 \geq 0 \end{cases}$$

In a word, when a mathematical model is constructed from a real problem, its standardization can be realized with the skill expressed in Table 9.6 (A compiling group for the Teaching Material of Operational Research, 1997; Shi, 1999).

9.4.3. Optimization of Exploration and Production Schemes Implemented by Linear Programming

9.4.3.1. Calculation Method

As mentioned, the original optimization can always be transformed to the standard pattern as expressed by Equations (9.61) and (9.62). The corresponding value coefficient of slack variable and surplus variable is zero in the object function. The big-M method is chosen to solve linear programming (A compiling group for the Teaching Material of Operational Research, 1997; Shi, 1999). For the sake of application and unification of the program, m artificial variables $(x_{n+1}, x_{n+2}, \cdots, x_{n+m})$ are introduced, of which the corresponding value coefficient is a negative big number in the object functions, which is called M (which can be taken as 10^{30} for the single-precision floating-point). Here Equations (9.61) and (9.62) should be transformed to the following standard pattern, which is applicable to solve linear programming by the big-M method.

Object function:

$$\max z = c_1 x_1 + c_2 x_2 + \cdots + c_n x_n$$

$$+ c_{n+1} x_{n+1} + c_{n+2} x_{n+2} + \cdots + c_{n+m} x_{n+m} \tag{9.63}$$

TABLE 9.6 Standardization Skills of Mathematical Models

Variable	$x_j \geq 0$		N/A
	$x_j \leq 0$		Let $x'_j = -x_j$, $x'_j \geq 0$
	x_j unconstraint		Let $x_j = x'_j - x''_j$, where x'_j and $x''_j \geq 0$
Constraint condition	$b \geq 0$		N/A
	$b < 0$		Two sides of constraint condition are multiplied with -1
	$\leq$		Add slack variable x_{si} as a base variable
	$=$		Add artificial variable x_{ai} as a base variable
	$\geq$		Subtract surplus variable x_{si}, add artificial variable x_{ai} as a base variable
Object function	max z		N/A
	min z		Let $z' = -z$, solve max z'
	Add coefficient of variable	Slack variable	0
		Artificial variable	$-M$

where $c_{n+1} = c_{n+2} = \cdots = c_{n+m} = -M$ and $x_{n+1} = x_{n+2} = \cdots = x_{n+m} = 0$ at the end of calculation.

Constraint condition:

$$\begin{cases} a_{11}x_1 + a_{12}x_2 + \cdots + a_{1n}x_n + x_{n+1} = b_1 \\ a_{21}x_1 + a_{22}x_2 + \cdots + a_{2n}x_n + x_{n+2} = b_2 \\ \qquad \cdots\cdots \\ a_{m1}x_1 + a_{m2}x_2 + \cdots + a_{mn}x_n + x_{n+m} = b_m \\ x_1, x_2, \cdots, x_n, x_{n+1}, x_{n+2}, \cdots, x_{n+m} \geq 0 \end{cases} \qquad (9.64)$$

Obviously, the last m columns in Equation (9.64) form an $(m \times m)$ unit matrix, and the artificial variables $x_{n+1}, x_{n+2}, \ldots, x_{n+m}$ are called *base variables*. Certainly, it is flexible in programming that the base variable do not always consist of only the artificial variable, but also the decision variable and the slack variable; and the columns in the $(m \times m)$ unit matrix that the base variables correspond to can also be spread in the location of any column. To make the following explanation easier, still assume that there are m artificial variables.

The calculation process of the big-M method consists of the following five steps:

1. Build an initial simplex table (Table 9.7).

In Table 9.7, the value coefficient of base variable is in column C_b, i.e., $c_{n+1}, c_{n+2}, \ldots, c_{n+m}$ at the initial; the base variable is in column X_b, i.e., $x_{n+1}, x_{n+2}, \ldots, x_{n+m}$ at the initial; the constant at the right-hand side of the constraint equations (resource coefficient) is in column B, i.e., $b_1, b_2, \ldots, b_m$ at the initial; the value coefficient of decision variable and base

TABLE 9.7   Initial Simplex Table

		c_j	c_1	...	c_n	$c_{n+1}=-M$	...	$c_{n+m}=-M$	θ_i
C_b	X_b	B	x_1	...	x_n	x_{n+1}	...	x_{n+m}	
c_{n+1}	x_{n+1}	b_1	a_{11}	...	a_{1n}	1	...	0	θ_1
c_{n+2}	x_{n+2}	b_2	a_{21}	...	a_{2n}	0	...	0	θ_2
...	...	...	...	...	...	...	...	...	...
c_{n+m}	x_{n+m}	b_m	a_{m1}	...	a_{mn}	0	...	1	θ_m
	σ_j	$-\sum\limits_{i=1}^{m} c_{n+i}b_i$	$c_1-\sum\limits_{i=1}^{m} c_{n+i}a_{i1}$	...	$c_n-\sum\limits_{i=1}^{m} c_{n+i}a_{in}$	0	...	0	

variable in row c_j, i.e., $c_1, c_2, \ldots, c_n, c_{n+1}, c_{n+2}, \ldots, c_{n+m}$; the number calculated with θ rule after the entering variable is determined is in column θ_i column; row σ_j is called the row of validation number, the validation number corresponding to each variable x_j, $\sigma_j = c_j - \sum_{i=1}^{m} c_{n+i}a_{ij}$, $j = 1, 2, \cdots, n, n+1, \cdots, n+m$; but the first number of the last row is $-\sum_{i=1}^{m} c_{n+i}b_i$; and the others are $m \times (n+m)$ matrix formed by $a_{ij}(i = 1, 2, \cdots, m; j = 1, 2, \cdots, n, n+1, n+m)$.

Table 9.7 is called an *initial simplex table*. A new simplex table is formed by every iteration.

It can be seized from Equation (9.64) or Table 9.7 that the initial feasible basis is $(x_{n+1}, x_{n+2}, \cdots, x_{n+m})$ and the initial basic feasible solution is $(0, \cdots, 0, b_1, \cdots, b_m)$ where there are n zeros.

2. Calculate the last row of simplex table. To calculate the validation number of each variable x_j, $\sigma_j = c_j - \sum_{i=1}^{m} c_{n+i}a_{ij}$, and the other "special σ," $-\sum_{i=1}^{m} c_{n+i}b_i = -z$, forming the last row in the simplex table for validating the solution.

If σ_j is not all ≤ 0, then turn to Step 3; otherwise, σ_j is all ≤ 0. It is possible to have the following three solutions. (1) If there is nonzero artificial variable in the base variable X_b, that means "no feasible solution" because there exist contrary constraint conditions, and calculation should be stopped; otherwise, turn to (2). (2) If the validation number of a decision variable is 0, then there are "infinite optimal solutions" because the object function and constraint condition are not applicable, and calculation should be stopped; otherwise, turn to (3). (3) There is a "unique optimal solution," which is formed by m numbers in column B and n zeros, and of which the order is determined by the base variable in column X_b. If it is not required that the final solution x_j should be an integer, then $x_j(j = 1, 2, \cdots, n, n+1, \cdots, n+m)$ and z will be output to end the whole calculation process. Otherwise, x_j is required to be an integer, which is the case with integer programming. It can be solved by using a branch-bound method or a cutting plane method (A compiling group for the teaching material of operational research, 1997; Shi, 1999). Nevertheless, the programming of the two methods is complicated. A simple and convenient method for integer programming has been presented, which does not need to reprogram.

Now this simple and convenient integer programming method is described in detail. Based on the obtained unique optimal solution of noninteger $x_j(j = 1, 2, \cdots, n)$, three integers adjacent to x_j are determined, composing 3^n groups in which each group includes n integers $x_j^*(j = 1, 2, \cdots, n)$. Then n integers in each group are substituted in the left-hand side of the constraint condition Equation (9.64), attaining $b_i^*(i = 1, 2, \cdots, m)$. The b_i^* subtracts b_i, respectively, to get a mean square error. Finally, the group of x_j^* with the minimum mean square error is selected as n solutions of the unique optimal solutions.

In these n solutions, x_j is noninteger. The value of the three integers for each x_j is defined as follows:

$$\left.\begin{array}{l} v_2 = \text{INT}[x_j + 0.5] \\ v_1 = v_2 - 1 \\ v_3 = v_2 + 1 \end{array}\right\} \tag{9.65}$$

where INT is an integral function. If $v_1 < 0$, take $v_1 = 0$. For instance, if $x_j = 8.25$, then $v_2 = 8, v_1 = 7, v_3 = 9$; if $x_j = 4.625$, then $v_2 = 5, v_1 - 4, v_3 - 6$. Obviously, these three integers are the closest in the integer set near x_j. The aforementioned x_j^* is taken from one of $v_1, v_2,$ and v_3 that every x_j corresponds to.

The aforementioned simple and convenient integer programming method has been merged to the noninteger programming, which integrates the noninteger programming with integer programming in linear programming. It is easy to make a uniform program. Obviously, this method is much more simple and convenient than the branch-bound method and the cutting plane method (A compiling group for the teaching material of operational research, 1997), which specially work for integer programming and also have less calculation than the exhaustive attack method (A compiling group for the teaching material of operational research, 1997).

3. Continue on the solution validation. In $\sigma_j > 0$, if a column vector $a_{ik}(i = 1, 2, \cdots, m)$ that σ_k corresponds to is all ≤ 0, that means there is an "unbounded solution" because of lack of necessary constraint conditions, and calculation should be stopped; otherwise, turn to Step 4.
4. Determine entering variable x_k and leaving variable x_l. If $\max(\sigma_j > 0) = \sigma_k$, then x_k is the entering variable, which is calculated in terms of th eθ rule as follows:

$$\theta = \min(b_i/a_{ik} > 0) = \frac{b_l}{a_{lk}}$$

Thus x_l is the leaving variable.
5. Modify the simplex table. There are two operations:
Operation 1 is to change c_l in column C_b to c_k, and x_l in column X_b to x_k.
Operation 2 is to perform Gauss elimination on the augmented matrix.
The $(m + 1) \times (n + m + 1)$ augmented matrix is composed of column B, a_{ij} and row σ_j. Taking a_{lk} as the principal element, Gauss elimination is performed to transform the column vector $(a_{1k}, a_{2k}, \cdots, a_{lk}, \cdots, a_{mk}, \sigma_k)^T$ that x_k corresponds to into $(0, 0, \ldots, 1, \ldots, 0, 0)^T$. The specific process is given here:
To the principal element row (the lth row),

$$a_{lk} \Rightarrow a^*; b_l/a^* \Rightarrow b_l; a_{lj}/a^* \Rightarrow a_{lj}$$

To the elements on the other columns and rows in the augmented matrix (except for the lth row and the kth column),

$$b_i - b_l \cdot a_{ik} \Rightarrow b_i; \quad a_{ij} - a_{lj} \cdot a_{ik} \Rightarrow a_{ij}$$

To the principal element column (the kth column),

$$1 \Rightarrow a_{lk}; \quad 0 \Rightarrow \text{the other elements}$$

Repeat Steps 2–5 till termination.

To help readers grasp the aforementioned big-M method with noninteger programming and integer programming, here an example of noninteger programming is introduced at first.

The problem of this noninteger programming is (A compiling group for the teaching material of operational research, 1997; Shi, 1999) is

$$\max \ z = 3x_1 - x_2 - x_3 \tag{9.66}$$

$$\begin{cases} x_1 - 2x_2 + x_3 \leq 11 \\ -4x_1 + x_2 + 2x_3 \geq 3 \\ -2x_1 \quad\quad + x_3 = 1 \\ x_1, x_2, x_3 \geq 0 \end{cases} \tag{9.67}$$

Try to solve by the big-M method.

Solution: To add slack variable x_4, surplus variable x_5, and artificial variables x_6 and x_7 in the preceding constraint conditions, attaining the standard pattern that follows. Object function:

$$\max \ z = 3x_1 - x_2 - x_3 - +0x_4 + 0x_5 - Mx_6 - Mx_7 \tag{9.68}$$

Constraint condition:

$$\begin{cases} x_1 - 2x_2 + x_3 + x_4 \quad\quad\quad\quad = 11 \\ -4x_1 + x_2 + 2x_3 \quad - x_5 + x_6 \quad\quad = 3 \\ -2x_1 \quad\quad + x_3 \quad\quad\quad\quad + x_7 = 1 \\ x_1, x_2, x_3, x_4, x_5, x_6, x_7 \geq 0 \end{cases} \tag{9.69}$$

Here M is an arbitrary big integer and can be taken as 10^{30}.

Obviously, x_4, x_6, and x_7 can be regarded as base variables and the coefficient column vector that they correspond to form (3×3) unit matrix. The results of the calculation process are listed in Table 9.8. In the first subtable, [1] denotes $a_{lk} = 1$, indicating the entering variable in next step is $x_3 (k = 3)$, leaving variable is $x_7 (l = 3)$ that is the third base variable in X_b, and the principal element of Gauss elimination is a_{33}. In the second subtable, [1] denotes $a_{lk} = 1$, indicating the entering variable in next step is $x_2 (k = 2)$, leaving variable is $x_6 (l = 2)$ that is the second base variable in X_b, and the principal element of Gauss elimination is a_{22} In the third subtable, [3] denotes $a_{lk} = 3$, indicating the entering variable in next step is $x_1 (k = 1)$, leaving variable is $x_4 (l = 1)$ that is the first base variable in X_b, and the principal element of Gauss elimination is a_{11}. In the fourth subtable (the last subtable), all $\sigma_j \leq 0$, indicating no case of "unbounded

TABLE 9.8 Simplex Table for Calculation Process of the Big-M Method

c_j			3	-1	-1	0	0	-M	-M	θ_i
C_b	X_b	B	x_1	x_2	x_3	x_4	x_5	x_6	x_7	
0	x_4	11	1	-2	1	1	0	0	0	11
-M	x_6	3	-4	1	2	0	-1	1	0	3/2
-M	x_7	1	-2	0	[1]	0	0	0	1	1
σ_j		4M	3-6M	M-1	3M-1	0	-M	0	0	
0	x_4	10	3	-2	0	1	0	0	-1	1
-M	x_6	1	0	[1]	0	0	-1	1	-2	
-1	x_3	1	-2	0	1	0	0	0	1	
σ_j		M+1	1	M-1	0	0	-M	0	1-3M	
0	x_4	12	[3]	0	0	1	-2	2	-5	4
-1	x_2	1	0	1	0	0	-1	1	-2	
-1	x_3	1	-2	0	1	0	0	0	1	
σ_j		2	1	0	0	0	-1	1-M	-M-1	
3	x_1	4	1	0	0	1/3	-2/3	2/3	-5/3	
-1	x_2	1	0	1	0	0	-1	1	-2	
-1	x_3	9	0	0	1	2/3	-4/3	4/3	-7/3	
σ_j		-2	0	0	0	-1/3	-1/3	1/3-M	2/3-M	

solution"; there is no nonzero artificial variable in X_b, indicating no case of "no feasible solution"; and there is no zero validation number of base variable (x_4, x_5, x_6, x_7), indicating no case of "infinite optimal solutions." So there exists only the case of "unique optimal solution." It is seen from column B that this unique optimal solution is

$$x_1 = 4, \ x_2 = 1, \ x_3 = 9, \ x_4 = x_5 = x_6 = x_7 = 0$$

and the maximum of the corresponding object function is

$$z = 2 \ (\text{due to} \ -z = -2)$$

Here one more example of noninteger programming is introduced.

The problem of this noninteger programming is (A compiling group for the teaching material of operational research, 1997; Shi, 1999) is

$$\max z = 2x_1 + 3x_2 \tag{9.70}$$

$$\begin{cases} x_1 + 2x_2 \leq 8 \\ 4x_1 \leq 16 \\ 4x_2 \leq 12 \\ x_1, x_2 \geq 0 \end{cases} \tag{9.71}$$

Solution: To add slack variable x_3, x_4, and x_5 in the preceding constraint condition, attaining the standard pattern that follows.

Object function:

$$\max \ z = 2x_1 + 3x_2 + 0x_3 + 0x_4 + 0x_5 \tag{9.72}$$

Constraint condition:

$$\begin{cases} x_1 + 2x_2 + x_3 \quad\quad\quad\quad = 8 \\ 4x_1 \quad\quad\quad\quad + x_4 \quad\quad = 16 \\ 4x_2 \quad\quad\quad\quad\quad + x_5 = 12 \\ x_1, x_2, x_3, x_4, x_5 \geq 0 \end{cases} \tag{9.73}$$

Obviously, x_3, x_4, and x_5 can be regarded as base variables, and the coefficient column vector that they correspond to form (3×3) unit matrix. The calculation process is omitted here. Finally, the unique optimal solution is obtained as follows:

$$x_1 = 4, \ x_2 = 2, \ x_3 = 0, \ x_4 = 0, \ x_5 = 4$$

and the maximum of the corresponding object function is:

$$z = 14$$

Though the unique optimal solution of the preceding aforementioned samples is all integers, yet the solution method still uses noninteger programming. The sample of integer programming is introduced in the following simple case study.

9.4.3.2. Simple Case Study: Optimal Profit of Exploration Drilling Plan

Assume that there are five oil-bearing traps in an exploration area of China. From risk analysis, the risk coefficients of exploration drilling for these five traps are found to be 0.10, 0.25, 0.40, 0.55, and 0.70, respectively, i.e., the success probability for exploration drilling is 0.90, 0.75, 0.60, 0.45, and 0.30, respectively. The oil revenue for each well in these five traps in two years after the exploration drilling is completed is estimated to be 600, 500, 400, 300, and 200 (10^4 yuan), which are multiplied by the success probability, respectively, resulting in the reliable probability revenue of 540, 375, 240, 135, and 60 (10^4 yuan). Let the number of wells drilled in these five traps be x_1, x_2, x_3, x_4, and x_5, respectively. Now the question is, how much $x_j (j = 1, 2, \cdots, 5)$ is under a given constraint condition when the total oil revenue is the maximum? Consequently, the object function is

$$\max \ z = 540x_1 + 375x_2 + 240x_3 + 135x_4 + 60x_5 \tag{9.74}$$

The following are the geological exploration requirements for this exploration drilling plan: (a) 31 wells to be drilled in total; (b) the number of wells to be drilled in the first trap is as much as double number in the second trap minus one well; (c) the number of wells to be drilled in the second trap is as much as double the number in the third trap minus one well; (d) the number of wells to be drilled in the third trap is as much as the number of wells to be drilled in the fourth trap plus the double number in the fifth trap and plus one well; and (e) there is only one well to be drilled in the fifth trap. So the constraint condition is

$$\begin{cases} x_1 + x_2 + x_3 + x_4 + x_5 & = 31 \\ -x_1 + 2x_2 & = 1 \\ \quad\quad -x_2 + 2x_3 & = 1 \\ \quad\quad\quad\quad x_3 - x_4 - 2x_5 & = 1 \\ \quad\quad\quad\quad\quad\quad x_5 & = 1 \\ x_1, x_2, x_3, x_4, x_5 \geq 0 \end{cases} \tag{9.75}$$

Solution: Since no sign of $<$ or $>$ is available in Equation (9.75), it is not necessary to introduce slack variables or surplus variables. Since no unconstraint decision variable exists, it is not necessary to introduce dual slack variables. Since the coefficient column vector of each decision variable is all not unit column vector, all decision variables cannot be used as base variables. Therefore, only five artificial variables x_6, x_7, x_8, x_9, and x_{10} can be introduced as base variables. As a result, Equations (9.74) and (9.75) is rewritten as:

Object function:

$$\max \ z = 540x_1 + 375x_2 + 240x_3 + 135x_4 + 60x_5$$
$$-Mx_6 - Mx_7 - Mx_8 - Mx_9 - Mx_{10} \tag{9.76}$$

Constraint condition:

$$\begin{cases} x_1 + x_2 + x_3 + x_4 + x_5 + x_6 & = 31 \\ -x_1 + 2x_2 \quad\quad\quad\quad + x_7 & = 1 \\ \quad\quad -x_2 + 2x_3 \quad\quad\quad\quad + x_8 & = 1 \\ \quad\quad\quad\quad x_3 - x_4 - 2x_5 \quad\quad\quad + x_9 & = 1 \\ \quad\quad\quad\quad\quad\quad x_5 \quad\quad\quad\quad + x_{10} & = 1 \\ x_1, x_2, x_3, x_4, x_5, x_6, x_7, x_8, x_9, x_{10} \geq 0 \end{cases} \tag{9.77}$$

using the big-M method to meet x_1, x_2, x_3, x_4, and x_5 in Equations (9.76) and (9.77).

Obviously, x_6, x_7, x_8, x_9 and x_{10} can be as regarded as base variables, and the coefficient column vector that they correspond to forms a (5×5) unit matrix.

If the noninteger programming is adopted, the unique optimal solution is obtained as follows:

$$x_1 = 15.5, \ x_2 = 8.25, \ x_3 = 4.625, \ x_4 = 1.625, \ x_5 = 1, \ x_6 = x_7 = x_8 = x_9 = x_{10} = 0$$

and the maximum of the corresponding object function is

$$z = 0.1285 \times 10^5 \ (10^4 \text{yuan}),$$

In practical application, the number of wells drilled can be but an integer, and it cannot be those noninteger values in the preceding unique optimal solution. So the integer programming must be adopted.

If the simple and convenient integer programming method is used, the unique optimal solution will be obtained as follows:

$$x_1 = 15, \ x_2 = 8, \ x_3 = 5, \ x_4 = 2, \ x_5 = 1, \ x_6 = x_7 = x_8 = x_9 = x_{10} = 0$$

and the maximum of the corresponding object function is:

$$z = 0.1263 \times 10^5 \ (10^4 \text{yuan})$$

Now the question is, is it possible not to apply this simple and convenient integer programming method but to use the round rule to directly transform the unique optimal solution of the noninteger programming $x_j = (j = 1, 2, \cdots, n, \ n+1, \cdots, n+m)$ to integer? The answer is that it does not always work, because the unique optimal solution then obtained is as follows:

$$x_1 = 16, \ x_2 = 8, \ x_3 = 5, \ x_4 = 2, \ x_5 = 1, \ x_6 = x_7 = x_8 = x_9 = x_{10} = 0$$

But this $x_1 = 16$ has been verified not to be the optimal with the simple and convenient integer programming method, for $x_1 = 15.5$, $v_2 = 16$, $v_1 = 15$, and $v_3 = 17$ are obtained by Equation (9.65). Three corresponding mean square errors are obtained by substituting these three numbers into the constraint condition expressed by Equation (9.77), and $v_1 = 15$ leads the mean square error of the constraint condition to be minimum, whereas $v_2 = 16$ and $v_3 = 17$ cannot make the mean square error of the constraint condition be minimum.

1. Calculation flowchart (Figure 9.9) (A compiling group for the teaching material of operational research, 1997; Shi, 1999). In this flowchart, $X_a = (x_1, \ x_2, \ \cdots, x_n)$.
2. Results and analyses. For the integer programming, it is impossible to ensure the unique optimal solution obtained always to completely satisfy the original constraint condition. When it does not satisfy the constraint condition, one of several possible optimal solutions is selected so that the mean square error of the constraint condition that it corresponds to reaches minimum. Let the right-hand side of the constraint condition after the unique optimal solution of the integer programming is substituted into it be designated as $b_i^* (i = 1, 2, \cdots, m)$, b_i^* is not surely equal to b_i. For example, $b_1^* = b_1$, $b_2^* = b_2$, $b_4^* = b_4$, $b_5^* = b_5$ in this exercise, only $b_3^* \neq b_3 (b_3^* = 2, b_3 = 1)$. At this time, it is possible to modify the original constraint condition, to rewrite Equation (9.75) to

$$\begin{cases} x_1 + x_2 + x_3 + x_4 + x_5 & = 31 \\ -x_1 + 2x_2 & = 1 \\ -x_2 + 2x_3 & = 2 \\ x_3 - x_4 - 2x_5 & = 1 \\ x_5 & = 1 \\ x_1, x_2, x_3, x_4, x_5 \geq 0 \end{cases} \tag{9.78}$$

The only difference between Equation (9.75) and Equation (9.78) is that for the right-hand side of the third constraint condition, the former is 1 ($b_3 = 1$) whereas the latter is 2 ($b_3^* = 2$), which is allowable in engineering. As a result, the unique optimal solution of the integer programming (15, 8, 5, 2, 1) completely satisfies the actual constraint condition expressed by Equation (9.78).

In summary, using the data of five oil-bearing traps in this area, including the success probability of exploration drilling for each oil-bearing trap, the oil revenue in two

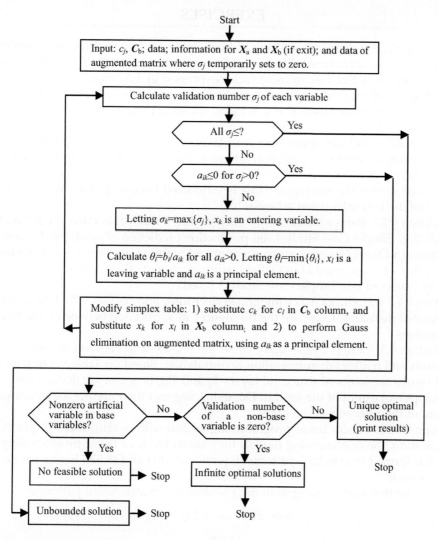

FIGURE 9.9 Calculation flowchart of linear programming.

years for each well as well as five geological exploration requirements, the object function (9.74) and its constraint condition (9.75) in this area are obtained by a data mining tool (linear programming). The results coincide with practicality. This object function and its constraint condition are called *mined knowledge*, which is applied to calculate the optimal number of wells to be drilled in each oil-bearing trap. Therefore, this method can be spread to the optimization for number of wells to be drilled in other areas.

EXERCISES

9-1. In Section 9.1.3, a simple case study of fuzzy integrated evaluation on an exploration drilling object is introduced. The fuzzy integrated decision value $E = 0.495$ is calculated by Equation (9.1). But the comment vector expressed by Equation (9.6) is $(2, 1, 0, -1, -2)$, denoting excellent, good, average, poor, and very poor, respectively. Is it possible to use the round rule to transform $E = 0.495$ to $E = 0$ so as to make it consistent with 0 in the comment vector?

9-2. Section 9.2.3 introduces space-time calculations conducted by gray prediction. In gray prediction there are two calculation schemes: Calculation Scheme 1, in which the original data are not processed; and Calculation Scheme 2, in which the original data are processed so that the minimum value is subtracted from each data. Which one is the better of the two calculation schemes?

9-3. In Section 9.2.3, there are two simple case studies for oil production prediction by gray prediction: Simple Case Study 1 (oil production prediction of an oil production zone in the Shengli Oilfield using gray prediction) and Simple Case Study 3 (oil production prediction of the Romashkino Oilfield using gray prediction). Why did Simple Case Study 1 succeed but Simple Case Study 3 failed?

9-4. Section 9.1.3 introduces fuzzy integrated decisions, and Section 9.2.4 introduces gray integrated decisions. Compare the differences in application between these two algorithms.

9-5. In Section 9.3, there is a case study about the relationship between the fractal features of faults and hydrocarbon migration accumulation. Based on Table 9.5 (fractal data and results for faults of structural layers T_2 and T_6), how can we deduce the development extent of the fault in Dongying Sag and then predict the development extent of the reservoir in the sag?

9-6. Section 9.4 introduces linear programming to deal with optimization. In practical application, the mathematical models for various kinds of linear programming should be transformed to standard patterns [Equations (9.61) and (9.62)] before the solution is worked out.

Assume that a linear programming problem in a nonstandard pattern is:

$$\min\ z = -x_1 + 2x_2 - 3x_3$$

$$\begin{cases} x_1 + x_2 + x_3 \leq 7 \\ x_1 - x_2 + x_3 \geq 2 \\ -3x_1 + x_2 + 2x_3 = 5 \\ x_1 \geq 0, x_2 \leq 0, x_3 \quad \text{is in unconstraint} \end{cases}$$

Write the process to transform the problem from a nonstandard pattern to a standard pattern.

9-7. Section 9.4.3 introduces optimization of exploration and production schemes implemented by linear programming, and the calculation method used is noninteger programming. Integer programming can be solved by using a branch-bound method or a cutting plane method. Nevertheless, the programming of the two methods is complicated. A simple and convenient method for integer programming has been presented, which does not need to reprogram. Now the question is, is it possible not to apply this simple and convenient integer programming method but to use the round rule to directly transform the unique optimal solution of the noninteger programming $x_j = (j = 1, 2, \cdots, n, n + 1, \cdots, n + m)$ to integer?

References

A compiling group for the Teaching Material of Operational Research, 1997. Operational Research. Tsinghua University Press, Beijing, China (in Chinese).

A compiling group in Chongqing University, 1991. Probability Theory and Statistics. Chongqing University Press, Chongqing, China (in Chinese).

Chen, Y., Chen, L., 1998. Fractal Geometry. Seismological Press, Beijing, China (in Chinese).

Deng, J., 1986. Gray Prediction and Decision. Central China University of Science and Technology Press, Wuhan, China (in Chinese).

Guo, S., Sun, C., 1992. An application of fuzzy identification in the evaluation of potential oil/gas reservoir by means of multiple geological, geophysical, and geochemical parameters. Petrol. Expl. Devel 19 (Suppl.), 106—111 (in Chinese with English abstract).

Qi, Z., 1996. Gray model prediction of depth and thickness of oil and gas formations during drilling. Petrol. Expl. Devel 23 (4), 75—77 (in Chinese with English abstract).

Shen, Z., Feng, Z., Zhou, G., Wang, J., Hong, Z., 1996. Oil-controlling fractal structure feature and its petroleum geology meaning of petroliferous basin. In: Chinese Mathematical Geology 7. Geological Publishing House, Beijing, China, pp. 215—222 (in Chinese).

Shi, G., 1999. New Computer Application Technologies in Earth Sciences. Petroleum Industry Press, Beijing, China (in Chinese).

Sun, B., Fan, H., 1996. Application of gray system theory in forecasting oil production. J. Xi'an Petro. Inst. 11 (4), 54—56 (in Chinese with English abstract).

Wang, J., Xie, J., 1996. Fundamentals and Applications of Systems Engineering. Geological Publishing House, Beijing, China (in Chinese).

Zhang, J., Tian, G., Liou, J., 1996. A fractal analysis on structural fractures of reservoirs. Petrol. Expl. Devel 23 (4), 65—67 (in Chinese with English abstract).

A Practical Software System of Data Mining and Knowledge Discovery for Geosciences

Data Mining and Knowledge Discovery for Geoscientists
http://dx.doi.org/10.1016/B978-0-12-410437-2.00010-2

This chapter presents a practical software system of data mining and knowledge discovery for geosciences. To check out the feasibility and validity of the automatic selection of regression and classification algorithms in this system, this chapter provides three typical case studies. To cite these three case studies in the instruction of this system, they are put at the beginning of this chapter.

As we know, there are three regression algorithms: multiple regression analysis (MRA), which we discussed in Chapter 2; error back-propagation neural network (BPNN), which we covered in Chapter 3; and regression of support vector machine (R-SVM), described in Chapter 4. Because of the complexities of geosciences rules, the studied problems are nonlinear in most of cases. Since MRA is a linear algorithm, it is not chosen in the automatic selection of regression algorithms. However, since the total mean absolute relative residual $\overline{R}*(\%)$ of MRA for a studied problem expresses the nonlinearity of this problem (Table 1.2), MRA should be used in applications. Thus, the automatic selection of regression algorithms, as presented by this chapter, contains two regression algorithms: BPNN and R-SVM.

As we also know, there are five classification algorithms: classification of support vector machine (C-SVM), which we learned about in Chapter 4; decision trees, discussed in Chapter 5; Naïve Bayesian (NBAY), covered in Chapter 6; Bayesian discrimination (BAYD), discussed in Chapter 6; and Bayesian successive discrimination (BAYSD), introduced in Chapter 6. Since the use of decision trees is quite complicated, that topic is not chosen in the automatic selection of classification algorithms; and since BAYSD is superior to NBAY and BAYD, NBAY and BAYD are not chosen in the automatic selection of classification algorithms. Thus, the automatic selection of classification algorithms, as presented in this chapter, contains two classification algorithms: C-SVM and BAYSD.

Section 10.1 (reservoir classification in the Keshang Formation) introduces Typical Case Study 1. Using data for the oil productivity index and oil layer classification in the Keshang Formation based on three parameters, i.e., the three parameters (porosity, permeability, resistivity) and an oil test result of 18 samples, of which 14 are taken as learning samples and four are taken as prediction samples, regression algorithms (R-SVM, BPNN, MRA) are adopted for the oil productivity index, and classification algorithms (C-SVM, BAYSD) are adopted for the oil layer classification. It is found that (a) since this oil productivity index is a very strong nonlinear problem, these regression algorithms are not applicable; and (b) since this oil layer classification is a moderate nonlinear problem, these classification algorithms are applicable.

Section 10.2 (reservoir classification in the lower H3 Formation) introduces Typical Case Study 2. Using data for the oil layer classification in the lower H3 Formation based on eight parameters, i.e., the eight parameters (elf-potential, micronormal, microinverse, acoustic-time, 0.45 m apparent resistivity, 4 m apparent resistivity, conductivity, caliper given) and an oil test result of 32 samples, of which 24 are taken as learning samples and eight are taken as prediction samples, classification algorithms (C-SVM, BAYSD) are adopted for the oil layer classification. It is found that since this oil layer classification is a moderate nonlinear problem, these classification algorithms are applicable.

Section 10.3 (reservoir classification in the Xiefengqiao Anticlinal) introduces Typical Case Study 3. Using data for the oil layer classification in the Xiefengqiao Anticlinal based on five parameters, i.e., the five parameters (true resistivity, acoustictime, porosity, oil saturation, permeability) and an oil test result of 27 samples, of which 24 are taken as learning samples and three are taken as prediction samples, classification algorithms (C-SVM, BAYSD) are adopted for the oil layer classification. It is found that since this oil layer classification is a strong nonlinear problem, only C-SVM is applicable, but BAYSD is not applicable.

The following three conclusions are gained from the aforementioned three typical case studies:

1. From Typical Case Study 1, it is concluded that when regression algorithms (R-SVM, BPNN, MRA) fail in a regression problem, this regression problem can be changed to a classification problem, then classification algorithms (C-SVM, BAYD) succeed in this classification problem.
2. From Typical Case Study 2, it is concluded that when the nonlinearity of a classification problem is very weak, weak, or moderate, classification algorithms (C-SVM, BAYD) succeed in this classification problem.
3. From Typical Case Study 3, it is concluded that when the nonlinearity of a classification problem is strong or very strong, only C-SVM succeeds in this classification problem, but BAYSD fails.

Section 10.4 (a practical system of data mining and knowledge discovery for geosciences) briefly presents this practical software system, including minable subsurface data and system components, proving that the feasibility and practicality of this system is based on the aforementioned three typical case studies as well as a large amount of other case studies in this book.

10.1. TYPICAL CASE STUDY 1: OIL LAYER CLASSIFICATION IN THE KESHANG FORMATION

10.1.1. Studied Problem

The objective of this case study is to conduct an oil layer classification in glutenite using porosity, permeability, and resistivity, which has practical value when oil test data are less limited.

Located in the District 8 of Kelamayi Oilfield in western China, the Keshang Formation is a monocline fault block oil layer with low porosity and middle-low permeability lying at depths of 1700–2200 m.

Using data of 21 samples from the Keshang Formation, with each sample containing three parameters (x_1 = porosity, x_2 = permeability, x_3 = resistivity) and an oil test result (y^* = reservoir type), Tan et al. (2004) adopted BPNN for the prediction of oil layer productivity index and classification. In this case study, among these 21 samples, 14 were taken as learning samples and 4 as prediction samples (Table 10.1).

TABLE 10.1   Input Data for Oil Layer Productivity Index and Classification of the Keshang Formation

				Well Logging[a]			Oil Test[b]	
				$x1$	$x2$	$x3$	y^*	
Sample Type	Sample No.	Well No.	Interval Depth (m)	ϕ (%)	k (mD)	RT ($\Omega \cdot$m)	OLPI (t/(MPa.m.d)	OLC[c]
Learning samples	1	150	1794.0–1810.0	15.7	32.0	20.0	0.208	3
	2	150	1832.0–1839.0	16.2	40.1	25.5	0.502	3
	3	150	1878.5–1880.5	10.3	3.4	21.5	0.010	4
	4	151	2101.0–2110.6	17.0	54.1	28.0	1.125	2
	5	152	2161.0–2165.0	11.2	4.9	13.0	0.030	4
	6	8239	2055.0–2073.0	16.1	37.2	9.0	0.464	3
	7	8252	2016.5–2022.5	15.7	32.0	5.0	0.466	3
	8	8259	2132.5–2138.5	15.7	32.0	26.0	0.470	3
	9	8227	2099.0–2104.0	15.7	32.0	17.0	0.665	3
	10	8271	1926.0–1930.0	14.8	22.0	25.0	0.376	3
	11	8235	2070.2–2074.8	16.6	46.6	24.0	1.163	2
	12	8247	2059.6–2062.8	8.5	1.6	20.0	0.006	4
	13	8241	1992.0–1994.2	18.4	98.4	40.0	1.551	1
	14	8201	2133.0–2134.8	17.2	58.3	30.0	1.094	2
Prediction samples	15	150	1926.0–1934.0	17.5	67.7	26.0	(1.226)	(2)
	16	152	2111.0–2112.5	15.7	32.0	23.0	(0.428)	(3)
	17	8220	2116.0–2118.0	18.1	84.7	40.0	(1.738)	(1)
	18	8228	2104.2–2107.2	10.1	3.2	28.0	(0.003)	(4)

[a]x_1 = porosity (ϕ); x_2 = permeability (k); and x_3 = resistivity (RT).
[b]y^* = oil-layer-productivity-index (OLPI) or oil-layer-classification (OLC) determined by the oil test. The numbers in parentheses are not input data but are used for calculating R(%).
[c]when y^* = OLC, it is oil layer classification (1—high-productivity oil layer, 2—intermediate-productivity oil layer, 3—low-productivity oil layer, 4—dry layer) determined by the oil test (Table 10.2).

TABLE 10.2 Oil Layer Classification Based on Productivity Index

Oil Layer Classification	Productivity Index (t/(MPa.m.d)	y^*
High-productivity oil layer	>1.40	1
Intermediate-productivity oil layer	(0.70, 1.40]	2
Low-productivity oil layer	[0.05, 0.70]	3
Dry layer	<0.05	4

10.1.2. Known Input Parameters

They are the values of the known variables x_i ($i = 1, 2, 3$) for 14 learning samples and four prediction samples, and the value of the prediction variable y^* for 14 learning samples (Table 10.1). Note: $y^* =$ productivity index for regression calculation; $y^* =$ oil layer classification for classification calculation.

10.1.3. Regression Calculation

R-SVM, BPNN, and MRA are adopted.

10.1.3.1. Learning Process

Using the 14 learning samples (Table 10.1) and by R-SVM, BPNN, and MRA, the three functions of oil-layer-productivity-index (y) with respect to three parameters (x_1, x_2, x_3) have been constructed, i.e., Equation (10.1) corresponding to SVM formula (4.1), Equation (10.2) corresponding to BPNN formula (3.1), and Equation (10.3) corresponding to MRA formula (2.14), respectively.

Using R-SVM, the result is an explicit nonlinear function:

$$y = \text{R-SVM}(x_1, x_2, x_3) \tag{10.1}$$

with $C = 1$, $\gamma = 0.333333$, and nine free vectors x_i.

The BPNN used consists of three input layer nodes, one output layer node, and seven hidden layer nodes. Setting the control parameter of t maximum $t_{max} = 100000$, the calculated optimal learning time count $t_{opt} = 94765$. The result is an implicit nonlinear function:

$$y = \text{BPNN}(x_1, x_2, x_3) \tag{10.2}$$

with the root mean square error $RMSE(\%) = 0.6124 \times 10^{-2}$.

Using MRA, the result is an explicit linear function:

$$y = -0.25509 + 0.012258x_1 + 0.015561x_2 + 0.0047419x_3 \tag{10.3}$$

Equation (10.3) yields a residual variance of 0.12794 and a multiple correlation coefficient of 0.93384. From the regression process, oil-layer-productivity-index (y) is shown to depend on the three parameters in decreasing order: permeability (x_2), resistivity (x_3), and porosity (x_1).

Substituting the values of porosity, permeability, and resistivity given by the 14 learning samples (Table 10.1) in Equations (10.1), (10.2), and (10.3), respectively, the oil-layer-productivity-index (y) of each learning sample is obtained. Table 10.3 shows the results of learning process by R-SVM, BPNN, and MRA.

10.1.3.2. Prediction Process

Substituting the values of porosity, permeability, and resistivity given by the four prediction samples (Table 10.1) in Equations (10.1), (10.2), and (10.3), respectively, the oil-layer-productivity-index (y) of each prediction sample is obtained. Table 10.3 shows the results of prediction process by R-SVM, BPNN, and MRA.

TABLE 10.3   Prediction Results of Oil Layer Productivity Index of the Keshang Formation

		Productivity Index (t/(MPa.m.d))						
					Regression Algorithm			
		Oil Test	R-SVM		BPNN		MRA	
Sample Type	Sample No.	y^*	y	$R(\%)$	y	$R(\%)$	y	$R(\%)$
Learning samples	1	0.208	0.520	150.07	0.257	23.60	0.530	154.88
	2	0.502	0.711	41.63	0.570	13.58	0.688	37.13
	3	0.010	0.073	627.15	0.006	40	0.026	160.32
	4	1.125	0.991	11.93	1.180	5.05	0.928	17.52
	5	0.030	0.125	318.08	0.006	79.99	0.020	33.00
	6	0.464	0.564	21.61	0.567	22.18	0.564	21.51
	7	0.466	0.509	9.23	0.547	17.40	0.459	1.50
	8	0.470	0.570	21.25	0.537	14.27	0.559	18.85
	9	0.665	0.505	24.12	0.745	11.99	0.516	22.42
	10	0.376	0.375	0.26	0.355	5.71	0.387	2.99
	11	1.163	0.813	30.10	1.241	6.72	0.787	32.30
	12	0.006	0.106	1670.77	0.006	0	−0.031	619.26
	13	1.551	1.451	6.47	1.550	0.09	1.691	9.05
	14	1.094	1.081	1.16	1.172	7.17	1.005	8.12
Prediction samples	15	1.226	1.160	5.39	1.547	26.20	1.136	7.32
	16	0.428	0.542	26.68	0.258	39.80	0.544	27.19
	17	1.738	1.407	19.05	1.490	14.28	1.475	15.16
	18	0.003	0.106	3419.47	0.006	100	0.051	1609.68

10.1.3.3. Application Comparisons Among R-SVM, BPNN, and MRA

It can been seen from $\overline{R}^*(\%) = 248.10$ of MRA (Table 10.4) that the nonlinearity of the relationship between the predicted value y and its relative parameters (x_1, x_2, x_3) is very strong from Table 1.2. $\overline{R}_1(\%)$ of the three regression algorithms (R-SVM, BPNN, and MRA) are big, and $\overline{R}_2(\%) > \overline{R}_1(\%)$.

10.1.3.4. Brief Summary

Since this case study is a very strong nonlinear problem, R-SVM, BPNN, and MRA are not applicable.

10.1.4. Classification Calculation

C-SVM and BAYSD are adopted. Since this case study is a classification problem, the regression algorithm is not adopted.

10.1.4.1. Learning Process

Using the 14 learning samples (Table 10.1) and by C-SVM, BAYSD, and MRA, the three functions of oil-layer-classification (y) with respect to three parameters (x_1, x_2, x_3) have been constructed, i.e., Equations (10.4) corresponding to SVM formula (4.1), Equation (10.5) corresponding to BAYSD discriminate function (6.30), and Equation (10.6) corresponding to MRA formula (2.14), respectively.

Using C-SVM, the result is an explicit nonlinear function:

$$y = \text{C-SVM}(x_1, x_2, x_3) \tag{10.4}$$

with $C = 512$, $\gamma = 0.03125$, 10 free vectors x_i, and the cross-validation accuracy CVA=92.8571%.

Using BAYSD, the result is a discriminate function:

$$\left. \begin{aligned}
B_1(x) &= \ln(0.071) - 407.034 + 36.474x_1 + 0.722x_2 + 1.769x_3 \\
B_2(x) &= \ln(0.214) - 347.655 + 43.648x_1 - 1.632x_2 + 1.562x_3 \\
B_3(x) &= \ln(0.500) - 322.547 + 44.749x_1 - 2.523x_2 + 1.343x_3 \\
B_4(x) &= \ln(0.214) - 174.608 + 33.896x_1 - 2.705x_2 + 1.056x_3
\end{aligned} \right\} \tag{10.5}$$

TABLE 10.4 Comparisons Among the Applications of Regression Algorithms (R-SVM, BPNN, and MRA) to the Oil Layer Productivity Index of the Keshang Formation

Algorithm	Fitting Formula	Mean Absolute Relative Residual			Dependence of the Predicted Value (y) on Parameters (x_1, x_2, x_3), in Decreasing Order	Time Consumed on PC (Intel Core 2)	Solution Accuracy
		$\overline{R}1(\%)$	$\overline{R}2(\%)$	$\overline{R}^*(\%)$			
R-SVM	Nonlinear, explicit	209.56	867.65	538.61	N/A	3 s	Very low
BPNN	Nonlinear, implicit	17.70	45.07	31.39	N/A	30 s	Very low
MRA	Linear, explicit	81.35	414.84	248.10	x_2, x_3, x_1	<1 s	Very low

TABLE 10.5 Prediction Results from Oil Layer Classification of the Keshang Formation

		Oil Layer Classification						
			Classification Algorithm					
		Oil Test[a]	**C-SVM**		**BAYD**		**MRA**	
Sample Type	Sample No.	y^*	y	$R(\%)$	y	$R(\%)$	y	$R(\%)$
Learning samples	1	3	3	0	3	0	2.913	2.90
	2	3	3	0	3	0	2.621	12.64
	3	4	4	0	4	0	3.931	1.73
	4	2	2	0	2	0	2.197	9.87
	5	4	4	0	4	0	3.934	1.66
	6	3	3	0	3	0	2.882	3.94
	7	3	3	0	3	0	3.079	2.64
	8	3	3	0	3	0	2.846	5.12
	9	3	3	0	3	0	2.946	1.79
	10	3	3	0	3	0	3.160	5.33
	11	2	2	0	2	0	2.452	22.60
	12	4	4	0	4	0	4.100	2.50
	13	1	1	0	1	0	0.880	12.00
	14	2	2	0	2	0	2.059	2.94
Prediction samples	15	2	2	0	2	0	1.852	7.41
	16	3	3	0	3	0	2.880	4.01
	17	1	1	0	1	0	1.238	23.83
	18	4	4	0	4	0	3.876	3.11

[a] $y^* =$ oil-layer-classification (1—high-productivity oil layer, 2—intermediate-productivity oil layer, 3—low-productivity oil layer, 4—dry layer) determined by the oil test (Table 10.2).

From the discriminate process of BAYSD, oil-layer-classification (y) is shown to depend on the three parameters in decreasing order: permeability (x_2), porosity (x_1), and resistivity (x_3). Using MRA, the result is an explicit linear function:

$$y = 4.8716 - 0.060006x_1 - 0.024835x_2 - 0.011092x_3 \tag{10.6}$$

Equation (10.6) yields a residual variance of 0.051881 and a multiple correlation coefficient of 0.97371. From the regression process, oil-layer-classification (y) is shown to depend on the three parameters in decreasing order: permeability (x_2), porosity (x_1), and resistivity (x_3). This order is consistent with the aforementioned one by BAYSD due to the fact that although MRA is a linear algorithm, whereas BAYSD is a nonlinear algorithm, the nonlinearity of the studied problem is moderate, whereas the nonlinearity ability of BAYSD is weak.

Substituting the values of porosity, permeability, and resistivity given by the 14 learning samples (Table 10.1) in Equations (10.4), (10.5) [and then use Equation (6.34)], and (10.6) respectively, the oil-layer-classification (y) of each learning sample is obtained. Table 10.5 shows the results of the learning process by C-SVM, BAYSD, and MRA.

10.1.4.2. Prediction Process

Substituting the values of porosity, permeability, and resistivity given by the four prediction samples (Table 10.1) in Equations (10.4), (10.5) [and then use Equation (6.34)], and (10.6), respectively, the oil-layer-classification (y) of each prediction sample is obtained. Table 10.5 shows the results of the prediction process by C-SVM, BAYSD, and MRA.

10.1.4.3. Application Comparisons Between C-SVM and BAYSD

It can been seen from $\overline{R}^*(\%) = 7.93$ of MRA (Table 10.6) that the nonlinearity of the relationship between the predicted value y and its relative parameters (x_1, x_2, x_3) is moderate from Table 1.2. The results of the two classification algorithms (C-SVM and BAYSD) are the same, i.e., not only the $\overline{R}_1(\%) = 0$, but also $\overline{R}_2(\%) = 0$, and thus the total mean absolute relative residual $\overline{R}^*(\%) = 0$, which coincides with practicality.

10.1.4.4. Brief Summary

Since this case study is a moderate nonlinear problem, C-SVM and BAYSD are applicable.

10.1.5. Summary and Conclusions

In summary, using data for the oil productivity index and oil layer classification in the Keshang Formation based on three parameters, i.e., the three parameters (porosity, permeability, resistivity) and an oil test result of 18 samples, of which 14 are taken as learning samples and four are taken as prediction samples, regression algorithms (R-SVM, BPNN, MRA) are adopted for the oil productivity index, and classification algorithms (C-SVM, BAYSD) are adopted for the oil layer classification. It is found that (a) since this oil productivity index is a

TABLE 10.6   Comparisons Among the Applications of Classification Algorithms (C-SVM and BAYD) to Oil Layer Classification of the Keshang Formation

| Algorithm | Fitting Formula | Mean Absolute Relative Residual | | | Dependence of the Predicted Value (y) on Parameters (x_1, x_2, x_3), in Decreasing Order | Time Consumed on PC (Intel Core 2) | Solution Accuracy |
		$\overline{R}1(\%)$	$\overline{R}2(\%)$	$\overline{R}^*(\%)$			
C-SVM	Nonlinear, explicit	0	0	0	N/A	3 s	Very high
BAYD	Nonlinear, explicit	0	0	0	x_2, x_1, x_3	3 s	Very high
MRA	Linear, explicit	6.26	9.59	7.93	x_2, x_1, x_3	<1 s	Problem's nonlinearity is moderate

very strong nonlinear problem, these regression algorithms are not applicable; and (b) since this oil layer classification is a moderate nonlinear problem, these classification algorithms are applicable.

10.2. TYPICAL CASE STUDY 2: OIL LAYER CLASSIFICATION IN THE LOWER H3 FORMATION

10.2.1. Studied Problem

The objective of this case study is to conduct an oil layer classification in clayey sandstone using eight well-logging data, which has practical value when oil test data are less limited.

Located in the Xia'ermen Oilfield of Henan oil province in central China, the lower H3 Formation is an oil layer with low resistivity lying at depths of 2500–3500 m.

Using data of 32 samples from the lower H3 Formation, with each sample containing eight well-logging data ($x_1 = SP$, $x_2 = R_{xo}1$, $x_3 = R_{xo}2$, $x_4 = \Delta t$, $x_5 = R_a0.45$, $x_6 = R_a4$, $x_7 = Cond$, $x_8 = Cal$) and an oil test result ($y^* = $ reservoir type), Yang et al. (2001) adopted BPNN for the oil layer classification. In this case study, among these 32 samples, 24 are taken as learning samples and eight as prediction samples (Table 10.7).

10.2.2. Known Input Parameters

These input parameters are the values of the known variables x_i ($i = 1, 2, ..., 8$) for 24 learning samples and eight prediction samples and the value of the prediction variable y^* for 24 learning samples (Table 10.7).

TABLE 10.7 Input Data for Oil Layer Classification of the Lower H3 Formation

Sample Type	Sample No.	x_1 SP (mv)	x_2 $R_{xo}1$ ($\Omega\cdot$m)	x_3 $R_{xo}2$ ($\Omega\cdot$m)	x_4 Δt (µs/m)	x_5 $R_a0.45$ ($\Omega\cdot$m)	x_6 R_a4 ($\Omega\cdot$m)	x_7 Cond [10^{-3} /($\Omega\cdot$m)]	x_8 Cal (cm)	Oil Test[b] y^*
					Well Logging[a]					
Learning samples	1	−68.526	1.881	2.453	273.022	8.731	5.419	176.956	23.331	3
	2	−69.165	2.175	2.633	265.584	9.060	5.545	175.115	22.896	3
	3	−69.046	2.333	2.910	257.446	9.447	5.708	173.925	22.415	3
	4	−68.622	2.138	3.206	249.330	9.871	5.870	174.479	22.052	3
	5	−67.953	2.177	2.930	240.054	9.939	6.040	175.085	21.990	3
	6	−67.012	2.229	2.735	231.610	9.772	6.210	177.295	21.984	3
	7	−65.995	2.493	2.831	225.396	9.438	6.408	179.505	22.125	3

(Continued)

TABLE 10.7   Input Data for Oil Layer Classification of the Lower H3 Formation—cont'd

Sample Type	Sample No.	Well Logging[a]								Oil Test[b]
		x_1	x_2	x_3	x_4	x_5	x_6	x_7	x_8	
		SP (mv)	$R_{xo}1$ ($\Omega\cdot$m)	$R_{xo}2$ ($\Omega\cdot$m)	$\triangle t$ (μs/m)	$R_a0.45$ ($\Omega\cdot$m)	R_a4 ($\Omega\cdot$m)	Cond [10^{-3} /($\Omega\cdot$m)]	Cal (cm)	y^*
	8	−64.934	2.287	2.788	228.311	9.211	6.613	180.699	22.258	3
	9	−63.484	1.792	2.552	234.416	9.347	6.797	181.612	22.173	3
	10	−61.957	1.398	1.971	245.015	9.681	6.968	177.258	22.088	3
	11	−60.451	1.481	1.622	255.382	10.415	7.140	169.957	21.935	3
	12	−63.926	1.843	1.808	262.898	11.168	7.314	161.975	21.764	3
	13	−83.106	2.097	2.032	243.210	4.543	3.180	226.574	22.474	2
	14	−82.941	2.250	2.250	245.319	4.554	3.302	223.415	22.339	2
	15	−82.776	1.719	1.602	248.132	4.561	3.433	222.170	22.226	2
	16	−82.631	1.014	1.071	250.823	4.561	3.566	222.200	22.040	2
	17	−82.503	0.875	0.906	253.298	4.558	3.708	222.488	21.760	2
	18	−82.369	0.881	0.956	255.029	4.523	3.851	222.776	21.498	2
	19	−82.138	0.824	0.958	255.288	4.489	4.036	223.063	21.313	2
	20	−81.908	0.759	0.907	255.547	4.420	4.222	224.067	21.141	2
	21	−81.380	1.240	0.740	256.740	4.020	4.850	244.530	21.990	2
	22	−82.680	0.760	1.070	257.500	4.040	3.030	224.910	19.470	1
	23	−82.540	0.810	1.080	255.960	4.150	3.050	212.910	19.350	1
	24	−82.450	0.870	1.030	254.440	4.200	3.100	203.440	19.240	1
Prediction samples	25	−67.550	1.960	2.210	270.410	11.840	7.490	154.600	21.590	(3)
	26	−70.070	1.870	2.520	276.510	12.750	7.660	148.750	21.420	(3)
	27	−71.740	1.890	2.500	282.210	13.310	7.890	144.470	21.310	(3)
	28	−72.020	1.920	2.470	287.920	13.710	8.190	142.240	21.310	(3)
	29	−81.330	1.010	0.770	258.410	3.980	4.910	248.400	22.180	(2)
	30	−81.330	1.130	0.980	258.980	3.940	4.980	253.430	22.280	(2)
	31	−81.480	1.570	1.310	258.780	3.900	5.040	255.970	21.900	(2)
	32	−82.370	0.880	1.010	253.300	4.260	3.160	195.170	19.280	(1)

[a] x_1 = self-potential (SP); x_2 = micronormal ($R_{xo}1$); x_3 = microinverse ($R_{xo}2$); x_4 = acoustictime ($\triangle t$); x_5 = 0.45 m apparent resistivity ($R_a0.45$); x_6 = 4 m apparent resistivity (R_a4); x_7 = conductivity (Cond); and x_8 = caliper (Cal).
[b] y^* = oil layer classification (1—oil layer, 2—water layer, 3—dry layer) determined by the oil test. The numbers in parentheses are not input data but are used for calculating R(%).

10.2.3. Classification Calculation

C-SVM and BAYSD are adopted.

10.2.3.1. Learning Process

Using the 24 learning samples (Table 10.7) and by C-SVM, BAYSD, and MRA, the three functions of oil-layer-classification (y) with respect to eight parameters ($x_1, x_2, ..., x_8$) have been constructed, i.e., Equation (10.7) corresponding to SVM formula (4.1), Equation (10.8) corresponding to BAYSD discriminate function (6.30), and Equation (10.9) corresponding to MRA formula (2.14), respectively.

Using C-SVM, the result is an explicit nonlinear function:

$$y = C\text{-}SVM(x_1, x_2, ..., x_8) \tag{10.7}$$

with $C = 32$, $\gamma = 0.03125$, eight free vectors x_i, and CVA = 100%.

Using BAYSD, the result is a discriminate function:

$$
\left.
\begin{aligned}
B_1(x) &= \ln(0.125) - 22558.460 - 243.408x_1 - 1296.484x_2 + 734.070x_3 - 19.501x_4 \\
&\quad + 515.456x_5 + 1218.567x_6 - 13.246x_7 + 1408.733x_8 \\
B_2(x) &= \ln(0.375) - 26792.570 - 263.680x_1 - 1422.524x_2 + 810.842x_3 - 21.583x_4 \\
&\quad + 556.603x_5 + 1339.537x_6 - 15.380x_7 + 1554.840x_8 \\
B_3(x) &= \ln(0.500) - 30575.700 - 271.657x_1 - 1573.476x_2 + 984.037x_3 - 21.867x_4 \\
&\quad + 593.776x_5 + 1476.612x_6 - 18.647x_7 + 1685.443x_8
\end{aligned}
\right\} \tag{10.8}
$$

From the discriminate process of BAYSD, oil layer classification (y) is shown to depend on the eight well-logging data in decreasing order: 0.45 m apparent resistivity (x_5), caliper (x_8), 4 m apparent resistivity (x_6), self-potential (x_1), microinverse (x_3), micronormal (x_2), acoustic-time (x_4), and conductivity (x_7).

Using MRA, the result is an explicit linear function:

$$
\begin{aligned}
y &= -8.549 - 0.0245925x_1 - 0.31546x_2 + 0.21615x_3 - 0.0045599x_4 \\
&\quad + 0.099409x_5 + 0.30047x_6 - 0.0055501x_7 + 0.35258x_8
\end{aligned} \tag{10.9}
$$

Equation (10.9) yields a residual variance of 0.0044158 and a multiple correlation coefficient of 0.99779. From the regression process, oil layer classification (y) is shown to depend on the eight well-logging data in decreasing order: 0.45 m apparent resistivity (x_5), caliper (x_8), 4 m apparent resistivity (x_6), self-potential (x_1), acoustictime (x_4), micronormal (x_2), microinverse (x_3), and conductivity (x_7). This order slightly differs from the aforementioned one by BAYSD in the last four parameters because MRA is a linear algorithm, whereas BAYSD is a nonlinear algorithm, and the nonlinearity of the studied problem is moderate.

Substituting the values of self-potential, micronormal, microinverse, acoustictime, 0.45 m apparent resistivity, 4 m apparent resistivity, conductivity, and caliper given by the 24 learning samples (Table 10.7) in Equations (10.7), (10.8) [and then use Equation (6.34)], and (10.9), respectively, the oil layer classification (y) of each learning sample is obtained. Table 10.8 shows the results of learning process by C-SVM, BAYSD, and MRA.

TABLE 10.8   Prediction Results from Oil Layer Classification of the Lower H3 Formation

Sample Type	Sample No.	Oil Test[a]	C-SVM		BAYD		MRA	
		y^*	y	$R(\%)$	y	$R(\%)$	y	$R(\%)$
Learning samples	1	3	3	0	3	0	3.030	1.00
	2	3	3	0	3	0	2.967	1.10
	3	3	3	0	3	0	2.933	2.23
	4	3	3	0	3	0	3.036	1.19
	5	3	3	0	3	0	3.008	0.27
	6	3	3	0	3	0	2.965	1.17
	7	3	3	0	3	0	2.948	1.74
	8	3	3	0	3	0	3.021	0.69
	9	3	3	0	3	0	3.065	2.17
	10	3	3	0	3	0	3.024	0.81
	11	3	3	0	3	0	2.917	2.75
	12	3	3	0	3	0	3.080	2.66
	13	2	2	0	2	0	2.010	0.49
	14	2	2	0	2	0	1.999	0.04
	15	2	2	0	2	0	2.013	0.67
	16	2	2	0	2	0	2.076	3.81
	17	2	2	0	2	0	2.009	0.47
	18	2	2	0	2	0	1.950	2.51
	19	2	2	0	2	0	1.942	2.91
	20	2	2	0	2	0	1.922	3.89
	21	2	2	0	2	0	2.039	1.97
	22	1	1	0	1	0	0.994	0.60
	23	1	1	0	1	0	1.022	2.22
	24	1	1	0	1	0	1.029	2.90
Prediction samples	25	(3)	3	0	3	0	3.361	12.04
	26	(3)	3	0	3	0	3.659	21.96
	27	(3)	3	0	3	0	3.809	26.95
	28	(3)	3	0	3	0	3.922	30.72

(Continued)

TABLE 10.8   Prediction Results from Oil Layer Classification of the Lower H3 Formation—cont'd

| | | Oil Layer Classification | | | | | | |
| | | Oil Test[a] | C-SVM | | BAYD | | MRA | |
Sample Type	Sample No.	y^*	y	$R(\%)$	y	$R(\%)$	y	$R(\%)$
	29	(2)	2	0	2	0	2.168	8.41
	30	(2)	2	0	2	0	2.198	9.87
	31	(2)	2	0	2	0	2.004	0.19
	32	(1)	1	0	1	0	1.107	10.70

[a]y^* = oil layer classification (1—oil layer, 2—water layer, 3—dry layer) determined by the oil test.

10.2.3.2. Prediction Process

Substituting the values of self-potential, micronormal, microinverse, acoustictime, 0.45 m apparent resistivity, 4 m apparent resistivity, conductivity, and caliper given by the eight prediction samples (Table 10.7) in Equations (10.7), (10.8) [and then use Equation (6.34)], and (10.9), respectively, the oil layer classification (y) of each prediction sample is obtained. Table 10.8 shows the results of the prediction process by C-SVM, BAYSD, and MRA.

10.2.3.3. Comparisons Between the Applications of C-SVM and BAYSD

It can been seen from $\overline{R}^*(\%) = 8.40$ of MRA (Table 10.9) that the nonlinearity of the relationship between the predicted value y and its relative parameters ($x_1, x_2, ..., x_8$) is moderate from Table 1.2. The results of the two classification algorithms (C-SVM and BAYSD) are the same, i.e., not only the $\overline{R}_1(\%) = 0$, but also $\overline{R}_2(\%) = 0$, and thus $\overline{R}^*(\%) = 0$, which coincides with practicality.

TABLE 10.9   Comparisons Among the Applications of Classification Algorithms (C-SVM and BAYD) to Oil Layer Classification of the Lower H3 Formation

| Algorithm | Fitting Formula | Mean Absolute Relative Residual | | | Dependence of the Predicted Value (y) on Parameters ($x_1, x_2, ..., x_8$), in Decreasing Order | Time Consumed on PC (Intel Core 2) | Solution Accuracy |
		$\overline{R}1(\%)$	$\overline{R}2(\%)$	$\overline{R}^*(\%)$			
C-SVM	Nonlinear, explicit	0	0	0	N/A	3 s	Very high
BAYD	Nonlinear, explicit	0	0	0	$x_5, x_8, x_6, x_1, x_3, x_2,$ x_4, x_7	3 s	Very high
MRA	Linear, explicit	1.68	15.11	8.40	$x_5, x_8, x_6, x_1, x_4, x_2,$ x_3, x_7	<1 s	Problem's nonlinearity is moderate

10.2.4. Summary and Conclusions

In summary, using data for the oil layer classification in the lower H3 Formation based on eight parameters, i.e., the eight parameters (elf-potential, micronormal, microinverse, acoustictime, 0.45 m apparent resistivity, 4 m apparent resistivity, conductivity, caliper given) and an oil test result of 32 samples, of which 24 are taken as learning samples and eight are taken as prediction samples, classification algorithms (C-SVM, BAYSD) are adopted for the oil layer classification. It is found that since this oil layer classification is a moderate nonlinear problem, these classification algorithms are applicable.

10.3. TYPICAL CASE STUDY 3: OIL LAYER CLASSIFICATION IN THE XIEFENGQIAO ANTICLINE

In Sections 4.3 and 6.4, the data in Table 4.4 were used, and C-SVM, MRA, and BAYSD were adopted to construct the function of oil layer classification (y) with respect to five relative parameters (x_1, x_2, x_3, x_4, x_5) in the Xiefengqiao Anticlinal, and the calculation results were filled into Tables 4.5, 4.6, 6.10, and 6.11. Now Table 4.4 is used as input data for Typical Case Study 3 of this chapter. For the sake of algorithm comparisons, Tables 4.5 and 6.10 are accordingly copied into Table 10.10, and Tables 4.6 and 6.11 are accordingly copied into Table 10.11.

It can been seen from $\overline{R}^*(\%) = 19.33$ of MRA (Table 10.11) that the nonlinearity of the relationship between the predicted value y and its relative parameters (x_1, x_2, x_3) is strong from Table 1.2. Only C-SVM is applicable, i.e., not only $\overline{R}_1(\%) = 0$, but also $\overline{R}_2(\%) = 0$, and thus $\overline{R}^*(\%) = 0$, which coincide with practicality; as for BAYSD and MRA, the solution accuracy is low, and $\overline{R}_2(\%) > \overline{R}_1(\%)$.

In summary, using data for the oil layer classification in the Xiefengqiao Anticlinal based on five parameters, i.e., the five parameters (true resistivity, acoustictime, porosity, oil saturation, permeability) and an oil test result of 27 samples, of which 24 are taken as learning samples and three are taken as prediction samples, classification algorithms (C-SVM, BAYSD) are adopted for the oil layer classification. It is found that since this oil layer classification is a strong nonlinear problem, only C-SVM is applicable, but BAYSD is not applicable.

10.4. A PRACTICAL SYSTEM OF DATA MINING AND KNOWLEDGE DISCOVERY FOR GEOSCIENCES

The data of geosciences discussed in this book are limited in the subsurface of the earth excluding the air, and further, the data discussed are minable subsurface data. Therefore, Section 1.1.4 discusses major issues in data mining for geosciences; Section 1.2 briefly introduces data systems usable by data mining, including database, data warehouse, and data bank. Practically speaking, the usable data also include file systems and any dataset organized in any form. Section 1.4 briefly introduces the idea of a *data mining system*. Aiming at major issues in data mining for geosciences, this chapter presents a practical *software system* of data mining and knowledge discovery for geosciences (Figures 10.1 and 10.2).

TABLE 10.10 Prediction Results from Oil Layer Classification of the Xiefengqiao Anticlinal

Sample Type	Sample No.	Well No.	Layer No.	Oil Test[a] y^*	C-SVM y	C-SVM $R(\%)$	BAYD y	BAYD $R(\%)$	MRA y	MRA $R(\%)$
Learning samples	1	ES4	5	2	2	0	2	0	2.194	9.67
	2		6	3	3	0	3	0	3.242	8.05
	3		7	2	2	0	2	0	2.189	9.45
	4		8	3	3	0	3	0	3.251	8.36
	5		9	1	1	0	1	0	0.909	9.14
	6		10	2	2	0	2	0	1.727	13.65
	7		11	2	2	0	2	0	2.097	4.83
	8		12	3	3	0	3	0	2.578	14.06
	9		13	1	1	0	1	0	1.276	27.60
	10	ES6	4	2	2	0	2	0	1.908	4.60
	11		5	3	3	0	3	0	2.479	17.38
	12		6	3	3	0	2	33.33	2.299	23.37
	13		8	3	3	0	3	0	3.580	19.33
	14	ES8	5_1	2	2	0	2	0	1.742	12.91
	15		5_2	3	3	0	2	33.33	2.542	15.28
	16		5_3	2	2	0	2	0	1.904	4.80
	17		6	3	3	0	3	0	2.851	4.96
	18		11	2	2	0	2	0	1.936	3.22
	19		12_1	1	1	0	1	0	1.371	37.08
	20		12_2	1	1	0	1	0	0.905	9.51
	21		12_3	1	1	0	1	0	1.113	11.29
	22		12_4	1	1	0	2	100	1.494	49.40
	23		13_1	1	1	0	1	0	0.883	11.73
	24		13_2	1	1	0	2	100	1.534	53.43
Prediction samples	25	ES8	13_4	1	1	0	2	100	1.511	51.14
	26		13_5	2	2	0	2	0	1.681	15.93
	27		7_2	3	3	0	3	0	2.969	1.03

[a]y^* = oil layer classification (1—oil layer, 2—poor oil layer, 3—dry layer) determined by the oil test.

TABLE 10.11   Comparisons Among the Applications of Classification Algorithms (C-SVM and BAYD) to Oil Layer Classification of the Xiefengqiao Anticlinal

Algorithm	Fitting Formula	Mean Absolute Relative Residual			Dependence of the Predicted Value (y) on Parameters (x_1, x_2, x_3, x_4, x_5), in Decreasing Order	Time Consumed on PC (Intel Core 2)	Integrated Evaluation
		$\overline{R}1(\%)$	$\overline{R}2(\%)$	$\overline{R}^*(\%)$			
C-SVM	Nonlinear, explicit	0	0	0	N/A	3 s	Very high
BAYD	Nonlinear, explicit	11.11	33.33	22.22	x_1, x_2, x_3, x_5, x_4	3 s	Very low
MRA	Linear, explicit	15.96	22.70	19.33	x_1, x_2, x_3, x_5, x_4	<1 s	Problem's nonlinearity is strong

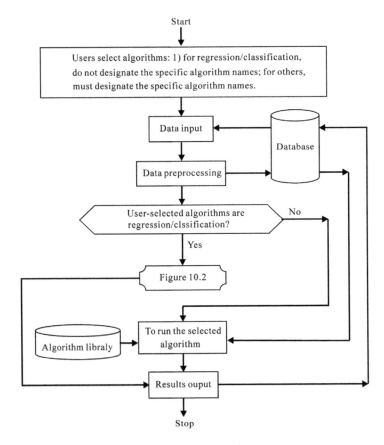

FIGURE 10.1   Logical chart of a practical data mining system for geosciences.

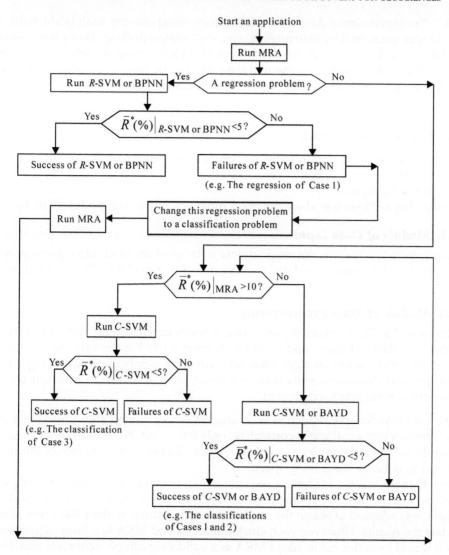

FIGURE 10.2   Automatic selection and calculation of regression and classification algorithms. [$i\,\overline{R}^*(\%)$, total mean absolute relative residual, calculated with Equation (1.6).]

10.4.1. Minable Subsurface Data

The subsurface data can be divided into two classes:

1. Class 1, the field-measured data, which are the interpretation results of the data indirectly measured by instruments at the Earth's surface or the air. Though the precision of these data is not high, they are of large amount and 3-D distributing, and thus they can be the basic data for 3-D data mining.

2. Class 2, the experimental data from a laboratory, which are the analytical results of the data directly measured by instruments, e.g., exploration drilling. Though the amount of this data is small and they are in a narrow view, their precision is high, and thus they can be used to calibrate the field-measured data and can be used as constraint conditions of data mining results. Both classes of subsurface data are minable.

10.4.2. System Components

The practical system of data mining for geosciences consists of five modules as follows: data input, data preprocessing, algorithm selection (Figure 10.2), running the selected algorithms, and results output.

The following sections introduce the main functions of the five modules one by one.

10.4.2.1. Module of Data Input

For a practical application, the original data to be used are read into a given project in a database, and in this project the data are reorganized to a readable file for algorithms to be adopted.

10.4.2.2. Module of Data Preprocessing

For the readable file mentioned previously, it needs data cleaning, data integration, data transformation, data reduction, and so on (see Section 1.4.3). For the data used by the regression algorithm or classification algorithm, data reduction is only discussed in regard to sample reduction and dimension reduction. In general, sample reduction is conducted at first, then dimension reduction is conducted.

1. *Sample reduction.* Sample reduction is conducted before mining algorithms start. Among those samples that are linearly correlative, only one is left, but others are deleted so as to reduce the number of samples as much as possible. Its benefits are to reduce the amount of data and to enhance the mining speed.

   Here Q-mode cluster analysis is recommended, to serve as a pioneering sample-reduction tool. For instance, in Case Studies 1, 2, 3, and 4 of Chapter 7, first Q-mode cluster analysis was adopted to obtain the results of sample reduction, then MRA was adopted to validate the results. However, each cluster analysis and MRA is a linear algorithm, and thus it would be better not to adopt MRA as a validating sample-reduction tool. From the application comparisons among algorithms in Chapters 3–6 and this chapter, it is found that (a) in regression algorithms, the application results of BPNN are best, and (b) in classification algorithms, the application results of C-SVM are best. Therefore, for regression problems, it is preferable to adopt BPNN as a validating sample-reduction tool, and for classification problems, it is preferable to adopt C-SVM as a validating sample-reduction tool.

2. *Dimension reduction.* Dimension reduction is conducted before mining algorithms start. It tries to reduce the number of x_i (independent variables or attributes) as much as possible. Its benefits are to reduce the amount of data that can enhance the mining speed, to reduce the attribute (variable) that can extend applying ranges, and to reduce misclassification ratio of prediction samples that can enhance mining quality.

Here MRA, BAYSD, and R-mode cluster analysis are recommended to serve as pioneering dimension-reduction tools, since these three algorithms can indicate the dependence of the predicted value (y) on parameters ($x_1, x_2, ..., x_m$).

a. *Adopt MRA as a pioneering dimension-reduction tool.* For instance, in a case study in Section 2.2.4, MRA can reduce an original 6-D problem ($x_1, x_2, x_3, x_4, x_5, y$) to a 4-D problem ($x_2, x_3, x_5, y$), which is validated by MRA itself. In a case study in Section 4.4.3, MRA also can reduce an original 8-D problem ($x_1, x_2, x_3, x_4, x_5, x_6, x_7, y$) to a 4-D problem ($x_2, x_4, x_5, y$), which is validated by C-SVM and BPNN.

b. *Adopt BAYSD as a pioneering dimension-reduction tool.* Though there are no calculations of dimension reduction in six case studies of Chapter 6, each case study gives the dependence of the predicted value (y) on parameters ($x_1, x_2, ..., x_m$) calculated by BAYSD and MRA, indicating that BAYSD also can serve as a pioneering dimension-reduction tool like MRA. Some dependences calculated by BAYSD and MRA are the same, e.g., Case Study 1 (Table 6.6), Case Study 3 (Table 6.11), and Case Study 4 (Table 6.13); other dependences are different, e.g., Case Study 2 (Table 6.9), Case Study 5 (Table 6.16), and Case Study 6 (Table 6.18). These differences stem from the fact that MRA is a linear algorithm, whereas BAYSD is a nonlinear algorithm. Therefore, the dependence obtained by BAYSD is relatively more reliable than that by MRA, and it is preferable to adopt BAYSD as a pioneering dimension-reduction tool.

c. *Adopt R-mode cluster analysis as a pioneering dimension-reduction tool.* Case Studies 1, 2, and 4 in Chapter 7 show that in R-mode cluster analysis, choosing an analog coefficient cannot be used for dimension reduction, whereas Case Study 3 shows that in R-mode cluster analysis, choosing a correlation coefficient can be used for dimension reduction. Therefore, when R-mode cluster analysis serves as a pioneering dimension-reduction tool, a correlation coefficient should be chosen.

Since R-mode cluster analysis cannot one by one indicate the dependence of the predicted value (y) on parameters ($x_1, x_2, ..., x_m$), it cannot be recommended. Between the remaining MRA and BAYSD, it is preferable to select nonlinear BAYSD.

Like the foregoing sample reduction, dimension reduction also should be validated. For regression problems, it is preferable to adopt BPNN to validate; for classification problems, it is preferable to adopt C-SVM to validate.

10.4.2.3. *Module of Algorithm Selection*

This book introduces 22 algorithms that can be summarized by six types: probability and statistics in Chapter 2; regression algorithms, including MRA, in Chapter 2, BPNN in Chapter 3, and R-SVM in Chapter 4; classification algorithms, including C-SVM, in Chapter 4, decision trees in Chapter 5, and Bayesian classification in Chapter 6; cluster analysis in Chapter 7; Kriging in Chapter 8; and other soft computing algorithms for geosciences in Chapter 9. The programs of these algorithms are stored in an algorithm library (Figure 10.1).

Besides regression algorithms and classification algorithms, each algorithm can be directly chosen by users. Moreover, regression algorithms and classification algorithms in applications can be chosen by automatic selection (Figure 10.2). In Figure 10.2, since the use of decision trees is quite complicated, decision trees are not chosen; and since BAYSD is superior to NBAY and BAYD in Bayesian classification, NBAY and BAYD are not chosen.

Hence, in the automatic selection, only R-SVM and BPNN are taken as regression algorithms, and only C-SVM and BAYSD are taken as classification algorithms. Since $\overline{R}^*(\%)$ of MRA for a studied problem expresses the nonlinearity of this problem (Table 1.2), this $\overline{R}^*(\%)$ is used for the selection of a classification algorithm, i.e., if >10, only C-SVM is taken; otherwise, C-SVM or BAYSD is taken (Figure 10.2).

10.4.2.4. Module of Running the Selected Algorithms

In database (Figure 10.1), Each practical application has its own corresponding project. To conduct a practical application, an algorithm selected from an algorithm library (Figure 10.1) points to its corresponding project, and then the algorithm runs in the project.

10.4.2.5. Module of Results Output

The results of the algorithm are also stored in the project. There are two kinds of results: (1) the results for the practical application, and (2) the knowledge expressed with a formula that can be spread to other areas, composing a knowledge library in a database.

EXERCISES

10-1. Section 10.1.3 (regression calculation) in Section 10.1 (Typical Case Study 1: oil layer classification in the Keshang Formation) introduces a calculation of oil productivity index. Why did this calculation fail?

10-2. Section 10.1.4 (classification calculation) in Section 10.1 (Typical Case Study 1: oil layer classification in the Keshang Formation) introduces a calculation of oil layer classification. Why did this calculation succeed?

10-3. From the answers to Questions 10-1 and 10-2, what conclusions do you draw?

10-4. Section 10.2 (Typical Case Study 2: oil layer classification in the lower H3 Formation) introduces a calculation of oil layer classification. Why did this calculation succeed?

10-5. From the answer to Question 10-4, what conclusions do you draw?

10-6. Section 10.3 (Typical Case Study 3: oil layer classification in the Xiefengqiao Anticline) introduces a calculation of oil layer classification. Why did this calculation succeed?

10-7. From the answer to Question 10-6, what conclusions do you draw?

10-8. What are two classes of minable subsurface data?

10-9. Why are not decision trees, NBAY, and BAYD included in Figure 10.2?

10-10. What algorithms can serve as pioneering sample-reduction tools? What algorithms can serve as pioneering dimension-reduction tools? What algorithms can serve as their better validating tools?

References

Tan, C., Ma, N., Su, C., 2004. Model and method for oil and gas productivity prediction of reservoir. J. Earth Sci. Env. 26 (2), 42–46 (in Chinese with English abstract).

Yang, Q., Deng, C., Yang, Y., Lu, G., 2001. Application of nerve network on oil-bearing formation with low resistivity. Special Oil Gas Reserv. 8 (2), 8–10 (in Chinese with English abstract).

Table of Unit Conversion*

Ordinal Numbers	Unit Symbol	Explanation	Physical Meaning	Conversion
1	a	year	Time	$1\ a = 365\ d$
2	atm	standard atmosphere	Pressure	$1\ atm = 1.01 \times 10^6\ dyn/cm^2$ $= 14.7\ psi$ $= 14.7\ lbf/in^2$ $1\ atm = 1.01 \times 10^5\ kg/(s^2 \cdot m)$ $= 1.01 \times 10^5\ Pa$ $\approx 10^5\ Pa = 0.1\ MPa$
3	bbl	American barrel	Volume	$1\ bbl = 5.6146\ ft^3$ $= 0.158988\ m^3$
4	bbl/d	American barrel/day	Flow rate	
5	bbl/(d·psi)	American barrel/(day·pound/inch^2)	Transmissibility	
6	°C	centigrade degree	Temperature	$°C = K - 273.15 = \frac{5}{9}(°F - 32)$
7	°C^{-1}	1/centigrade degree	Thermal expansion coefficient	
8	°C/100m	centigrade degree/100 meter	Geothermal gradient	
9	cal	calorie	Heat	$1\ cal = 4.1868\ J$
10	cal/(g·°C)	calorie/(gram·centigrade degree)	Specific heat capacity	
11	cm	centimeter	Length	$1\ cm = 10^{-2}\ m$
12	cm/s	centimeter/second	Velocity	
13	cm/s^2	centimeter/second^2	Acceleration	
14	cm^2	centimeter^2	Area	

(Continued)

*In alphabetical order

Data Mining and Knowledge Discovery for Geoscientists
http://dx.doi.org/10.1016/B978-0-12-410437-2.15001-5

Continued

Ordinal Numbers	Unit Symbol	Explanation	Physical Meaning	Conversion
15	cm^2/s	centimeter^2/second	Thermal diffusibility	
16	cm^3	centimeter^3	Volume	
17	cP	centipoise	Viscosity	$1\ cP = 1\ mPa\cdot s$
18	D	Darcy	Permeability	$1\ D = 9.87 \times 10^{-13}\ m^2$ $\approx 10^{-12}\ m^2$ $= 1\ \mu m^2$ $= 10^{-6}\ mm^2$
19	d	day	Time	
20	dyn	dyne	Force	$1\ dyn = 1\ g\cdot cm/s^2$ $= 10^{-5}\ N$
21	dyn/cm3	dyne/centimeter3	Stress gradient	
22	°F	Fahrenheit degree	Temperature	$°F = \frac{9}{5}°C + 32$
23	ft	foot	Length	$1\ ft = 0.3048\ m$
24	ft^2	foot^2	Area	$1\ ft^2 = 0.0929\ m^2$
25	ft^3	foot^3	Volume	$1\ ft^3 = 0.02832\ m^3$
26	g	gram	Mass	$1\ g = 10^{-3}\ kg$
27	g/cm^3	gram/centimeter^3	Density	$1\ g/cm^3 = 1\ t/m^3$
28	g/g(TOC)	gram(HC)/gram(TOC)	Hydrocarbon content in TOC	
29	[g/g(TOC)]/Ma	[gram(HC)/ gram(TOC)]/ million-year	Hydrocarbon generation velocity	
30	HFU	heatflow unit	Heat flow	$1\ HFU = 1\ \mu cal/(cm^2\cdot s)$ $= 41.868\ mW/m^2$
31	in	inch	Length	$1\ in = 2.54\ cm$
32	in^2	inch^2	Area	
33	in^3	inch^3	Volume	
34	J	Joule	Heat, power	$1\ J = (4.1868)^{-1}\ cal,$ $1\ J = 1\ N\cdot m = 1\ kg\cdot m^2/s^2$
35	J/kg	Joule/kilogram	Fluid potential	$1\ J/kg = 1\ m^2/s^2$
36	K	Kelvin	Thermodynamics temperature	$K = °C + 273.15$
37	kcal	kilocalorie	Heat	$1\ kcal = 10^3\ cal$
38	kcal/mol	kilocalorie/mole	Activation energy	$1\ kcal/mol = 10^3\ cal/mol$
39	kg	kilogram	Mass	$1\ kg = 10^3\ g$
40	kg/(s^2·m)	kilogram/ (second^2·meter)	Pressure	$1\ kg/(s^2\cdot m) = 1\ Pa$ $= (1.01 \times 10^5)^{-1}\ atm$
41	kg/t(TOC)	kilogram(HC)/ ton(TOC)	Oil-generation rate	
42	kJ	kilojoule	Heat	$1\ kJ = (4.1868)^{-1} \times kcal$

(Continued)

Continued

Ordinal Numbers	Unit Symbol	Explanation	Physical Meaning	Conversion
43	km	kilometer	Length	$1 \text{ km} = 10^3 \text{ m}$
44	km^2	kilometer^2	Area	$1 \text{ km}^2 = 10^6 \text{ m}^2$
45	km^3	kilometer^3	Volume	$1 \text{ km}^3 = 10^9 \text{ m}^3$
46	1bf	pound force	Force	
47	liter	liter	Volume	$1 \text{ liter} = 1000 \text{ cm}^3$
48	m	meter	Length	
49	Ma	million-year	Geological time, burial time	$1 \text{ Ma} = 10^6 \text{ a}$
50	Ma^{-1}	1/million-year	Reaction velocity, frequency factor	
51	m/Ma	meter/million-year	Depositional velocity	
52	mD	millidarcy	Permeability	$1 \text{ mD} = 10^{-3} \text{ D}$ $= 9.87 \times 10^{-16} \text{ m}^2$ $\approx 10^{-15} \text{ m}^2$ $= 10^{-3} \text{ μm}^2$
53	mm	millimeter	Length	$1 \text{ mm} = 10^{-1} \text{ cm}$ $= 10^{-3} \text{ m}$
54	mm^2	millimeter^2	Area, permeability	$1 \text{ mm}^2 = 10^6 \text{ μm}^2$ $\approx 10^6 \text{ D}$
55	Unit symbol	Explanation	Physical meaning	Conversion
56	m^2	meter^2	Area, permeability	$1 \text{ m}^2 = 10^6 \text{ mm}^2 = 10^{12} \text{ μm}^2$ $= 9.87^{-1} \times 10^{13} \text{ D}$ $\approx 10^{12} \text{ D}$
57	m^2/m^3	meter^2/meter^3	Specific surface area	
58	m^2/s^2	meter^2/second^2	Fluid potential	$1 \text{ m}^2/\text{s}^2 = 1 \text{ J/kg}$
59	m^3	meter^3	Volume	$1 \text{ m}^3 = 6.28979 \text{ bbl}$ $= 35.3147 \text{ ft}^3$
60	m^3/s	meter^3/second	Flow rate	
61	m^3/t	meter^3/ton	Oil-gas conversion equivalent	
62	m^3/t(TOC)	meter^3(gas)/ton(TOC)	Gas generation rate	
63	mol	mole	Mass	
64	mol/liter	mole/liter	Concentration of metal ion	
65	MPa	million pascal	Pressure	$1 \text{ MPa} = 10^6 \text{ Pa} \approx 10 \text{ atm}$
66	mPa	millipascal	Pressure	$1 \text{ mPa} = 10^{-3} \text{ Pa}$
67	mPa·s	millipascal·second	Viscosity	$1 \text{ mPa·s} = 1 \text{ cP}$
68	ms/m	millisiemens/meter	Conductivity	$1 \text{ ms/m} = 10^{-3}/(\Omega \cdot \text{m})$
69	Mt	million-ton	Mass	$1 \text{ Mt} = 10^6 \text{ t} = 10^9 \text{ kg}$
70	mv	millivolt	Voltage	
71	mW	milliwatt	Power	$1 \text{ mW} = 10^{-3} \text{ W}$

(Continued)

Continued

Ordinal Numbers	Unit Symbol	Explanation	Physical Meaning	Conversion
72	mW/m^2	milliwatt/meter^2	Heat flow	$1\ mW/m^2 = 41.868^{-1}\ HFU$
73	$mW/(m \cdot K)$	milliwatt/ (meter·Kelvin)	Thermal conductivity	$1\ mW/(m \cdot K) =$ $41.868^{-1}\ \mu cal/(s \cdot cm \cdot {}^\circ C)$
74	N	Newton	Force	$1\ N = 1\ kg \cdot m/s^2$ $= 10^5 dyn$
75	nm	nanometer (millimicron)	Length	$1\ nm = 10^{-9}\ m$
76	Pa	Pascal	Pressure	$1\ Pa = 1\ N/m^2$ $= 1\ kg/(s^2 \cdot m)$
77	psi	pound/inch^2	Pressure	$1\ psi = 1\ lbf/in^2$ $= 14.7^{-1}\ atm$
78	psi^{-1}	(pound/inch^2)$^{-1}$	Rock compressibility	
79	psi/ft	pound/(inch^2·foot)	Specific density of fluid	
80	s	second	Time	
81	t	ton	Mass	$1\ t = 10^3\ kg = 10^6\ g$
82	t/m^3	ton/meter^3	Density	$1\ t/m^3 = 1\ g/cm^3$
83	$t/(MPa \cdot m \cdot d)$	ton/(million pascal·meter·day)	Oil productivity index	
84	W	Watt	Power	
85	μcal	microcalorie	Heat	$1\ \mu cal = 10^{-6}\ cal$
86	$\mu cal/(cm^2 \cdot s)$	microcalorie/ (centimeter^2·second)	Heat flow	$1\ \mu cal/(cm^2 \cdot s) = 1\ HFU$ $= 41.868\ mW/m^2$
87	$\mu cal/(cm \cdot s \cdot {}^\circ C)$	microcalorie/ (centimeter·second· centigrade degree)	Thermal conductivity	$1\ \mu cal/(cm \cdot s \cdot {}^\circ C) =$ $0.41868\ mW/(m \cdot K)$
88	μm	micrometer	Length	$1\ \mu m = 10^{-6}\ m$
89	μm^2	micrometer^2	Area, permeability	$1\ \mu m^2 = 10^{-12}\ m^2$ $\approx 1\ D$
90	$\mu m/a$	micrometer/year	Migration velocity	
91	$\mu s/m$	microsecond/meter	Acoustictime	
92	$\Omega \cdot m$	ohm·meter	Resistivity	
93	$10^4 t$	10^4 ton	Annual oil production	$10^4 t$
94	$10^4 t/km^2$	10^4 ton/kilometer^2	Oil generation concentration, oil expulsion concentration	$10^4 t/km^2$
95	$10^8 m^3/km^2$	10^8 meter^3/kilometer^2	Gas generation concentration, gas expulsion concentration, proved reserve concentration	$10^8 m^3/km^2$

Note: Gravitational acceleration, $g \approx 981\ cm/s^2$; gas constant, $R \approx 1.986\ cal/(mol \cdot K)$; TOC is an abbreviation of *total organic carbon*; HC is an abbreviation of *hydrocarbon*.

2

Answers to Exercises

ANSWERS TO CHAPTER 1 EXERCISES

1-1. With over 20 years of development, data mining has applied to some scientific and technological as well as business fields, but its application in geosciences is still in the initial stage. Why?
Answer: This is attributed to the particularity that geoscience is different from other fields, with miscellaneous data types, huge quantities, different measuring precision, and many uncertainties as to data mining results.

1-2. The attributes, variables, and parameters mentioned in this book are the same terminology. But which area, such as datalogy, mathematics, or applications, do the three terminologies fall into?
Answer: *Attribute* is the terminology for datalogy, *variable* for mathematics, and *parameter* for applications.

1-3. Data are divided into two types: continuous and discrete in datalogy, whereas real type and integer type in software. The question is, are continuous data equal to real data and discrete data equal to integer data?
Answer: Yes.

1-4. What are the five most popular types of knowledge discovered by data mining?
Answer: Generalization, association, classification and clustering, prediction, and deviation.

1-5. What are the five problems that DM is facing compared with the data mining in other fields?
Answer: There are five special problems: (1) encrypted local area networks should be used due to expensive data; (2) data cleaning should follow specific physical and chemical rules since data come from field measuring and laboratory experimentation; (3) the space DM technique and visualization technique should be used from point to line, from line to plane, and from plane to volume because the study object is under ground; (4) the corresponding nonlinear DM algorithm should be used since most correlations between the factors underground are nonlinear, with different strengths; and (5) the knowledge obtained from data mining is of probability risk due to many uncertainties underground.

1-6. What data systems are usable by data mining?

Answer: In general, the data system usable by data mining includes database, data warehouse, and data bank as well as file system or any dataset with another organization.

1-7. The absolute relative residual $R(\%)$ is defined in Equation (1.4), where the denominator is an observed value y_i^*. How can we avoid $y_i^* = 0$?

Answer: To avoid floating-point overflow, zero must not be taken as a value of y_i^*. Therefore, for regression algorithms, delete the sample if its $y_i^* = 0$; for classification algorithms, positive integers are taken as values of y_i^*.

1-8. (a) What is the basic difference between a regression algorithm and a classification algorithm? (b) What is the sameness in application? (c) Is it possible to approximately regard a regression algorithm as a classification algorithm? (d) What are the regression algorithms and classification algorithms introduced in this book? (e) Among them, which one is a linear algorithm?

Answer: (a) The essential difference between classification and regression algorithms is represented by the data type of calculation results y. For regression algorithms, y is a real-type value and generally differs from y^* given in the corresponding learning sample; whereas for classification algorithms, y is an integer-type value and must be one of y^* defined in the learning samples. In the view of datalogy, the real-type value is called a *continuous attribute*, whereas the integer-type value is called a *discrete attribute*. (b) In application, whether regression algorithm or classification algorithm, the same known parameter is used for the study, the same unknown value to be predicted, but different calculation results. (c) To approximately regard a regression algorithm as a classification algorithm, the results y of regression algorithms are converted from real numbers to integer numbers using the round rule. Certainly, it is possible that some y after the conversion are not equal to any y^* in all learning samples. (d) This book introduces six algorithms for regression and classification: MRA, BPNN, R-SVM, C-SVM, DTR, and BAC. Among the six algorithms, MRA and R-SVM are regression algorithms, whereas others are classification algorithms. (e) Only MRA is a linear algorithm because MRA constructs a linear function, whereas the other five are nonlinear algorithms because they construct nonlinear functions.

1-9. What are the two uses of the total mean absolute relative residual $\overline{R}^*(\%)$ expressed in Equation (1.6)?

Answer: There are two major uses:

1. To express nonlinearity of a studied problem. Since MRA is a linear algorithm, its $\overline{R}^*(\%)$ for a studied problem expresses the nonlinearity of $y = y(x)$ to be solved, i.e., the nonlinearity of the studied problem. This nonlinearity can be divided into five classes: very weak, weak, moderate, strong, and very strong (Table 1.2).

2. Express the solution accuracy of a studied problem. Whether linear algorithm (MRA) or nonlinear algorithms (BPNN, C-SVM, R-SVM, DTR, and BAC), their $\overline{R}^*(\%)$ of a studied problem expresses the accuracy of $y = y(x)$ obtained by each algorithm, i.e., solution accuracy of the studied problem solved by each algorithm. This solution accuracy can be divided into five classes: very high, high, moderate, low, and very low (Table 1.2).

1-10. What is the definition of a data mining system? What are the three major steps of data mining?

Answer: A data mining system (DMS) is a kind of information system that effectively integrates one or a number of DM algorithms on a software platform and combines corresponding data sources to complete specific mining applications or common mining tasks, to extract the modes, rules, and knowledge that are useful to users. Research and development have been made to a specific DMS for a specific application area.

Three steps for DM are illustrated in Figure 1.4: (1) Data preprocessing, such as data selection, cleaning, handling missed ones, misclassification recognition, outlier recognition, transformation, normalization. (2) Knowledge discovery, first to make sure which factors are enough to support DM by using dimension-reduction tools, then to select an appropriate DM algorithm base on the characteristics of learning samples so as to obtain new knowledge. (3) Knowledge application, to deduce the target value for each prediction sample according to new knowledge.

ANSWERS TO CHAPTER 2 EXERCISES

2-1. For the various exploration degrees, what are different ways to construct the probability density function?

Answer: To predict undiscovered resources by probability density function, the chart of resource discovery density is drawn at first, then the probability density function is constructed by two different ways, according to the different exploration degrees. (1) For a more explored area (Figure 2.3), since the known data are enough, an appropriate function (e.g., a logarithmic normal distribution) is chosen as a probability density function, and this function is determined by the known data. (2) For a less or moderately explored area (Figure 2.4), since the known data are not enough, a probability density function (e.g., a logarithmic normal distribution) of another area is borrowed by analogy method.

2-2. In the simple case study for the pore volume of trap calculated by the Monte Carlo method, for parameters φ and H, logarithmic normal distribution is employed as a probability density function $p(x)$ to perform the cumulative frequency distribution as a probability distribution function $F(x)$, i.e., $F(x) = \int p(x)dx$. Can the data of φ and H be directly used to perform the cumulative frequency distribution rather than the logarithmic normal distribution?

Answer: Yes. Because the numbers of φ and H values are large enough (25 and 40, respectively), and φ and H are all uniformly distributed, the result V_ϕ is reliable.

2-3. The least-squares method can be adopted to construct a function of an unknown number with respect to a parameter. When the general polynomial fitting is used, does the case that the higher the order of polynomial, the better result in practical applications?

Answer: No. It is seen from the results of Simple Case Study 2 (polynomial of porosity with respect to acoustictime) that the larger m, i.e., the higher the order of a polynomial, the smaller the mean square error. But that does not mean that the higher the order of the polynomial, the better the results in practical applications . In general, $m = 2$ is taken when the relationship between an unknown number and a parameter is linear; $m > 2$ is taken when the relationship is nonlinear, but $m < 6$. In practical applications, when a fitting formula with large m is employed to calculate y from an x, if the value of the x is outside the range of n measured values, the calculated y would be probably far from the real value.

2-4. Multiple regression analysis (MRA) can be adopted to construct a function of an unknown number with respect to multiple parameters. When $F_1 = F_2 = 0$, the successive regression and the classical successive regression are coincident so that two additional benefits are obtained. What are the two benefits?

Answer: The two additional benefits are (1) all variables can be introduced in a regression equation, and the unknown number (y) can be shown to depend on all parameters ($x_1, x_2, \ldots, x_m$) in decreasing order; and (2) based on this order, MRA can serve as a pioneering dimension-reduction tool in data mining.

For example, in the case study (linear function of proved reserve with respect to its relative parameters), (a) the order of introduced x_k is x_3, x_5, x_2, x_4, x_1; and (b) through the observation on residual variance Q and multiple *correlation* coefficient R, the 6-D problem ($x_1, x_2, x_3, x_4, x_5, y$) can be reduced to a 4-D problem (x_2, x_3, x_5, y).

ANSWERS TO CHAPTER 3 EXERCISES

3-1. In applying the conditions of BPNN, the number of known variables for samples must be greater than one, i.e., the number of input layer nodes must be greater than one. Why?

Answer: If the number of known variables for samples, m, is equal to one, this is a problem of "to construct a function of a parameter with respect to another parameter" and should be solved by the least-squares method introduced in Section 2.2.3. As with MRA, introduced in Section 2.2.4, BPNN is adopted to a problem of "to construct a function of a parameter with respect to multiple parameters." For a special time-series prediction, however, special processing should be conducted to change $m = 1$ to $m > 1$, e.g., Simple Case Study 2 (prediction of oil production), introduced in Section 3.1.5; then BPNN can be applied.

3-2. Why does BPNN as introduced in this book use only one hidden layer?

Answer: In general, BPNN consists of one input layer, one or more hidden layers, and one output layer. There is no theory yet to determine how many hidden layers are needed for any given case, but in the case of an output layer with only one node, it is enough to define one hidden layer (Figure 3.2).

3-3. Why is the formula employed for the number of hidden nodes $N_{hidden} = 2(N_{input} + N_{output}) - 1$?

Answer: It is also difficult to determine how many nodes a hidden layer should have. For solving the local minima problem, it is suggested to use the large $N_{hidden} = 2(N_{input} + N_{output}) - 1$ estimate, where N_{hidden} is the number of hidden nodes, N_{input} is the number of input nodes, and N_{output} is the number of output nodes.

3-4. Since each of Case Studies 1 and 2 is a strong nonlinear problem, the data mining tool can adopt BPNN but not MRA. How can we determine whether these two case studies are strong nonlinear problems?

Answer: Since the total mean absolute relative residuals $\overline{R}^*$ (%) of MRA for Case Studies 1 and 2 are 15.16 (Table 3.9) and 19.50 (Table 3.12), respectively, it is seen from Table 1.2 that the nonlinearity of each case study is strong.

ANSWERS TO CHAPTER 4 EXERCISES

4-1. What are the differences between C-SVM and R-SVM in terms of use?

Answer: C-SVM is a classification algorithm in SVM, whereas R-SVM is a regression algorithm in SVM. In one word, when we desire the results of a real problem be integer numbers (discrete attributes), C-SVM is adopted to solve this classification problem; and when we desire the results of a real problem be real numbers (continuous attributes), R-SVM is adopted to solve this regression problem.

4-2. What is the advantage of SVM compared with existing methods of statistics?

Answer: The SVM procedure basically does not refer to the definition of probability measure and the law of great numbers, so SVM defers from existing methods of statistics. The final decision-making function of SVM [e.g., Equation (4.7) for C-SVM, Equation (4.11) for R-SVM] is defined by only few support vectors (free vectors) so that the complexity of calculation depends on the number of support vectors but not on the number of dimensions of sample space, which can avoid the dimension disaster to some extent and has better robustness. That is a distinguishing advantage of SVM.

4-3. To have the comparability of calculation results among various applications in this book, what operation options does SVM take?

Answer: RBF is taken as a kernel function, and the termination of calculation accuracy (TCA) is fixed to 0.001. The insensitive function ε in R-SVM is fixed to 0.1.

4-4. In Case Studies 1 and 2, C-SVM, R-SVM, BPNN, and MRA are adopted, and the results of the four algorithms are compared in Tables 4.3 and 4.6. From the two tables, review the results of each algorithm.

Answer: Since Case Studies 1 and 2 are classification problems and C-SVM is a classification algorithm, the used algorithm is reasonable and the calculation results are usable. Since the other three algorithms are regression algorithms, their calculation results are converted from real numbers to integer numbers by using the round rule so as to do the classification action. The calculation results of R-SVM and BPNN are basically usable. Table 4.3 shows that BPNN is better than R-SVM; contrarily, Table 4.6 shows that R-SVM is better than BPNN. Since the nonlinearity of Case Studies 1 and 2 are respectively very strong and strong, the calculation results of MRA are unusable, but $\overline{R}^*$ (%) of MRA can expresses the nonlinearity of the studied problem and can establish the order of dependence between prediction variable y and its relative known variables $(x_1, x_2, ..., x_m)$. This order can be used for dimension reduction.

4-5. What are the definitions and actions of dimension reduction?

Answer: The definition of dimension reduction is to reduce the number of dimensions of a data space as much as possible, but the results of the studied problem are unchanged. The action of dimension reduction is to reduce the amount of data so as to speed up the data mining process and especially to extend up the applying ranges.

ANSWERS TO CHAPTER 5 EXERCISES

5-1. For the classification algorithm, the dependent variable y^* in learning samples must be an integer number, but the independent variables $(x_1, x_2, ..., x_m)$ may be real numbers or integer numbers, e.g., for the two classification algorithms of C-SVM in Chapter 4 and BAC in Chapter 6. For the classification algorithm DTR in this chapter, can the independent variables $(x_1, x_2, ..., x_m)$ be real numbers?

Answer: No. It is known from Equation (5.3) that the independent variables $(x_1, x_2, ..., x_m)$ in each learning sample used by DTR must be nonzero integer numbers and then can be used in tree generation.

5-2. What are the advantages and shortcomings of ID3?

Answer: The advantages of ID3 are the clear basic theory, simple method, and strong learning ability. However, ID3 has some shortcomings. For instance, (a) it is sensitive to noise; (b) it is available only when the number of learning samples is small, but the tree changes as the number of learning samples increases, and then regenerating a new tree causes big expenditures; and (c) like BPNN, the local optimal solution occurs, but the global optimal solution does not.

5-3. C4.5 has inherited the advantages of ID3 and made four major improvements. What are the four improvements?

Answer: The four major improvements are (1) to substitute the information gain ratio for the information gain so as to overcome the deflection in gain calculation; (2) to use the pruning in the tree generation procedure so as to eliminate the noises or abnormal data; (3) to discrete the continuous variables; and (4) to process the nonintegral data.

5-4. From Tables 5.3, 5.6, and 5.9, $\overline{R}^*(\%) = 0$ when DTR and C-SVM are applied to the simple case study in Section 5.1 and Case Study 1 in Section 5.2; $\overline{R}^*(\%) = 0$ and $\overline{R}^*(\%) = 8.78$ when DTR and C-SVM are applied to Case Study 2 in Section 5.3, respectively. Does that mean DTR is superior to C-SVM?

Answer: No. The reason that $\overline{R}^*(\%) = 8.78$ when C-SVM is applied to Case Study 2 is that there is no big enough number of learning samples. Doing a comparison, the number of learning samples in Case Study 1 is 29 (Table 5.5), but in Case Study 2 the number is 26 (Table 5.7).

ANSWERS TO CHAPTER 6 EXERCISES

6-1. This chapter introduces naïve Bayesian (NBAY), Bayesian discrimination (BAYD), and Bayesian successive discrimination (BAYSD). What are the differences in methods and application values of these three algorithms?

Answer: At first, the three algorithms all employ Bayesian theorems expressed with (6.2). For both BAYD and BAYSD, it is assumed that the learning samples of each class have a normal distribution and the discriminate function is constructed in the learning process, expressed with Equations (6.10) and (6.30), respectively. The form of the two equations is the same, but their corresponding coefficients are different due to the fact that BAYD uses a covariance matrix (6.8), whereas BAYSD uses a deviation matrix (6.18) and a total deviation matrix (6.20). In the applications, the results of BAYD and BAYSD are the same, e.g., in Case Study 5 (Table 6.16). Comparing BAYD and BAYSD with NBAY, it is shown that different algorithms are applicable to different applications, e.g., NBAY is superior to BAYSD in a simple case study (Table 6.3). Especially since the prediction variable y is shown to depend on the known variables x_i ($i = 1, 2, ..., m$) in decreasing order given only by BAYSD (Tables 6.3, 6.6, 6.9, 6.11, 6.13, 6.16, 6.18), BAYSD can serve as a pioneering dimension-reduction tool.

6-2. In seven case studies from Section 6.1 to Section 6.7, BAYSD and C-SVM in Chapter 4 are applied. What are the solution accuracies of the two classification algorithms?

Answer: It is found from Table 6.19 that the solution accuracies of BAYSD and C-SVM are (a) high when the nonlinearity of case studies is very weak or moderate, but (b) different when the nonlinearity of case studies is strong, except for Case Study 5, and only one of the two algorithms is applicable.

6-3. Either BAYSD or MRA in Chapter 4 can serve as a pioneering dimension-reduction tool. What are the advantages and shortcomings of the two dimension-reduction tools?

Answer: Although no dimension-reduction calculations are conducted in seven case studies from Section 6.1 to Section 6.7, the prediction variable y is shown to depend on the known variables x_i ($i = 1, 2, ..., m$) in decreasing order given by BAYSD and MRA, indicating BAYSD or MRA can serve as a pioneering dimension-reduction tool. For the dependencies in each case study obtained by BAYSD and MRA, some are the same, e.g., in a simple case study, Case Studies 1, 3, and 4; but the others are not same, e.g., in Case Studies 2, 5, and 6. The different dependencies result from the fact that MRA is a linear algorithm whereas BAYSD is a nonlinear algorithm. Hence, the dependence obtained by BAYSD is more reliable than that of MRA, i.e., it is better to adopt BAYSD as a pioneering dimension-reduction tool than to adopt MRA. However, MRA can give the multiple correlation coefficient for y and introduced x_i so as to determine the reducible dimension (see Section 4.4). Certainly, whether MRA or BAYSD is adopted for dimension reduction, the results are required to be validated by a nonlinear tool (C-SVM or BPNN). Strictly and concretely, since MRA is a regression algorithm, it can serve as a pioneering dimension-reduction tool for regression problems, and the dimension-reduction results are validated by regression algorithm BPNN; whereas BAYSD is a classification algorithm that can serve as a pioneering dimension-reduction tool for classification problems, and the dimension-reduction results are validated by classification algorithm C-SVM.

ANSWERS TO CHAPTER 7 EXERCISES

7-1. Are there any similarities or differences among the data used by a regression/ classification algorithm and a cluster analysis algorithm?

Answer: This chapter introduced two algorithms of cluster analysis: Q-mode cluster analysis and R-mode cluster analysis. The data used by the two algorithms can be taken from the data used by regression algorithms and classification algorithms. First, the data used in regression algorithms and classification algorithms cover learning samples and prediction samples, whereas the data used by cluster analysis cover only samples, without the concept of learning samples and prediction samples. Taking the learning samples in regression algorithms and classification algorithms as an example, the data used is the x_i vector expressed by Equation (1.1) that includes y_i^*. But the data used by Q-mode cluster analysis is the sample vector x_i expressed by Equation (7.1), which does not include y_i^*, and the data used by R-mode cluster analysis is the parameter vector x_i expressed by Equation (7.2) that does not include y_i^* either. y_i^* is used only for comparison between the calculation results of cluster analysis and y_i^* if y_i^* exist. Therefore, Q-mode cluster analysis is a cluster analysis to n sample vectors, whereas R-mode cluster analysis is a cluster analysis to m parameter vectors. Obviously, the matrix defined by Equation (7.1) and the matrix defined by Equation (7.2) are mutually transposed.

7-2. What are the different purposes of Q-Mode cluster analysis and R-mode cluster analysis?

Answer: By adopting Q-mode cluster analysis, a cluster pedigree chart about n samples will be achieved, indicating the order of dependence between samples. Therefore, it possibly serves as a pioneering sample-reduction tool. By adopting R-mode cluster analysis, a cluster pedigree chart about m parameters will be achieved, indicating the order of dependence between parameters. Therefore, it possibly serves as a pioneering parameter-reduction (i.e., dimension-reduction) tool.

7-3. In this book, there are three kinds of data normalization (standard difference standardization, maximum difference standardization, maximum difference normalization), three cluster statistics (distance coefficient, analog coefficient, correlation coefficient), and three class-distance methods (the shortest-distance method, the longest-distance method, and the weighted mean method) for each cluster statistic. Hence, there are nine clustering methods from among which users can choose. How can we choose the best one in each application?

Answer: A suitable one will be chosen by depending on the practical application and specific data. For instance, this chapter provides four case studies extending from Chapter 6. Each case study uses the weighted mean method to perform Q-mode and R-mode cluster analyses. But the original data normalization method to be used and cluster statistics are not the same. The reason is that each case study tries to use nine methods (i.e., three methods for data normalization of standard difference standardization, maximum difference standardization, and maximum difference normalization, as well as the three methods for calculating cluster statistics of distance coefficients, analog coefficients, and correlation coefficients) for calculation, and then compares nine calculation results y by Q-mode cluster analysis with y^* of the corresponding case studies in Chapter 6, to select one result that is closest. The method obtaining the closest result is the method that is should be used for this case study. For example: Case Study 1 uses standard difference standardization and analog coefficients; Case Study 2 uses maximum difference normalization and analog coefficients; Case Study 3 uses maximum difference standardization and correlation coefficients, and Case Study 4 uses maximum difference standardization and analog coefficients. Therefore, the optimal selection of methods depends on applications, i.e., different applications have different optimal selections.

7-4. Q-mode cluster analysis and R-mode cluster analysis are applied to Case Studies 1, 2, 3, and 4. What are the solution accuracies of the two algorithms?

Answer: It can be seen from Table 7.13 that (a) Q-mode cluster analysis provides a cluster pedigree chart for samples, and the classification coincidence rate is passable; (b) if MRA is passed for validation, Q-mode cluster analysis can also serve as a pioneering sample-reduction tool; (c) R-mode cluster analysis provides a cluster pedigree chart for parameters; and (d) if MRA is passed for validation, R-mode cluster analysis (using correlation coefficients) can also serve as a pioneering dimension-reduction tool. The so-called "pioneering tool" is whether it succeeds or does not need nonlinear tool (C-SVM, R-SVM, or BPNN) for the second validation so as to determine how many samples and how many parameters can be reduced. Why does it need a second validation? Because of the complexities of geoscience rules, the *correlations between different classes* of geosciences data are nonlinear in most cases. In general, therefore, C-SVM or R-SVM is adopted when nonlinearity is strong, whereas BPNN, DTR, or BAC is adopted when nonlinearity is not strong. The two linear algorithms of cluster analysis and MRA can serve as auxiliary tools, cooperating with major tools (BPNN, C-SVM, R-SVM, DTR, and BAC) for data mining.

ANSWERS TO CHAPTER 8 EXERCISES

8-1. What are the advantages of geostatistics compared with conventional statistics?

Answer: Kriging is an approach of geostatistics. The differences between geostatistics and conventional statistics are (a) from methodology, conventional statistics uses conventional mathematics, whereas geostatistics uses a new combination of conventional statistics and geology; (b) from applying ranges, conventional statistics is applicable to multidisciplines, including geology, whereas geostatistics is applicable only to geology; and (c) from the space structure and randomicity of studied parameters, geostatistics is much superior to conventional statistics because geostatistics can furthest utilize various information in geosciences, evaluate in both whole and local terms, and gain accurate results.

Briefly, Kriging is a marginal discipline, taking the variogram as a basic tool, selecting various appropriate methods, and performing optimal linear unbiased interpolation estimates on the space structure and randomicity of parameters.

8-2. How is Kriging organically composed of five operation steps?

Answer: Figure 8.1 illustrates the order of the five steps of the Kriging operation, i.e., preprocessing, experimental variogram, optimal fitting of experimental variogram, cross-validation of Kriging, and applications of Kriging. In fact, the first three steps are run, then the fourth step, achieving the Kriging estimated value on the known point. If the Kriging estimated values on these known points are qualified in the fourth step, the fifth step is run, achieving the Kriging estimated value at each grid center. Here the following should be stressed: (a) the results from the first step will be used in Steps 2, 4, and 5, (b) the results from the second step will be used in Step 3, and (c) the results from the third step will be used in Steps 4 and 5.

8-3. Why is the coordinate system transformation required in the preprocessing of geostatistics (see Section 8.1)?

Answer: The study object of Kriging is the parameter with location coordinates (Figure 8.3, Table 8.1). For example, in 2-D space, let a parameter value be v_n, which is located at (x_n, y_n), $n = 1, 2, \ldots, N_{total}$. The study method of Kriging is of orientation, i.e., to conduct the study in a designated direction. Usually one study is carried out in the direction parallel to the x axis, and one is carried out in the direction parallel to the y axis, thus achieving geostatistical results in these two directions. If statistics is made in other directions, the old coordinate system (x, y) should be rotated in an angle to a new coordinate system (x^*, y^*), on which geostatistics is made in the directions of the x^* axis and the y^* axis. For example, let α be an included angle (units of angle) of this direction with x positive axis; new coordinate system (x^*, y^*) can be formed by anticlockwise rotation of the old coordinate system (x, y) in an angle of α, with the following Equations (8.1) and (8.2). To carry out geostatistics in the directions parallel to x^* and y^*, respectively, on this new coordinate system, which is definitely consistent with the geostatistics carried out in the direction parallel to α direction and $\alpha + 90$ direction on the old coordinate system (Figure 8.2).

8-4. What are the actions of the experimental variogram (see Section 8.2)?

Answer: The calculation of an experimental variogram is performed to a coordinate-transformed parameter (see Section 8.1), achieving the important properties of the affecting range, space distribution continuity, and anisotropy for the parameter (Table 8.2, Figure 8.4) and preparing for the optimal fitting of the experimental variogram (see Section 8.3).

8-5. What are the actions of the optimal fitting of the experimental variogram (see Section 8.3)?

Answer: When the discrete values of an experimental variogram for a parameter have been calculated (see Section 8.2), curve fitting (Figure 8.4) will be made to these discrete values (Table 8.2 or Figure 8.4) so as to intensify the important understanding to the affecting range, space distribution continuity, and anisotropy of this parameter and to prepare for the Kriging calculation (see Sections 8.4 and 8.5).

8-6. Why is the optimal fitting curve rather than the experimental variogram curve (Figure 8.4) provided for Kriging (see Sections 8.4 and 8.5)?

Answer: Two defects of the experimental variogram (Figure 8.4) are (1) some discrete values decrease as the lag distance increases, which does not accord with the whole rule; and (2) a straight line links the discrete points, which leads to decreased accuracy.

To avoid the first defect, it is required to delete those discrete points that decrease as the lag distance increases before fitting, which will be realized by the program (Tables 8.2 and 8.3). To avoid the second defect, the optimal fitting curve is selected (Figure 8.4).

8-7. Is it possible to adopt the least-squares method mentioned in Section 2.2.3 to realize the optimal fitting of the experimental variogram?

Answer: The answer is no. The reason is that the study object is the parameter with a location coordinate. It is impossible to fully consider the geological characteristics that the first three or four points in the experimental variogram curve are much more important than the last few points by simply adopting the least-squares method (Table 8.3).

8-8. What are the actions of the cross-validation of Kriging (see Section 8.4)?

Answer: When the known value of a coordinate-transformed parameter (see Section 8.1) and the optimal fitting of an experimental variogram (see Section 8.3) have been obtained, Kriging is used to calculate the estimation value of this parameter. Analysis is made to the variance and deviation between the estimation value and known value so as to check the validity that Kriging is used to this parameter (Tables 8.7 and 8.8) and confirm whether Kriging will be applied (see Section 8.5).

8-9. What are the actions of the applications of Kriging (see Section 8.5)?

Answer: When the known value of a coordinate-transformed parameter (see Section 8.1) and the optimal fitting of an experimental variogram (see Section 8.3) have been obtained, Kriging is used to calculate the estimate value of this parameter at each grid center on the plane (Table 8.9).

ANSWERS TO CHAPTER 9 EXERCISES

9-1. In Section 9.1.3, a simple case study of fuzzy integrated evaluation on an exploration drilling object is introduced. The fuzzy integrated decision value $E = 0.495$ is calculated by Equation (9.1). But the comment vector expressed by Equation (9.6) is $(2, 1, 0, -1, -2)$, denoting excellent, good, average, poor, and very poor, respectively. Is it possible to use the round rule to transform $E = 0.495$ to $E = 0$ so as to make it consistent with 0 in the comment vector?

Answer: Yes. In fact, E ranges between 1 and 0, and slightly leaning to 0, so the integrated evaluation result on this exploration drilling object approaches the average.

9-2. Section 9.2.3 introduces space-time calculations conducted by gray prediction. In gray prediction there are two calculation schemes: Calculation Scheme 1, in which the original data are not processed; and Calculation Scheme 2, in which the original data are processed so that the minimum value is subtracted from each data. Which one is the better of the two calculation schemes?

Answer: No conclusions can be drawn. It depends on practical applications, e.g., for Simple Case Study 1 (oil production prediction of an oil production zone in the Shengli Oilfield using gray prediction), Calculation Scheme 1 is better than Calculation Scheme 2, whereas for Simple Case Study 2 (reservoir prediction of a well in the Liaohe Oilfield using gray prediction), Calculation Scheme 2 is better than Calculation Scheme 1.

9-3. In Section 9.2.3, there are two simple case studies for oil production prediction by gray prediction: Simple Case Study 1 (oil production prediction of an oil production zone in the Shengli Oilfield using gray prediction) and Simple Case Study 3 (oil production prediction of the Romashkino Oilfield using gray prediction). Why did Simple Case Study 1 succeed but Simple Case Study 3 failed?
Answer: This is attributed to the annual oil production gradient variation of the learning sample: The annual oil production of that oil production zone is 428.8250–451.7581 (10^4t) for five years [see Equation (9.22)], whereas the annual oil production for 39 years of the Romashkino Oilfield is 200–8150 (10^4t) (Table 9.2). Obviously, it is impossible to adopt gray prediction but adopt BPNN to predict annual oil and gas production of an oilfield when annual production gradient is large, since the solution accuracy of BPNN is not restricted by the gradient (see Figure 9.3).

9-4. Section 9.1.3 introduces fuzzy integrated decisions, and Section 9.2.4 introduces gray integrated decisions. Compare the differences in application between these two algorithms.
Answer: These two algorithms all belong to integrated decisions. However, gray integrated decisions can determine the quality of the several geological objects and then select the optimal one, whereas fuzzy integrated decisions can determine the quality of only one exploration drilling object. If there are several exploration drilling objects, the fuzzy integrated decision can be made under the same evaluation standard (weight vector W and comment vector V), respectively, achieving their fuzzy integrated decision values, which will be ranked according to quality so as to reach the aim of optimal selection.

9-5. In Section 9.3, there is a case study about the relationship between the fractal features of faults and hydrocarbon migration accumulation. Based on Table 9.5 (fractal data and results for faults of structural layers T_2 and T_6), how can we deduce the development extent of the fault in Dongying Sag and then predict the development extent of the reservoir in the sag?
Answer: It is obvious from Table 9.5 that $r_{LN} > r_a$ in each area, indicating that the linear correlation between $\ln(L)$ and $\ln(N)$ in each area is highly significant and it is feasible to describe the fault feature in this sag using fractal dimension D value.
  It can be seen from Table 9.5 that for layer T_6, D value (1.2023) in the north area is larger than D value (1.1034) in the south area, and D value (1.1363) in the east area is smaller than D value (1.1496) in the west area. For layer T_2, D value (1.1187) in the north area is also larger than D value (1.0980) in the south area, and D value (1.1232) in the east area is larger than D value (1.0812) in the west area, which indicates that the tectonic movement in this sag leads the faults to be high in the north and low in the south. Since faults tend to play an important role in hydrocarbon migration accumulation, the reservoir is predicted to be high in the north and low in the south in this sag, which is proved by reality. The size and productivity of reservoirs in this sag are higher in the north area than in the south area.

9-6. Section 9.4 introduces linear programming to deal with optimization. In practical application, the mathematical models for various kinds of linear programming should be transformed to standard patterns [Equations (9.61) and (9.62)] before the solution is worked out.

Assume that a linear programming problem in a nonstandard pattern is:

$$\min z = -x_1 + 2x_2 - 3x_3$$

$$\begin{cases} x_1 + x_2 + x_3 \leq 7 \\ x_1 - x_2 + x_3 \geq 2 \\ -3x_1 + x_2 + 2x_3 = 5 \\ x_1 \geq 0,\ x_2 \leq 0,\ x_3 \quad \text{is in unconstraint} \end{cases}$$

Write the process to transform the problem from a nonstandard pattern to a standard pattern.

Answer: Using the standardization skills of a mathematical model shown in Table 9.6, the process to transform the problem from a nonstandard pattern to a standard pattern is, let $z' = -z$; let $x'_2 = -x_2$; let $x_3 = x_4 - x_5$, of which the slack variables $x_4 \geq 0$ and $x_5 \geq 0$; and add slack variable x_6 and subtract surplus variable x_7 at the left-hand side of the first and the second constraint inequality, respectively, of which $x_6 \geq 0$ and $x_7 \geq 0$.

Thus, this linear programming has been in a standard pattern:

$$\max z' = x_1 + 2x'_2 + 3(x_4 - x_5) + 0x_6 + 0x_7$$

$$\begin{cases} x_1 - x'_2 + x_4 - x_5 + x_6 & = 7 \\ x_1 + x'_2 + x_4 - x_5 - x_7 & = 2 \\ -3x_1 - x'_2 + 2x_4 - 2x_5 & = 5 \\ x_1,\ x'_2,\ x_4,\ x_5,\ x_6,\ x_7 \geq 0 \end{cases}$$

9-7. Section 9.4.3 introduces optimization of exploration and production schemes implemented by linear programming, and the calculation method used is noninteger programming. Integer programming can be solved by using a branch-bound method or a cutting plane method. Nevertheless, the programming of the two methods is complicated. A simple and convenient method for integer programming has been presented, which does not need to reprogram. Now the question is, is it possible not to apply this simple and convenient integer programming method but to use the round rule to directly transform the unique optimal solution of the noninteger programming $x_j = (j = 1, 2, \cdots, n,\ n+1, \cdots, n+m)$ to integer?

Answer: The answer is that it does not always work. It is discussed here with an example of a simple case study (optimal profit of exploration drilling plan) in Section 9.4.3:

If the noninteger programming is adopted, the unique optimal solution is obtained as follows:

$$x_1 = 15.5, \; x_2 = 8.25, \; x_3 = 4.625, \; x_4 = 1.625, \; x_5 = 1, \; x_6 = x_7 = x_8 = x_9 = x_{10} = 0$$

and the maximum of the corresponding object function is:

$$z = 0.1285 \times 10^5 \; (10^4 \; \text{yuan}),$$

In practical application, the number of wells drilled can be but an integer, and it cannot be those noninteger values in the unique optimal solution. So, integer programming must be adopted.

If the simple and convenient integer programming method is used, the unique optimal solution will be obtained as follows:

$$x_1 = 15, \; x_2 = 8, \; x_3 = 5, \; x_4 = 2, \; x_5 = 1, \; x_6 = x_7 = x_8 = x_9 = x_{10} = 0$$

and the maximum of the corresponding object function is:

$$z = 0.1263 \times 10^5 \; (10^4 \; \text{yuan})$$

Now the question is, is it possible not to apply this simple and convenient integer programming method but to use the round rule to directly transform the unique optimal solution of the noninteger programming $x_j = (j = 1, 2, \cdots, n, \; n+1, \cdots, n+m)$ to integer? The answer is that it does not always work, because the unique optimal solution then obtained is as follows:

$$x_1 = 16, \; x_2 = 8, \; x_3 = 5, \; x_4 = 2, \; x_5 = 1, \; x_6 = x_7 = x_8 = x_9 = x_{10} = 0$$

But this $x_1 = 16$ has been verified not to be the optimal with the simple and convenient integer programming method, for $x_1 = 15.5, v_2 = 16, v_1 = 15$ and $v_3 = 17$ are obtained by Equation (9.65). Three corresponding mean square errors are obtained by substituting these three numbers into the constraint condition expressed by Equation (9.77), and $v_1 = 15$ leads the mean square error of the constraint condition to be minimum, whereas $v_2 = 16$ and $v_3 = 17$ cannot make the mean square error of the constraint condition be minimum.

ANSWERS TO CHAPTER 10 EXERCISES

10-1. Section 10.1.3 (regression calculation) in Section 10.1 (Typical Case Study 1: oil layer classification in the Keshang Formation) introduces a calculation of oil productivity index. Why did this calculation fail?
Answer: From $\overline{R}^*(\%) = 248.10$ of MRA (Table 10.4), the nonlinearity of this calculation of oil productivity index (Tables 10.1 and 10.3) is very strong, so the three regression algorithms (R-SVM, BPNN, and MRA) are not applicable (Table 10.4).

10-2. Section 10.1.4 (classification calculation) in Section 10.1 (Typical Case Study 1: oil layer classification in the Keshang Formation) introduces a calculation of oil layer classification. Why did this calculation succeed?

Answer: From $\overline{R}^*(\%) = 7.93$ of MRA (Table 10.6), the nonlinearity of this calculation of oil layer classification (Tables 10.1, 10.2 and 10.5) is moderate, so the two classification algorithms (C-SVM and BAYSD) are applicable (Table 10.6).

10-3. From the answers to Questions 10-1 and 10-2, what conclusions do you draw?

Answer: From Typical Case Study 1, it is concluded that when regression algorithms (R-SVM, BPNN, MRA) fail in a regression problem, this regression problem can be changed to a classification problem, then classification algorithms (C-SVM, BAYD) succeed in this classification problem.

10-4. Section 10.2 (Typical Case Study 2: oil layer classification in the lower H3 Formation) introduces a calculation of oil layer classification. Why did this calculation succeed?

Answer: From $\overline{R}^*(\%) = 8.40$ of MRA (Table 10.9), the nonlinearity of this calculation of oil layer classification (Tables 10.7 and 10.8) is moderate, so the two classification algorithms (C-SVM and BAYSD) are applicable (Table 10.9).

10-5. From the answer to Question 10-4, what conclusions do you draw?

Answer: From Typical Case Study 2, it is concluded that when the nonlinearity of a classification problem is very weak, weak, or moderate, classification algorithms (C-SVM, BAYD) succeed in this classification problem.

10-6. Section 10.3 (Typical Case Study 3: oil layer classification in the Xiefengqiao Anticline) introduces a calculation of oil layer classification. Why did this calculation succeed?

Answer: From $\overline{R}^*(\%) = 19.33$ of MRA (Table 10.11), the nonlinearity of this calculation of oil layer classification (Tables 4.4 and 10.10) is strong, so only C-SVM succeeds in this classification problem, but BAYSD fails (Table 10.11).

10-7. From the answer to Question 10-6, what conclusions do you draw?

Answer: From Typical Case Study 3, it is concluded that when the nonlinearity of a classification problem is strong or very strong, only C-SVM succeeds in this classification problem, but BAYSD fails.

10-8. What are two classes of minable subsurface data?

Answer: The subsurface data can be divided to two classes: (1) Class 1, the field-measured data, which are the interpretation results of the data indirectly measured by instruments at the Earth's surface or in the air. Though the precision of this data is not high, they are of large amount and 3-D distribution, and thus they can be the basic data for 3-D data mining. (2) Class 2, the experimental data from a laboratory, which are the analytical results of the data directly measured by instruments, e.g., exploration drilling. Though the amount of this data is small and they are in a narrow view, their precision is high, and thus they can be used to calibrate the field-measured data and can be used as constraint conditions of data mining results. Both classes of subsurface data are minable.

10-9. Why are not decision trees, NBAY, and BAYD included in Figure 10.2?

Answer: Since the use of decision trees is quite complicated, decision trees are not chosen in the automatic selection of classification algorithms; and since BAYSD is superior to NBAY and BAYD, NBAY and BAYD are not chosen in the automatic selection of classification algorithms. Thus, the automatic selection of classification algorithms as presented by this chapter contains two classification algorithms of C-SVM and BAYSD.

10-10. What algorithms can serve as pioneering sample-reduction tools? What algorithms can serve as pioneering dimension-reduction tools? What algorithms can serve as their better validating tools?

Answer: Q-mode cluster analysis can serve as a pioneering sample-reduction tool. Each of MRA, BAYSD, and R-mode cluster analysis can serve as a pioneering sample-reduction tool. As for their better-validating tools, it is preferable to adopt BPNN as a validating sample-reduction tool for regression problems, and it is preferable to adopt C-SVM as a validating sample-reduction tool for classification problems.

Index

Note: Page numbers followed by "f" denote figures; "t" tables.

Printed and bound by CPI Group (UK) Ltd, Croydon, CR0 4YY

08/05/2025

01864871-0005